Understanding Inorganic Chemistry

Understanding Inorganic Chemistry

Remi Dalton

MURPHY & MOORE
www.murphy-moorepublishing.com

Published by Murphy & Moore Publishing,
1 Rockefeller Plaza,
New York City, NY 10020, USA

ISBN: 978-1-63987-553-5

Cataloging-in-Publication Data

Understanding inorganic chemistry / Remi Dalton.
p. cm.
Includes bibliographical references and index.
ISBN 978-1-63987-553-5
1. Chemistry, Inorganic. 2. Chemistry. I. Dalton, Remi.
QD151.3 .U53 2021
546--dc23

For information on all Murphy & Moore Publications
visit our website at www.murphy-moorepublishing.com

MURPHY & MOORE

Contents

Preface

The branch of chemistry which is concerned with the synthesis and analysis of inorganic and organometallic compounds is known as inorganic chemistry. The subject is further divided into organometallic chemistry, cluster chemistry and bioinorganic chemistry. The key feature of inorganic compounds is the absence of carbon-hydrogen bond in them. Inorganic compounds are generally classified into cluster compounds, transition metal compounds, coordination compounds and bioinorganic compounds. Some common inorganic compounds are ammonia, chlorine, aluminium sulphate, ammonium nitrate, etc. Some common features of inorganic compounds are high melting point, ease of crystallization, high specific heat capacity and poor electrical conductivity. Applications of inorganic chemistry are widespread ranging from agriculture and medications to fuels and catalysis. The topics included in this book on inorganic chemistry are of utmost significance and bound to provide incredible insights to readers. While understanding the long-term perspectives of the topics, the book makes an effort in highlighting their impact as a modern tool for the growth of the discipline. Those in search of information to further their knowledge will be greatly assisted by this book.

A detailed account of the significant topics covered in this book is provided below:

Chapter 1- Inorganic chemistry deals with the properties and behavior of inorganic compounds. Many metals, minerals and organometallic compounds fall under this category. Inorganic chemistry is a vast subject which includes coordination chemistry, magnetochemistry, bioinorganic chemistry, solid state chemistry, organolanthanide chemistry, etc. These topics have been extensively discussed in this chapter.

Chapter 2- Substances that do not contain both carbon and hydrogen are called inorganic compounds. They lack a carbon-hydrogen bond. Inorganic compounds can be further divided into various groups like inorganic carbon compounds, inorganic chlorine compounds, inorganic nitrogen compounds, inorganic phosphorus compounds, etc. The diverse aspects of inorganic compounds have been thoroughly discussed in this chapter.

Chapter 3- Bioinorganic chemistry deals with the role of inorganic metals and non-metals in biological systems. It includes the study of behaviors and effects of the various metalloproteins on biological systems. Biological processes like photosynthesis, respiration, metal ion transport, enzymatic actions, etc. all fall under its umbrella. This chapter has been carefully written to provide the reader with a thorough understanding of bioinorganic chemistry and its related processes.

Chapter 4- Coordination compounds are the compounds that have a central atom, which is often metallic and is bounded by ligands. The study of such compounds is known as coordination chemistry. Some of the significant topics of this field includes coordination number, coordination complex, coordination sphere, trans and cis effect, Jahn-Teller effect, etc. These topics have been closely examined in this chapter.

Chapter 5- Chemical bonding can be defined as the bond formed between molecules, atoms or ions which results in the formation of a compound. Inorganic compounds undergo chemical bonding with electrovalent or ionic bonding being the most common type of bond. This chapter covers all the topics related to chemical bonding in inorganic chemistry in a comprehensive manner.

Chapter 6- Inorganic chemistry finds its use in a wide variety of applications. Due to the high melting points and specific high or low electrical conductivity of inorganic compounds, they are used as catalysts and pigments in many industrial processes. Materials science is another field where inorganic chemistry is applied. This chapter deals with the various applications of inorganic chemistry to provide an extensive understanding of the subjects.

I would like to make a special mention of my publisher who considered me worthy of this opportunity and also supported me throughout the process. I would also like to thank the editing team at the back-end who extended their help whenever required.

Remi Dalton

1

Inorganic Chemistry and its Branches

Inorganic chemistry deals with the properties and behavior of inorganic compounds. Many metals, minerals and organometallic compounds fall under this category. Inorganic chemistry is a vast subject which includes coordination chemistry, magnetochemistry, bioinorganic chemistry, solid state chemistry, organolanthanide chemistry, etc. These topics have been extensively discussed in this chapter.

The word organic refers to the compounds which contain the carbon atoms in it. So the branch of chemistry that deals with the study of compounds, which does not consist of carbon-hydrogen atoms in it, is called 'Inorganic Chemistry.' In simple words, it is opposite to that of Organic Chemistry. The substances which do not have carbon-hydrogen bonding are the metals, salts, chemical substances, etc.

On this planet, there are known to exist about 100,000 numbers of Inorganic compounds. Inorganic chemistry studies the behaviour of these compounds along with their properties, their physical and chemical characteristics too. The elements of the periodic table except for carbon and hydrogen come in the lists of Inorganic compounds.

Many of the elements are technologically important: titanium, iron, nickel and copper, for example, are used structurally and electrically. Second, the transition metals form several useful alloys, with each other and with other metallic elements.

Organometallic Chemistry

Organometallic Chemistry, an interdisciplinary science in Inorganic Chemistry, has grown at a phenomenal pace during the last three to four decades. On the academic plane, efforts to elucidate the nature of bonds in the ever increasing list of exciting organometallic compounds have led to a clearer understanding of the nature and variety of chemical bonds. Organometallic compounds are primarily used as homogeneous catalysis agents in industries.

Organometallic chemistry is an organometallic compound study. Because many compounds without these bonds are chemically identical, an alternative may be compounds containing metallic bonds of a mostly covalent nature. Organometallic chemistry blends elements of inorganic chemistry with organic chemistry.

Transition Elements

A transition element may be defined as one which possesses partially filled d-orbitals in its penultimate shell. This conceptual definition is useful as it enables us to recognize a transition element merely by looking at its electronic configuration. This definition excludes zinc, cadmium and mercury from the transition elements as they do not have a partially filled d-orbital. However, they are also considered as transition elements, because their properties are an extension of the properties of transition elements in inorganic chemistry. In fact, the zinc group serves as a bridge between the transition elements and the representative elements.

The most notable characteristics shared by the 24 elements concerned are that they are all metals and that most of them are hard, solid and lustrous, have high melting and boiling points and are good conductors of heat and electricity. The range in these properties is considerable; hence, the statements are comparable to the general properties of all the other elements.

Coordination Chemistry

Coordination compounds found their applications long before the establishment of inorganic chemistry. A systematic investigation of structure and bonding in coordination chemistry began with the inquisitiveness of Tassaert which was extended by distinguished chemists like Wilhelm Blomstrand, Jorgensen and Alfred Werner until the end of the nineteenth century. In the events, Werners coordination theory became the base of modern coordination chemistry.

P-Block Elements

The elements placed in group 13 to group 18 of the periodic table constitute the p-block. The properties of inorganic chemistry p block elements like that of other block elements are greatly influenced by their atomic size, ionization enthalpy, electron gain enthalpy and electronegativity. The absence of d–orbitals in the second period and presence of d- or f-orbitals in heavier elements has a significant effect on the properties of the elements and therefore, heavier p-block elements differ from their lighter congeners.

Classification of Inorganic Compounds

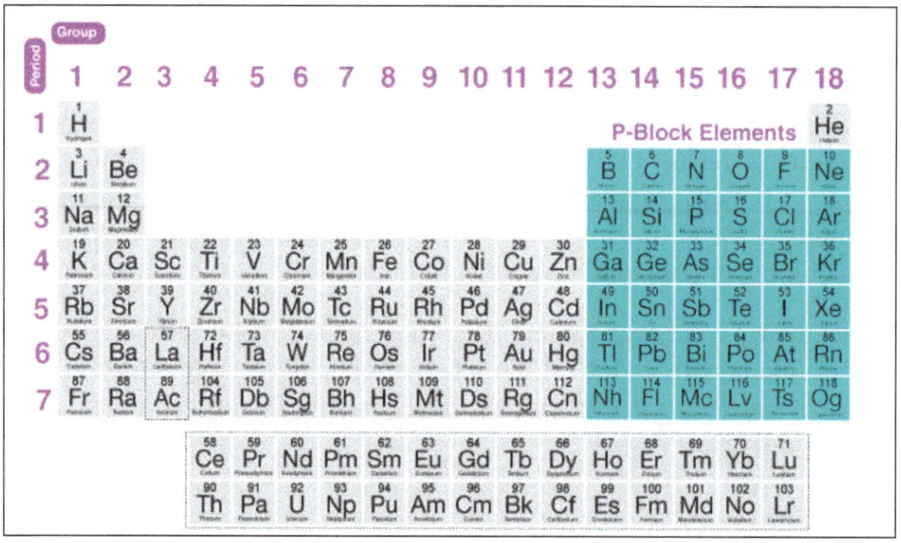

The organic compounds that are classified under Inorganic chemistry are:

- Acids: Acids are those compounds that dissolve in water and generate hydrogen ions or H+ Ions. The examples of acids include Hydrochloric acid, citric acid, sulphuric acid, vinegar, etc. One example of the acidic reaction is shown below-Hydrochloric acid + water → H+ + Cl.

- Bases: A base is a type of substance or a compound that produces hydroxyl ions when kept in water. The bases like potassium hydroxide, calcium hydroxide, ammonia, sodium hydroxide produce OH- ions when dissolved in water. Potassium Hydroxide + $H_2O \rightarrow K+ + OH-$.
- Salts: As you might be familiar with the word 'Salt'. The substances obtained as a result of the reaction between an acid and bases are called Salts. The table salt of sodium hydroxide is one of the typical examples of salts.
- Oxides: The compounds which consist of one oxygen atom called Oxides.

Types of Reactions and Examples

There are about four types of chemical reactions of Inorganic chemistry namely combination, decomposition, single displacement and double displacement reactions.

- Combination Reactions: As it is in the name 'Combination', here two or more substances combine to form a product which is called a Combination reaction. For example: Barium + $F_2 \rightarrow BaF_2$
- Decomposition Reaction: It is a type of reaction where a single element splits up or decomposes into two products. For example: FeS → Fe + S.
- Single Displacement Reactions: A reaction where a single atom of one element replaces another atom of one more element. For example:Zn (s) + CuSO4 (aq) → Cu (s) + ZnSO4 (aq).
- Double Displacement Reactions: This type of reaction is also called 'metathesis reactions'. Here two elements of two different compounds displace each other to form two new compounds. For example:CaCl2 (aq) + 2AgNO3 (aq) → Ca(NO3)2 (aq) + 2 AgCl (s).

Applications of Inorganic Chemistry

Inorganic chemistry finds its high number of applications in various fields such as Biology, chemical, engineering, etc.

- It is applied in the field of medicine and also in healthcare facilities.
- The most common application is the use of common salt or the compound Sodium hydroxide in our daily lives.
- Baking soda is used in the preparation of cakes and other foodstuffs.
- Many inorganic compounds are utilized in ceramic industries.
- In the electrical field, it is applied to the electric circuits as silicon in the computers, etc.

Branches of Inorganic Chemistry

Branches of inorganic chemistry include applications in organic chemistry, bioinorganic chemistry, coordination chemistry, geochemistry, inorganic technology, nuclear science and energy, organometallic compounds, reaction kinetics and mechanisms, solid-state chemistry, and synthetic inorganic chemistry.

Coordination Chemistry

Coordination chemistry is the study of compounds that have a central atom (often metallic) surrounded by molecules or anions, known as ligands. The ligands are attached to the central atom by dative bonds, also known as coordinate bonds, in which both electrons in the bond are supplied by the same atom on the ligand.

Complex formation is a fundamental chemical process in natural waters - complexes modify metal species in solution, reducing activity of free ion, changing solubility, mobility, and toxicity. Surface chemistry can also be interpreted from a coordination chemistry approach, and processes such as ion exchange, sorption, and weathering are really ligand exchange reactions.

Fundamentals

- Coordination compounds are complexes that consist of one or more central atoms or ions with one or more attached molecules.
- The central atom is called a metal, and is a Lewis Acid.
- The attached ions are ligands, and are Lewis Bases.
- The total number of attachments between metal and ligands is the coordination number.
- Ligands attach to the metal via coordinate covalent bonds, i.e., the electrons of the bond is entirely derived from the ligand.
    - Must involve electron pair not otherwise utilized in bonding withing the ligand. The electron pair is equally shared, but controlled by the ligand.
    - The ligand is a lewis base, and acts as an electron pair donor. The metal is the electron acceptor and a lewis acid.
- Most metals form cationic ions in solutions (electron acceptors), and participate in coordination complexes.
- The strength of the coordination interaction is a function of the ability of a ligand to donate or share their available pair(s) of electrons.
- Coordination number of a metal ion may vary from one complex to another.

$$Fe(H_2O)_6^{+3} \ ; \ Fe(CN)_6^{-3} \ ;' \ Fe(Cl)_4^{-}$$

- Water is also a (weak) Lewis Base, and can act as a ligand. Aqueous ions are almost always initially coordinated with water, and any other complex actually forms by ligand exchange between water and the new ligand, i.e.:

$$Fe(H_2O)_6^{2+} + Cl^- <----> Fe(H_2O)_5Cl^+ + H_2O$$

- Ligands that attach at a single point are monodentate. Examples are Cl^-, CN^-, OH^-.
    - Multidentate ligands chelate the metal, Ex. SO_4^{2-}.

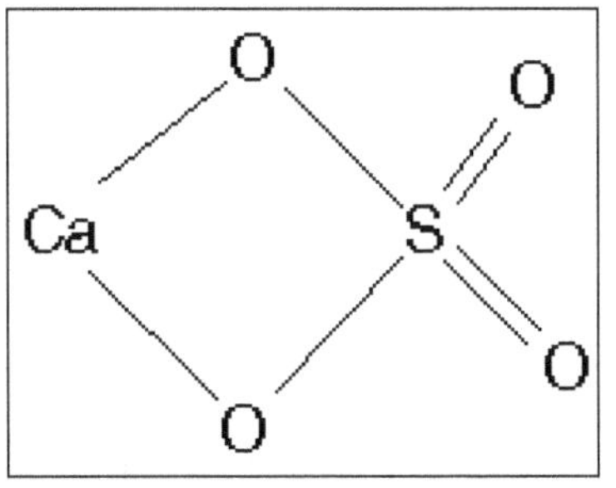

    - EDTA, for example,. forms 6-dentate complexes:

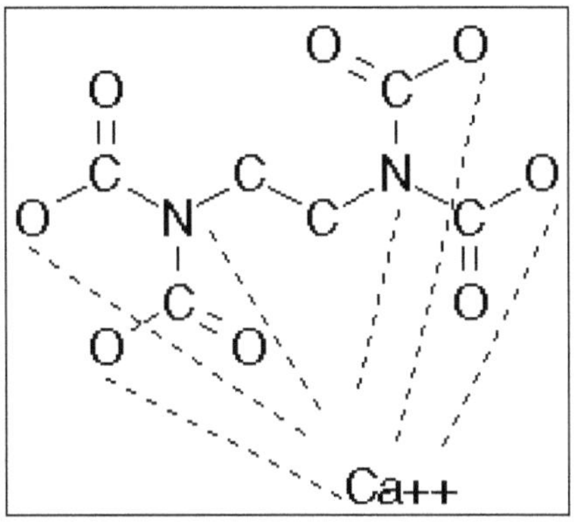

    - Polynuclear complexes form when there is more than one central ion or molecule:

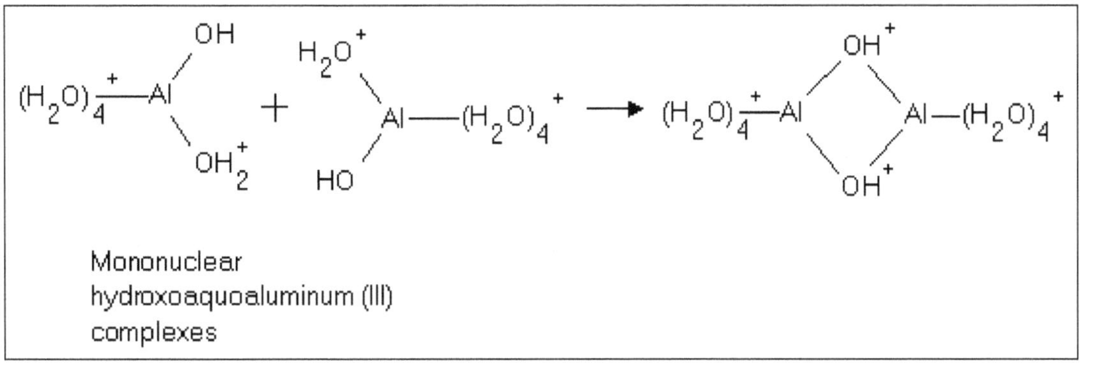

Ion Pairs

- Temporary electrostatic complexes where each ion retains their coordinated waters, but they are still close enough to effectively form a complex between the metal and the ligand.
- Usually form beween Hard cations and anions.
- Another term is outer sphere complex. These complexes are difficult to analyze directly, and are normally estimated from coulombic attraction and Bejerrum's ion association model.

- The log Kstab of ion pair complexes range from ~0 to 4, depending on charge density.
- There can be ion pair or coordinative complexes for the same pair of ions, with an equilibrium established between them. The major ion pairs in water are:

  $CaHCO_3^+$, ; $CaSO_4^\circ$; $CaOH^+$; $MgCO_3^\circ$; $MgSO_4^\circ$

- The degree of interaction is only slightly greater than the electrostatic interactions accounted for by the DeBye-Huckel theory, but this is enough to change activities.
- Ion pairs become more important in saline waters and brines, where the interactions must be specifically accounted for using the Pitzer Equations.

Reaction Rate

- The rates of coordination reactions are characterized by the terms labile (very fast reactions) and inert (very slow reactions).
- These terms do not necessarily indicate the stability of the complex, only the kinetics.
- Ligand exchange involving water is usually fast, on the order of seconds.

Complex Stability

- Equilibrium constants for complexes are usually stated for reactions written in the direction of complex formation, that is a stability constant:

  Ligand + central metal ion -----> Complex

- The Keq of the reaction is the stability or formation constant. Large values of stability constants indicate stable complexes. If the equilibrium constant is stated for the dissociation of the complex, it is called a dissociation or and instability constant.

HSAB

- Complex stability can be predicted in a general sense by the HSAB appraoch, or Hard soft acid bases rules. The basic premise is that an individual cation "prefers" to to bond with one type of ligand over another.

Type A metals tend to be sperically symetrical, with intert gas electron configurations, and are known as hard spheres, or hard acids. Examples are:

- Alkalis; Na^+, K^+ (Weakest interaction, rarely considered).
- Alkaline earths; Ca^{+2}, Mg^{2+}.
- Al+3; Th+4; Si+4.

These cations tend to complex F- and ligands containing oxygen:

CO_3^{2}-; OH-; Borate

B-type cations have more easily polarized electron shells, form covalent complexes, and are termed soft acids, e.g.:

Ag^{+}; Zn^{+2}; Hg^{+2}; Pb^{+2}, Sn^{+2}

These metals tend to form complexes with soft ligands containing I, S, P, and N.

NH_4^{+}; S-; PO_3^{-3} (Phosphite)

Within these groups of metal ions complex stability is proportional to the charge of the metal ion - alkali metals are the weakest, and only complex with water, while transition metals complex with a variety of ligands.

Irving Williams Series

Mn2+ < Fe2+ < Co2+ < Ni2+ < Cu2+

- ClO4- and NO3- rarely complex metals, while phosphate, hydroxide, and carbonate are potent complex formers.
- Complexes containing 5 and 6 membered rings are the most stable, while four or 7+ membered rings are less stable.
- Stability increases with the number of attachment points, so multiligand complexes are more stable than monodentate, called the chelate effect.

When a ligand forms complexes with two metal ions at equal concentration in a solution, the ligand will form complexes preferentially with the metal ion that produces the complex with the larger stability constant. Example: EDTA titration of hardness:

$Ca^{2+} + EDTA^{4-}$ <----> $(EDTA\text{-}Ca)^{2-}$: $K = 10^{+10.7}$

$Mg^{2+} + EDTA^{-2}$ <----> $(EDTA\text{-}Mg)^{-2}$: $K = 10^{+8.7}$

Since the stability constant of the calcium complex is significantly larger than the Mg complex, the reaction with calcium proceeds to completion before any reaction with magnesium:

Remember, as metal is complexed its concentration decreases, changing the releation to the equilibrium constant.

When a ligand that forms a very stable complex with a metal ion is added to a solution of the metal ion complexed with another ligand in a less stable complex, the stronger complexing agent (ligand) will tend to decompose the existing complex and withdraw the metal ion into a complex with itself (assuming the weaker complex is labile): Example, EDTA Hardness titration:

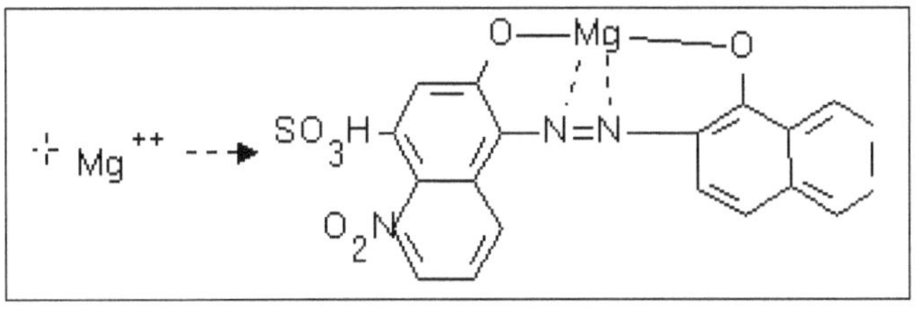

A solution containing both Mg and Ca is dosed with the complexing agent Erichrome Black T (EBT), which forms a complex with Mg2+. The EBT-Mg complex (K = 10+7) is weaker than the EDTA-Mg complex (K=10+8.7).

The color of the EBT-Mg complex is red, and the color of uncomplexedd EBT is blue. A solution with excess EBT is therefore purple. Adding EDTA, the color of the solution will change from Purple to pure blue as the Red EBT-Mg complex is consumed.

Stability Constant Expressions

- For complexes that contain more than one ligand or central metal ion, there are two ways of writing stability constants.
    - Stepwise formation constants are equilibrium constants for the reaction in which the central ion consecutively adds one ligand.

$$Hg^{2+} + Cl^- <\!\!-\!\!-\!\!> HgCl^+; \quad \log K_1 = 7.15$$
$$HgCl^+ + Cl^- <\!\!-\!\!-\!\!-\!\!-\!\!> HgCl_2{}^\circ: \quad \log K_2 = 6.9$$
$$HgCl_2{}^\circ + Cl^- <\!\!-\!\!-\!\!-\!\!-\!\!> HgCl_3{}^- \quad \log K_3 = 2.0$$

Where Ki is the equilibrium constant for the reaction in which the complex containing i ligands is formed from the complex containing i-1 ligands.

- An Overall formation constant is the equilibrium constant for the reaction in which the central metal ion combines with all of the ligands necessary to form a specific complex:

$$Hg^{2+} + Cl^- <\!\!-\!\!-\!\!-\!\!-\!\!> HgCl \quad \log b_1 = 7.15$$
$$Hg^{2+} + 2Cl^- <\!\!-\!\!-\!\!-\!\!-\!\!> HgCl_2 \quad \log b_1 = 14.05$$
$$Hg^{2+} + 3Cl^- <\!\!-\!\!-\!\!-\!\!-\!\!> HgCl_3 \quad \log b_1 = 16.05$$
$$Hg^{2+} + 4Cl^- <\!\!-\!\!-\!\!-\!\!-\!\!> HgCl_4 \quad \log b_1 = 16.75$$

Where bi is the equilibrium constant for the reaction in which the complex containing i ligands is formed from the metal ion and i ligands.

- Either type of constant can be used, but not together. It is probably safer to use the stepwise formation constants, but the books will usually have the overall.
- The general equation for these two then are:

$$K_i = \frac{(ML_i)}{(ML_i)_{i-1}(L)}; \beta = \frac{(ML_i)}{(M)(L)^i}$$

pH Effect

The addition of a protonated ligand introduces the complexity of what to do with the proton, as the ligand usually loses the proton during complexation (the proton often occupies the site of metal chelation...). This produces a more complex form of the equation:

$$K_i = \frac{(ML_i)(H^+)}{(ML_i)_{i-1}(HL)}; \beta = \frac{(ML_i)(H^+)}{(M)(HL)^i}$$

Remember that this reaction is reversible, so if the pH decreases (Proton activity increases), this type of complex will decompose.

- General Rule 1: The tendency of metal-ion solutions to protolyze (hydrolyze) increases with dilution and with decreasing pH.
- General Rule 2: The fraction of polynuclear complexes in a solution decreases with dilution.

Organic Complexes

- These complexes are very difficult to predict, because of the complexity of the natural organic compounds.
- These complexes can act as independent ligands, forming multi-dentate complexes, or they can form "ternary complexes, especially on surfaces".
- Iron, in particular, can be associated with organics as Fe (III), with the apparent solubility of iron exceeding the equilibrium by 3 orders of magnitude. Aluminum as well.

Hydrolysis of Metal Ions

- In aqueous solution, free metal ions are complexed with water (aquated complexes), forming hydration shells. The metal carries the original charge however, since water is a neutral ligand that does not have to deprotonate to chelate.
- The interaction of these hydrated ions with acids and bases is a ligand exchange reaction called hydrolysis.

$$Al(H_2O)_6^{+3} + H_2O - \ <----> \ Al(H_2O)_5OH^{2+} + H_3O^+$$

- So in a stepwise replacement of water molecules with hydroxyls, the loss of a proton into solution means that aluminum acts as both a Lewis and Bronsted acid.
- Hydrolysis is when water as a reactant is split up into proton and hydroxide, vs hydration, where water maintains itself. A similar reaction would be:

$$Al(H_2O)_6^{+3} + OH^- <----> \ Al(H_2O)_5OH^{2+} + H_2O$$

- So the complete reaction sequence of aluminum in water would be:

$$Al\left(H_2O\right)_6^{+3} + H_2O \ <----> \ Al\left(H_2O\right)_5 OH^{2+} + H_3O^+ \, K_{1,Al} =$$

$$Al(H_2O)_5 OH^{2+} + H_2O <----> Al(H_2O)_4 (OH)_2^{\ +} + H_3O^+ K_{2,Al} =$$

$$Al(H_2O)_4 (OH)_2^{\ +} + H_2O <----> Al(H_2O)_3 (OH)_3 + H_3O^+ K_{3,Al} =$$

$$Al(H_2O)_3 (OH)_3^{\ -} + H_2O <----> *Al(OH)_4^{\ -} + H_3O^+ K_{4,Al} =$$

There is some evidence that Al (OH) 4- assumes a tetrahedral geometry, without additional waters of hydration.

Magnetochemistry

Magnetochemistry is the study of the magnetic properties of materials. By "magnetic properties" we mean not only whether a material will make a good bar magnet, but whether it will be attracted or repelled by a magnet. This includes synthesis, analysis and understanding. This short description is meant to give a basic understanding before you delve into a more complex treatment.

Magnetism arises from moving charges, such as an electric current in a coil of wire. In a material which does not have a current present, there are still magnetic interactions. Atoms are made of charged particles (protons and electrons) which are moving constantly. The processes which create magnetic fields in an atom are

- Nuclear spin. Some nuclei, such as a hydrogen atom, have a net spin which creates a magnetic field.
- Electron spin. An electron has two intrinsic spin states (similar to a top spinning) which we call up and down or alpha and beta.
- Electron orbital motion. There is a magnetic field due to the electron moving around the nucleus.

Each of these magnetic fields interacts with one another and with external magnetic fields. However, some of these interactions are strong and others are negligible.

Measurement of interactions with nuclear spins are used to analyze compounds in nuclear magnetic resonance (NMR) and electron spin resonance (ESR) spectroscopy. In most other situations, interaction with nuclear spins is a very minor effect.

Interactions between the intrinsic spin of one electron and the intrinsic spin of another electron are strongest for very heavy elements such as the actinides. This is called spin-spin coupling. For these elements this coupling can shift the electron orbital energy levels.

The interaction between an electron's intrinsic spin and it's orbital motion is called spin-orbit coupling. Spin-orbit coupling has a significant effect on the energy levels of the orbitals in many inorganic compounds.

Macroscopic effects, such as the attraction of a piece of iron to a bar magnet are primarily due to the number of unpaired electrons in the compound and their arrangement. The various possible cases are called magnetic states of matter.

Magnetic States of Matter

- Diamagnetic - A diamagnetic compound has all of it's electron spins paired giving a net spin of zero. Diamagnetic compounds are weakly repelled by a magnet.
- Paramagnet - A paramagnetic compound will have some electrons with unpaired spins. Paramagnetic compounds are attracted by a magnet.
- Ferromagnet - In a ferromagnetic substance there are unpaired electron spins, which are held in alignment by a process, known as ferromagnetic coupling. Ferromagnetic compounds, such as iron, are strongly attracted to magnets.
- Ferrimagnet - Ferrimagnetic compounds have unpaired electron spins, which are held in an pattern with some up and some down. This is known as ferrimagnetic coupling. In a ferrimagnetic compound, there are more spins held in one direction, so the compound is attracted to a magnet.
- Antiferromagnetic - When unpaired electrons are held in an alignment with an equal number of spins in each direction, the substance is strongly repelled by a magnet. This is referred to as an antiferromagnet.
- Superconductor - Superconductors are repelled by magnetic fields because the magnetic field is excluded from passing through them. This property of superconductors, called the Meissner effect, is used to test for the presence of a superconducting state. The underlying theory of how superconductivity arises is still a matter of much research and debate at the time of this writing. It does appear that the mechanism behind the magnetic properties of superconductors is significantly different from the other classes of compounds discussed here.

Interaction with an External Magnetic Field

A magnetic field is given the symbol "H" which is a vector since the field has both a direction and a magnitude. We will consider only interactions in one dimension making H and many other quantities we will define scalars. This gives us results for a homogeneous magnetic field and is a very good approximation for the way that most magnetic property measurements are performed. The magnitude of the magnetic field is usually given in units of gauss (G) or tesla (T) where 1 tesla = 10000 gauss.

When a material is placed in a magnetic field, the magnetic field inside the material will be the sum of the external magnetic field and the magnetic field generated by the material itself. The magnetic field in a material is called the magnetic induction and given the symbol "B". The formula for this is:

B = H + 4 Pi M

Where,

B = magnetic induction

H = external magnetic field

Pi = 3.14159

M = magnetization (a property of the material)

For mathematical and experimental convenience this equation if often written as:

$$\frac{B}{H} = 1 + \frac{4 \text{ Pi } M}{H} = 1 + 4 \text{ Pi Xv}$$

Where,

Xv = M/H = volume magnetic susceptibility

The volume magnetic susceptibility is so named because B, H and M are defined per unit volume. However this results in Xv being unitless. It is convenient to use the magnetic susceptibility instead of the magnetization because the magnetic susceptibility is independent of the magnitude of the external magnetic field, H, for diamagnetic and paramagnetic materials.

Many studies are done using Xg, magnetic susceptibility per gram, which is Xv divided by the density. This gives units of cm cubed per gram. Another useful form is Xm, molar magnetic susceptibility, which is Xg times the molecular weight. This gives units of cm cubed per mole. Another measure of magnetic interaction that is often used is an effective magnetic moment, mu, where,

mu = 2.828 (Xm T)$^{1/2}$

where,

mu = effective magnetic moment

Xm = molar magnetic susceptibility

T = temperature

The numeric factor puts mu in units of Bohr magnetons (BM). Where one BM equals 9.274 x 10^-24 joules per tesla. The effective magnetic moment is a convenient measure of a material's magnetic properties because it is independent of temperature as well as external field strength for diamagnetic and paramagnetic materials. This said, we would now like to examine how the magnetization, magnetic susceptibility and effective magnetic moment depend on molecular structure.

Diamagnetism

Diamagnetism can be described by electrons forming circular currents, orbiting the nucleus, in the presence of a magnetic field. As such, a diamagnetic contribution can be calculated for any atom. However, the magnitude of the diamagnetic contribution is so much smaller than the magnitude of paramagnetic and other effects that it is usually ignored for any other type of materials.

In this orbital model, the diamagnetic susceptibility from a given electron is proportional to the square of its mean distance from the nucleus. Thus larger atoms are expected to have a larger diamagnetic interaction than smaller atoms. Often, the contributions for common atoms are tabulated along with corrections for multiple bonds. Thus a magnetic susceptibility can be predicted merely by adding together the contributions from all of the atoms and bonds in the molecule.

Paramagnetism

The structural feature most prominent in determining paramagnetic behavior is the number of unpaired electrons in the compound. A spin only formula for the magnetic moment of a paramagnetic compound is:

$$mu = g\{ S (S + 1) \}^{1/2}$$

where,

mu = effective magnetic moment

g = 2.0023

S = 1/2 for one unpaired electron

1 for two unpaired electrons

3/2 for three unpaired electrons, etc.

This equation is sometimes written with g=2. This does not introduce a significant error since this simple spin only treatment is a decent approximation but is often not accurate even to two significant digits. An equation which takes into account both spin and orbital motion of the electrons is,

$$mu = \{ 4 S (S + 1) + L (L + 1) \}^{1/2}$$

where,

mu = effective magnetic moment

S = 1/2 for one unpaired electron, 1 for two, etc.

L = total orbital angular momentum

This equation is derived for atoms. It is applicable only to molecules with very high symmetry where the energies of the orbitals containing unpaired electrons are degenerate. If the splitting of orbital energy levels is large relative to k T (k is the Boltzman constant) then the applicable formula is,

$$mu = g\{ J (J + 1) \}^{1/2}$$

Where,

$$g = 1 + \frac{S (S + 1) - L (L + 1) + J (J + 1)}{2 J (J + 1)}$$

Where,

$$J = S + L$$

This formula is usually used for the lanthanide and actinide elements. If the splitting of orbital energy levels is comparable in magnitude to k T then the expression for magnetic properties must incorporate a Boltzman distribution. This is often the case with high spin transition metal complexes. The worst case, where this procedure is absolutely imperative, is the description of spin cross overs such as exhibited by some iron coordination compounds. Examples of this type of treatment are given in both the Drago and Selwood texts.

For all of the cases of paramagnetic behavior the spin only formula is often used as a first rough approximation. If the only purpose for measuring the magnetic susceptibility is to determine the number of unpaired electrons this is often all that is done.

Ferromagnetism, Antiferromagnetism and Ferrimagnetism

The advantage of using effective magnetic moments for describing paramagnetic behavior is that it is a measure of the materials magnetic behavior which is not dependent upon either the temperature or the magnitude of the external field. It is not possible to set up such a convention for ferromagnetic, antiferromagnetic and ferrimagnetic materials.

All three of these classes of materials can be considered a special case of paramagnetic behavior. The description of paramagnetic behavior is based on the assumption that every molecule behaves independently. The materials discussed here result from a situation in which the direction of the magnetic field produced by one molecule is affected by the direction of the magnetic field produced by an adjacent molecule, in other words their behavior is coupled. If this occurs in a way in which the magnetic fields all tend to align in the same direction, a ferromagnetic material results and the phenomenon is called ferromagnetic coupling. Antiferromagnetic coupling gives an equal number of magnetic fields in opposite directions. Ferrimagnetic coupling gives magnetic fields in two opposite orientations with more in one direction than in the other.

With a few exceptions, the magnetic moments are not aligned throughout the entire material. Typically regions, called domains, will form with different orientations. The existence of domains of coupled molecules gives rise to a number of types of behavior as described in the following paragraphs.

The tendency of molecules to align themselves to one another enhances the magnetization of the material due to the presence of an external magnetic field. This is why ferromagnetic and ferrimagnetic materials can have magnetic susceptibilities several orders of magnitude large than paramagnetic materials. This also gives rise to the fact that the magnetic susceptibility of these materials is not independent of the magnitude of the external magnetic field as was the case for diamagnetic and paramagnetic materials.

For a ferromagnetic material, the actual field acting on a given magnetic dipole (unpaired electron) is designated Ht and given by an equation similar to the equation for magnetic induction given above.

$$Ht = H + Nw\ M$$

where,

Ht = magnetic field felt by an electron

H = external magnetic field

Nw = molecular field constant, approximately 10000

M = magnetization

This equation is used because it allows a mathematical treatment of a ferromagnetic substance similar to that used for paramagnetic substances. In this form the molecular field constant, Nw, is typically defined empirically in order to take the ferromagnetic coupling into account. To obtain the molecular field constant in a rigorous way would require a quantum mechanical calculation that takes into account the elements, their arrangement in the solid, kinetic energy of the electrons, coulombic attraction of electrons to the nucleus and repulsion with other electrons as well as spin interactions. What is most often done is a computer simulation using the Ising model, which is not truly rigorous but is based on quantum mechanics. This is a spin only quantum mechanical treatment assuming that the values of neighboring spins can be replaced by their average over time.

Vibrational motion of the molecules, which increases with temperature, can disrupt the domain structure. Thus the magnetic properties of all three of these types of materials are strongest at low temperatures. At sufficiently high temperatures, no domain structure is able to form so all of these materials become paramagnetic at high temperatures. The temperature at which paramagnetic behavior is seen called the Curie temperature for ferromagnetic and ferrimagnetic materials and called the Neel temperature for antiferromagnetic materials. This is why a temperature independent effective magnetic moment cannot be defined for these materials.

The alignment of the magnetic moments of the domains may give the material a net magnetic moment even in the absence of an external field. This gives a permanent magnet, such as a bar magnet. A material with no net moment prior to being exposed to an external magnetic field may retain a net moment after being exposed to an external magnetic field. This is how cassette and video tapes and computer disks store information. The magnitude of this memory effect can be quantified by plotting magnetization vs field strength as the external field intensity is varied from one polarity to the other and back again. A strong memory effect will be indicated by a wide hysteresis loop.

Over a period of time magnetic domains tend to return to a random orientation. This makes the kinetics of this relaxation process another factor in the magnetic behavior of these materials. This is also responsible for the limited life span of magnetically stored music, video and computer data.

Variation with Temperature

The source of variation of magnetic properties with temperature is the disruption of the alignment of molecular magnetic moments due to the thermal motion of the atoms. As such, it should come as no surprise that diamagnetic behavior shows no variation with temperature.

Paramagnetism

As temperature increases, the magnetic susceptibility of a paramagnetic substance decreases. In some paramagnetic compounds the magnetic susceptibility is inversely proportional to the temperature. These are called “normal paramagnets” and have magnetic properties arising primarily due to the presence of permanent magnetic dipoles. This is referred to as the Curie Law and is expressed in mathematical form as:

$$X = C / T$$

Where,

$$C = \frac{Na\, g^2\, b^2}{4\,k}$$

Where,

X = magnetic susceptibility

C = the Curie constant

T = temperature

Na = Avogadro's number

g = the electron g factor

b = the Bohr magneton

k = the Boltzman constant

In most paramagnetic compounds, an inverse relationship is observed, but the extrapolation to zero temperature does not obey the Curie Law. These compounds obey the Curie-Weiss Law which is:

$$X = \frac{C}{T - \text{theta}}$$

Where theta is a constant referred to as the Weiss constant. The Weiss constant can have a large range of values from -70 K to 3000 K. Most often it is positive.

Ferromagnetism & Ferrimagnetism

Ferromagnetic and ferrimagnetic compounds also show a decrease in magnetic susceptibility with increasing temperature. However, a plot of magnetic susceptibility vs. temperature shows a different line shape for these compounds than for paramagnetic compounds. This plot would have a positive curvature for paramagnetic compounds and a negative curvature for ferromagnetic compounds. A rough sketch of the shapes of these curves is as follows:

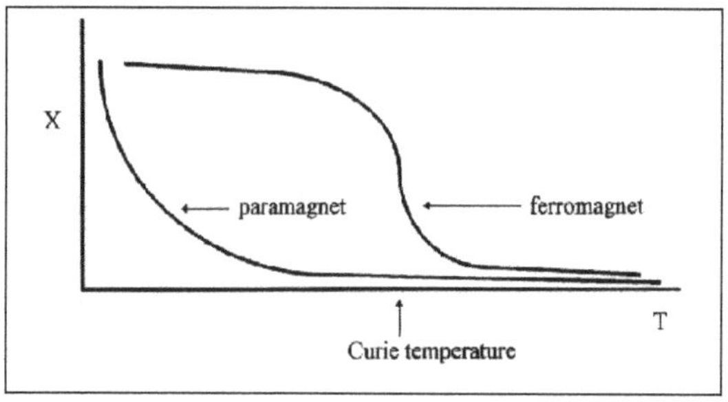

When a critical temperature (called the Curie temperature) is reached, the curvature of the plot changes. At the Curie temperature, ferromagnetic and ferrimagnetic compounds become paramagnetic. Curie temperatures range from 16 C for Gd to 1131 C for Co.

For ferromagnetic substances a universal temperature curve can be constructed, meaning that all substances with the same total spin follow the same curve. This is done by plotting M(T)/M vs T/Tc where M(T) is the magnetization at a given temperature, M is the magnetization at absolute zero, T is the temperature and Tc is the Curie temperature. For more information, see Morrish.

Antiferromagnetism

Antiferromagnetic compounds show an increase in magnetic susceptibility until their critical temperature, called the Neel temperature, is reached. Above the Neel temperature these compounds also become paramagnetic. Neel temperature range from 1.66 K for $MnCl_2*4H_2O$ to 953 K for alpha-Fe_2O_3.

As with ferromagnetic substances, a universal temperature curve can be constructed that all substances with the same number of unpaired electrons follow. This is done by plotting X(T)/X(Tn) vs T/Tn where X(T) is the magnetic susceptibility at a given temperature, X(Tn) is the magnetic susceptibility at the Neel temperature, T is the temperature and Tn is the Neel temperature.

Organolanthanide Chemistry

An organolanthanide is a type of organometallic chemical compound in which the metal is of the lanthanide series in the periodic table. Presently, the term lanthanide is used to describe elements 57-71 (lanthanum to lutetium inclusively). "Ln" is widely used in chemical formulas to stand for any of these elements. Although both lanthanide and lanthanoid terminology has been used in the past, IUPAC currently recommends lanthanoid. An organolanthanoid consists of a lanthanioid bonded to at least one carbon atom through either single sigma-bonds and/or multiple pi-bonds. In contrast to d-block metals, organolanthanoids do not form complexes with CO under normal conditions and are also usually air- and moisture-sensitive. As with most organometallic compounds, organolanthanoids are able to act as effective catalysts for a variety of organic tranformations such as: hydrogenation, hydrosilylation, hydroboration, hydroamination reactions and the cyclization/polymerization of alkenes.

Organolanthanides as Catalysts

The availability of a range of different lanthanoid metals, coupled with a large amount of different ligands, provides an effective way to systematically alter the properties of the organometallic complex. This helps to control their catalytic behaviour, such as chemoselectivity, enantioselectivity and diasterioselectivity. Reactions involving organolanthanoid catalysts involve mild inert reaction conditions. The presence of an (n5-C5R5) ligand on an organolanthanoid complex is a common feature. However, when R = H, the complex tends to be poorly soluble in hydrocarbon solvents and the catalytic activity is generally low. Since hydrocarbon solvents are generally used for catalytic reactions and are able to bind to the Ln3+ center, bulkier R groups are required to increase the catalytic value of the complex. This may affect the catalytic activity due to the steric demands of the ligands, which can hinder the association of the lanthanoid with the desired organic substrate, but also allows specificity to be built into the catalyst. One effective strategy to optimize a catalyst is to tilt the angle between two (n5-C5R5) groups by attaching them together.

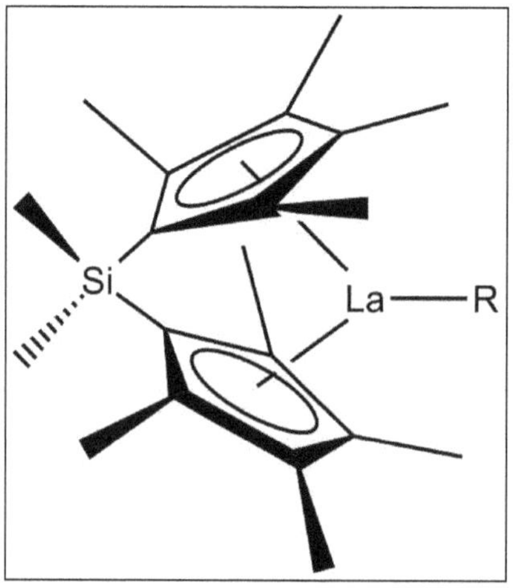

Example of increase the catalytic activity of an organolanthanide complex by tilting the cyclopentapentadienyl ligands.

Organolanthanides as Catalysts for the Polymerization of α-olefins

Group 3 organometallic and organolanthanide compounds are among the most active known catalysts for the Ziegler-Natta type polymerization processes. The absense of a rigid stereo-chemical arrangement around the lanthanide allows for the ligands to determine their stereochemistry; thus, many catalytic processes involving organolanthanides are highly stereospecific. The most active of these catalysts are those containing cyclopentadienyl ligands (Cp2MX2). These complexes are extremely active homogeneous ethylene polymerization catalysts with turnover frequencies exceeding 1800 S-1 (25 °C, 1 atm of ethylene pressure) for M = La. Lanthanide cyclopentadienyl sandwich complexes can be modified by substituting different lanthanide metals (due to their varied atomic size) or modifying the ligands to allow olefins better access to the catalytic center. Ligand modification can involve changing the substituents on the Cp, bridging the Cp's, or tethering a Cp to create a constrained geometry ligand system.

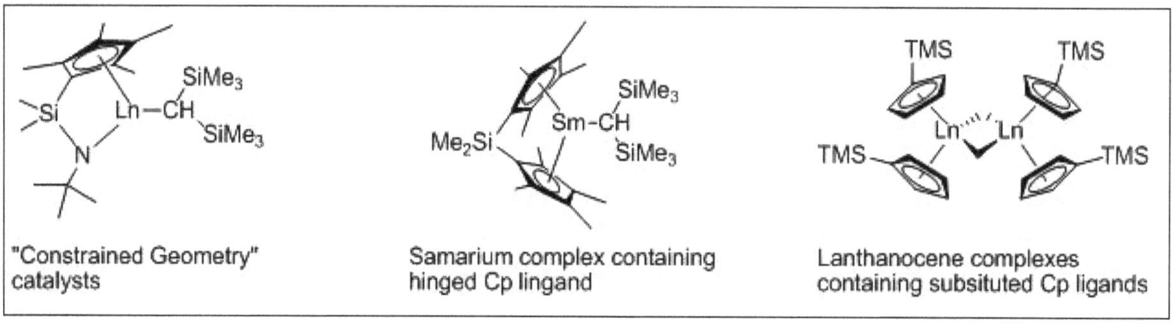

Examples of homogeneous catalysts for polymerization.

Organolanthanides as Catalysts for other for other Reactions

Selective Hydrogenation

Organolanthanides have been shown to be selective hydrogenation catalysts. Catalysts with the formula Cp2L2Ln are selective towards the hydrogenation of terminal alkenes. Recent work with these catalysts have been able to achieve enantioselective olefin hydrogenation.

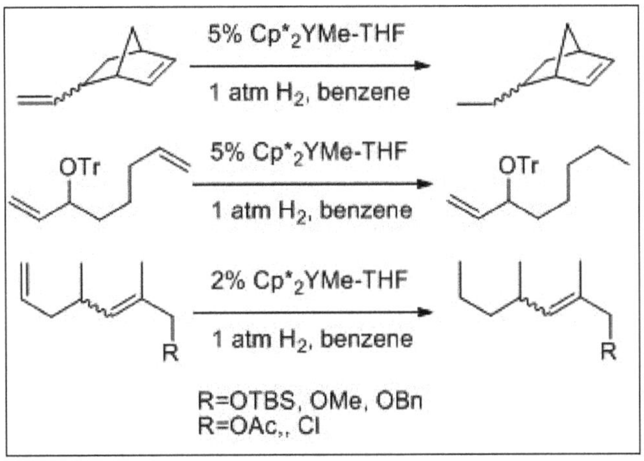

Examples of Cp*YMe-THF catalysts, reagents and products in Selective Hydrogenation.

Despite their activity and selectivity, these catalysts can be costly and complicated to prepare and handle, making them less attractive than many alternative transition-metal-based catalysts.

The mechanism for the hydrogenation of olefins begins with a sigma-bond metathesis of R' group of the precursor with H2, releasing R'H and the active catalyst. The olefin then coordinates to the catalyst through an agostic interaction, followed by the 1, 2-insertion of hydrogen onto the olefin. The product is then released from the catalyst after undergoing another sigma-bond methathesis with hydrogen, regenerating the active catalyst.

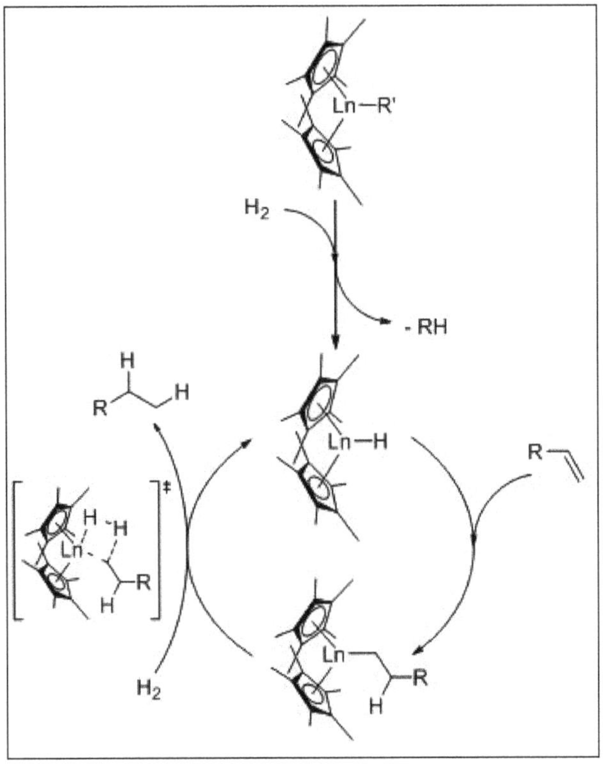

Catalytic cycle of olefin hydrogenation.

Hydroamination/Cyclization

A reaction which has generated much interest is the bonding of an amine onto an unsaturated carbon-carbon bond. This reaction is very useful in organic synthesis as it is atom economical, can be done inter- and/or intra-molecularly and, depending on the choice of amine and unsaturated bond, can infer chirality into the molecule. After many attempts at producing a transition metal catalyst for this reaction, success was found with lanthanide and actinide catalysts of the form Cp2LnR (where R= H, CH2SiMe3; Ln = La, Sm, YNd, Lu). These catalysts were found to be very kinetically labile, having high electrophilicity, high turnover frequencies, high stereoselectivtity and diastereoselectivity.

Since these metal centers only have one stable oxidation state (3+), they usually only undergo two types of reactions: olefin insertions and sigma-bond metathesis. The single oxidation state prevents other reaction types such as reductive elimination and oxidative addition. The catalyst precursor undergoes a sigma-bond metathesis with H2NR', transfering a hydrogen from the amine to R' while simultaneously forming a L-N sigma bond, the product of this substitution is the active catalyst in the hydroamination/cyclization cycle. The rate limiting formation of an agostic bond between the unsaturated bond and the metal center, is then followed by a 1,2-insertion of the amine across the unsaturated bond. In the case of intramolecular olefin insertion, this is the cyclization step. The product is then released via another sigma-bond metathesis with the starting substrate and the cycle repeats.

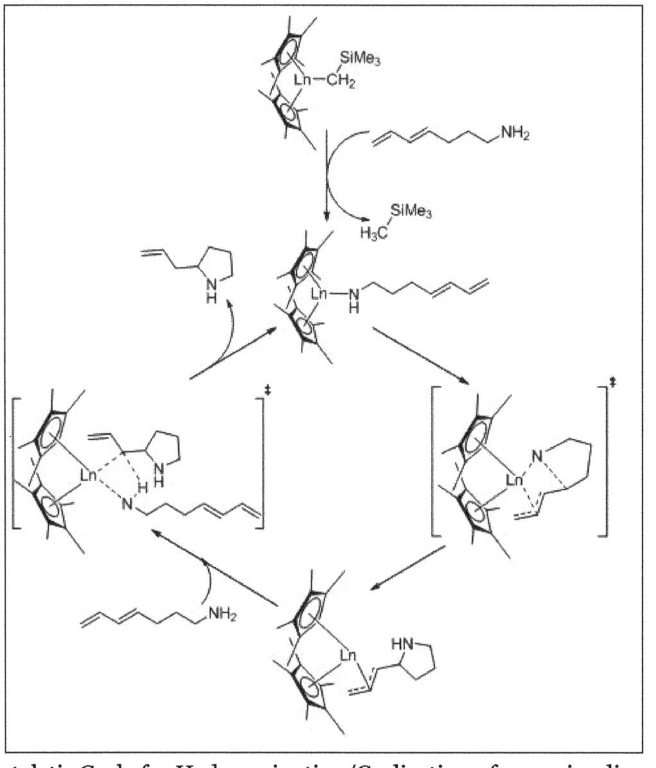

Catalytic Cycle for Hydroamination/Cyclization of an aminodiene.

In the case where R' is an alkene, the molecule can undergo an intramolecular amination reaction; further studies have found that when R' is an diene, allene of alkyne the reaction proceeds becomes more exergonic with a quicker with a higher turnover number.

Other Catalytic Activity

There are other less thoroughly investigated instances of lanthanoid catalysis. These include the hydrosilylation of alkynes, catalysis in the Mukaiyama reaction, the hydroboration of olefins, and hydrogenation of imines. The mechanisms for these reactions follow the same general principles as previously observed. Sigma-bond metathesis of the precursor to the activated form Cp*[2]LnH followed by the coordination of the unsaturated bond. This is proceeded by a 1,2-insertion of H, then release of product through another sigma-bond metathesis to regnerate the active catalyst.

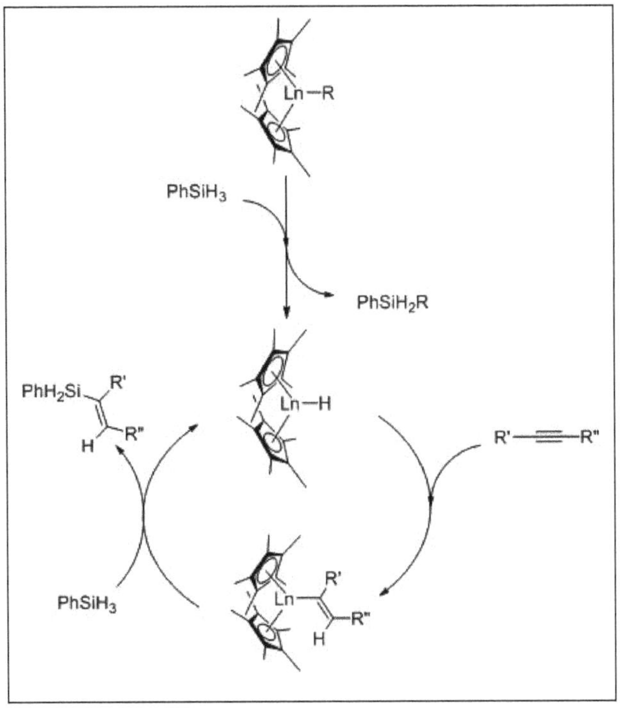

Typical Catalytic Cycle for Hydrosilylation: Hydroboration also undergoes similar process.

Organolanthanides in Materials Chemistry

Semi-Conductors

Lanthanide metals are suitable doping agents for III-V semiconductors such as semi-conductor lasers, optical amplifiers and light emitting diodes. Since all metals in the lanthanide series typically carry a +3 charge, this offers tunability with regards to metal choice, since all of the lanthanides will be suitable +3 doping agents. Each lanthanide metal varies in atomic size which tunes for incorporation in solid semi-conductor lattices. Each lanthanide also varies in its optical transmission spectrum which tunes for the optical properties of absorption or emission.

Incorporating lanthanides into semi-conductors presents an issue. The synthetic routes require the starting materials to be liquids or gases for a homogeneous distribution throughout the product material. Organolanthanide complexes are known to be highly air-sensitive, thermally very robust high-melting materials. High melting points in the range of 150-350 °C are common, and the design of volatile, low-melting precursors is a synthetic challenge.

The solution was found using organolanthanide complexes such as tris(cyclopentadienyl) lanthanides. These complexes are very volatile, allowing gas-phase incorporation into the semi-conductor. These compounds were first synthesized from LnCl3 and Na [C_5H_5] via a salt metathesis but other methods and derivatives are also possible. Other appropriate precursor compounds include mixed-sandwich lanthanide complexes containing n^5 cyclopentadienyl and n^8 cyclooctatetraenyl rings. These compounds also exhibit high volatility.

Magnetic Semi-Conductors

Rare earth monochalcogenides LnE (Ln = Yb, Eu, Sm; E = S, Se, Te) have interesting magnetic properties. These materials are normally obtained by high-temperature syntheses from the correspondent elements, but in this case the materials often contain impurities which are readily incorporated due to the high oxophilicity of the rare earth center. The use of organolanthanides could enable us to obtain the LnE materials with required purity by a relative low temperature synthesis.

2

Types of Inorganic Compounds

Substances that do not contain both carbon and hydrogen are called inorganic compounds. They lack a carbon-hydrogen bond. Inorganic compounds can be further divided into various groups like inorganic carbon compounds, inorganic chlorine compounds, inorganic nitrogen compounds, inorganic phosphorus compounds, etc. The diverse aspects of inorganic compounds have been thoroughly discussed in this chapter.

Inorganic compounds are typically compounds without carbon atoms. H_2O, O_2, and NaCl are examples of inorganic compounds. Water is the most abundant substance in the body. Its abundance is due partly to its unique chemical properties created by the influence of its hydrogen bonds. These properties include the following:

- Solvency: Water is an excellent solvent. Ionic substances are soluble in water (they dissolve) because the poles of the polar water molecules pull them apart, forming ions. Polar covalent substances are also water-soluble because they share the same hydrogen bonding as water shares with itself. For this reason, polar covalent substances are called hydrophilic (water loving). Because they lack charged poles, nonpolar covalent substances do not dissolve in water and are called hydrophobic (water fearing).
- Cohesion: Because water molecules are held together by hydrogen bonds, water molecules have a high degree of cohesion, or the ability to stick together. As a result, water has strong surface tension. This tension, in turn, gives water strong capillary action, allowing water to creep up narrow tubing. These qualities contribute to the movement of water through capillaries.
- Stability: The temperature of water is stable. You must add a relatively large amount of energy to warm (and boil) it and remove a large amount of energy to cool (and freeze) it. So, when sweat evaporates from your forehead, a large amount of heat is taken with it and you are cooled.

Types of Inorganic Compounds

A chemical compound is a substance consisting of atoms or ions of two or more elements that are chemically bonded together whereas a chemical element is a substance of only one type of atom. Most elements are inorganic but technically are not inorganic compound since they are comprised of only one type of atom. Thus, the classification of inorganic compounds entails grouping of substances comprised of more than one type of atom. Conversely, simple inorganic substances (not necessarily compounds) are typified as either metal or non-metal. However, there is no clear distinction between metals and non-metals.

Most inorganic compounds are ionic compounds. This means that the chemical bond that holds the atoms together is an ionic bond. Based on inorganic compound constituents, ionic compounds could be classified into bases, acids, and salts. The ionic bond is the bond where there is a complete transfer of an electron from one atom to another. It is an electrostatic attraction between oppositely charged ions, i.e. cation and anion. A cation is a positively charged ion whereas an anion is a negatively charged ion. For instance, sodium chloride is an ionic compound where the cation Na^+ and the anion Cl^- are held together by an ionic bond. An ionic compound that has hydrogen ions (H^+) is classified as an acid. Conversely, an ionic compound that has hydroxide (OH^-) or oxide (O_2^-) is classified as a base. An ionic compound formed by acid-base reactions and without those ions is called a salt.

Water is definitely one of the most important inorganic compounds to all living things. It is a compound comprised of two hydrogen atoms and one oxygen atom. However, it is not an ionic compound but a molecule held by a covalent bonding between hydrogen and oxygen.

Inorganic Carbon Compounds

Carbides

In chemistry, a carbide is a compound of carbon with a less electronegative element. Many carbides are important industrially; for example, calcium carbide (CaC_2) is a feedstock for the chemical industry and iron carbide (Fe_3C, cementite) is formed in steels to improve their properties. In general, carbides are classified according to the chemical bonding in the compounds, as follows:

- Salt-like ionic carbides;
- Covalent carbides;
- Interstitial carbides;
- "Intermediate" transition metal carbides. (In bonding terms, they sit between the salt-like and interstitial carbides.)

Calcium carbide.

Ionic Carbides

Salt-like, ionic carbides are formed by the following metals:

- Group one metals (the alkali metals);

- Group two metals (the alkaline earths);
- Group three metals (scandium, yttrium, and lanthanum);
- Group 11 metals (copper, silver, and gold);
- Group 12 metals (zinc, cadmium, and mercury);
- Only aluminum from group 13, (gallium, indium and thallium do not appear to form carbides);
- Lanthanides, when forming MC_2 and M_2C_3 carbides (where M is the metal);
- Actinides, when forming MC_2 and M2C_3 carbides.

Most commonly, they are salts of C_2^{2-} and are called acetylides, ethynides, acetylenediides, or (rarely) percarbides. Some ionic carbides contain other anionic species, such as:

- C^{4-}, sometimes called methanides (or methides) because they hydrolyze to give methane gas;
- C_3^{4-} ion, sometimes called sesquicarbides, which hydrolyze to give methylacetylene.

The naming of ionic carbides is not consistent and can be quite confusing.

Acetylides

Acetylides contain the polyatomic ion $C_{22}-$, in which there is a triple bond between the two carbon atoms (similar to acetylene). Examples are carbides of the alkali metals (such as Na_2C_2), some alkaline earths (such as CaC_2) and lanthanoids (such as LaC_2). The C-C bond distance ranges from 109.2 picometers (pm) in CaC_2 (similar to acetylene), to 130.3 pm in LaC_2 and 134pm in UC_2.

Methanides

Methanides contain the monatomic ion C^{4-}. Examples of methanides are Be2C and Al_4C_3. The C^{4-} ion is a very strong base and will combine with four protons to form methane. The reaction may be written as follows:

$$C^{4-} + 4H^+ \rightarrow CH_4$$

Methanides commonly react with water to form methane, but reactions with other substances are also common.

Sesquicarbides

The polyatomic ion C_3^{4-} is found in, for instance, Li_4C_3 and Mg_2C_3. The ion is linear and isoelectronic with CO_2. The C-C distance in Mg_2C_3 is 133.2 pm. Hydrolysis of Mg_2C_3 yields methylacetylene (CH_3CCH), which was the first indication that it may contain C_3^{4-}.

Covalent Carbides

Silicon and boron form covalent carbides. Silicon carbide (SiC, or carborundum) has two similar crystalline forms, both of which are related to the structure of diamond. Boron carbide (B_4C), on the other hand, has an unusual structure, which includes icosahedral boron units linked by carbon atoms. In this respect, boron carbide is similar to the boron-rich borides. Both silicon carbide and boron carbide are very hard, refractory (heat-resistant) materials. Both are important industrially. Boron also forms other covalent carbides, such as B_25_C.

Interstitial Carbides

Properties

The carbides of transition metals of groups four, five, and six (with the exception of chromium) are often described as interstitial compounds. They are chemically quite inert, have metallic properties, and are refractory. Some exhibit a range of stoichiometries (that is, the metal and carbon atoms can be combined in different proportions). Titanium carbide and tungsten carbide are important industrially and are used to coat metals in cutting tools.

Structure

The carbon atoms are thought to fit into octahedral interstices in the metal lattice, when the metal atom radius is greater than 135 pm. If the metal atoms are cubic close packed (face-centered cubic), then eventually all the interstices could be filled to give a 1:1 stoichiometry, with the rock salt structure, as in the case of tungsten carbide (WC). When the metal atoms are hexagonal close packed, then only half the interstices are filled, giving a stoichiometry of 2:1, as in the case of divanadium carbide (V_2C).

The following table shows actual structures of metals and their carbides. The notation "h/2" refers to the V_2C-type structure mentioned above, which is an approximate description of the actual structures. The simple view that the lattice of the pure metal "absorbs" carbon atoms is true only for the monocarbides of vanadium (VC) and niobium (NbC).

Metal	Structure	Metallic radius (pm)	MC structure	M_2C structure	Other carbides
titanium	hexagonal	147	rock salt		
zirconium	hexagonal	160	rock salt		
hafnium	hexagonal	159	rock salt		
vanadium	cubic body centered	134	rock salt	h/2	V_4C_3
niobium	cubic body centered	146	rock salt	h/2	Nb_4C_3
tantalum	cubic body centered	146	rock salt	h/2	Ta_4C_3
chromium	cubic body centered	128			$Cr_{23}C_6$, Cr_3C, Cr_7C_3, Cr_3C_2
molybdenum	cubic body centered	139	hexagonal	h/2	Mo_3C_2
tungsten	cubic body centered	139	hexagonal	h/2	

For a long time, the non-stoichiometric phases were believed to be disordered, with a random filling of the interstices. However, short and longer range ordering has been detected.

Intermediate Transition Metal Carbides

In these carbides, the transition metal ion is smaller than the critical 135 pm, and the structures are not interstitial but are more complex. Multiple stoichiometries are common. For example, iron forms a number of carbides: Fe_3C, Fe_7C_3, and Fe_2C. The best-known of these is cementite (Fe3C), which is present in steels.

These carbides are more reactive than the interstitial carbides. For example, the carbides of Cr, Mn, Fe, Co, and Ni are all hydrolyzed by dilute acids and sometimes by water, to give a mixture of hydrogen and hydrocarbons. These compounds share features with both the inert interstitials and the more reactive, salt-like carbides.

Some Carbide and their uses

- Calcium carbide (CaC_2): It is used for the production of acetylene and calcium cyanamide ($CaCN_2$), which is used in fertilizers. It is also important for the desulfurization of iron.
- Silicon carbide (SiC) or carborundum: It is used as an abrasive and in ceramics.
- Tungsten carbide (often called simply carbide): Being extremely hard, it is useful for the manufacture of cutting tools, abrasives, and bearings.
- Cementite (iron carbide; Fe_3C): It is an important constituent of steel.
- Boron carbide: It is used in cutting tools, dies, and anti-ballistic armor plating. It is an effective absorbent of neutron radiation in nuclear power plants.
- Tantalum carbide: It is an extremely hard, refractory, ceramic material, used in tool bits for cutting tools.
- Titanium carbide: Similar to tantalum carbide, it is an extremely hard, refractory, ceramic material.

Carbon Dioxide

Carbon dioxide (CO_2) is a naturally occurring gas, important to the carbon cycle for life and a by-product of many forms of energy production. It is also a greenhouse gas.

Chemical formula	CO_2
Molar mass	44.01 grams/mole

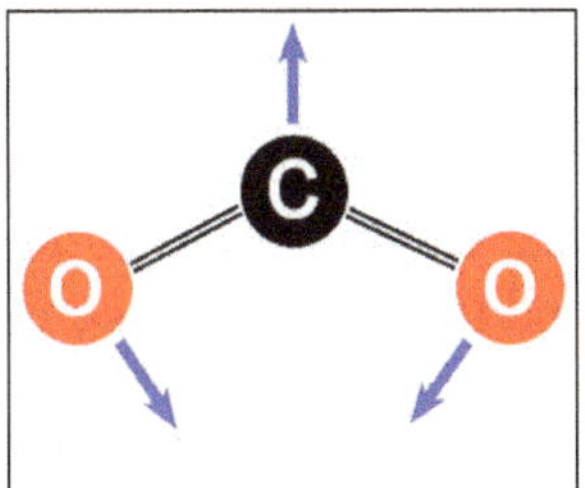

Carbon dioxide molecule. The arrows show how the molecule vibrates when it absorbs energy.

Carbon dioxide with its vibration modes for absorbing infrared radiation.

Carbon dioxide is taken out of the atmosphere by photosynthesis in plants to produce energy, and chemical uptake by the ocean (leading to ocean acidification). The total amount of carbon dioxide in the atmosphere has historically been in balance with the respiration of animals and plants. However, recently there has been a huge net increase in the levels of atmospheric and oceanic CO_2. This is mainly due to humans digging carbon (in the form of fossil fuels) out of the ground and releasing it into the atmosphere. These fuels undergo combustion—which produces carbon dioxide as a byproduct. Chemically, carbon dioxide is one of the two main outputs in any form of combustion (the other being water).

Global Warming Potential

The global warming potential (GWP) scale is a measurement of how much heat a greenhouse gas traps in the atmosphere. CO_2 is used as a benchmark for this scale; hence carbon dioxide has a GWP value of 1. It is not easy to calculate how long carbon dioxide remains in the atmosphere after emission.

Carbon dioxide lets visible light through but its ability to vibrate on a molecular scale causes CO_2 to absorb infrared radiation. This infrared radiation is what causes increased temperatures.

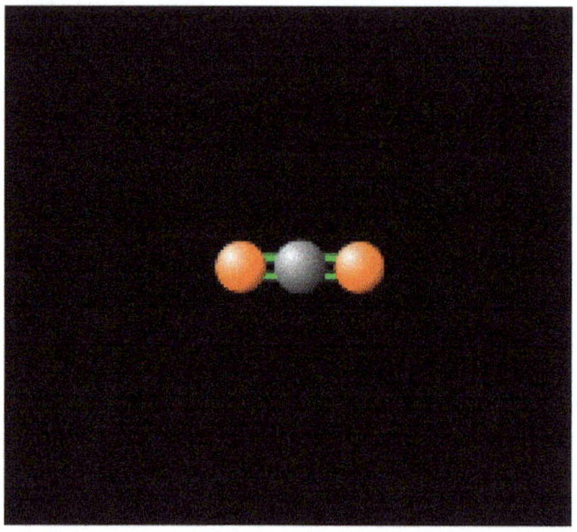

Carbon dioxide is able to interact with infrared radiation, leading to an imbalance of radiation entering and leaving the atmosphere.

Emissions

Carbon is the buzzword when talking about emissions, and because of this it is important to know the difference between carbon and carbon dioxide. Carbon is an element present in all organic material and is not, in itself, a bad thing. In fact, carbon is necessary to all life. Carbon dioxide, on the other hand, is what most people use when referring to "carbon emissions." It's a greenhouse gas and contributes to climate change because it helps to trap heat inside our atmosphere. Currently CO_2 levels in the oceans and atmosphere are on the rise, and this increase can mostly be attributed to human activities. This is also referred to as anthropogenic CO_2 emission.

Carbon dioxide emissions are often measured in Gigatonnes of carbon or carbon dioxide. To convert from gigatons of carbon to gigatons of carbon dioxide, multiply by 44.01/12.01, or 3.664. The interactive graph below shows Megatonnes (1000 Megatonnes = 1 Gigatonne) of carbon dioxide emissions by country (or region depending on personal preference).

Scientists around the world are currently researching ways of capturing carbon and burying it underground in a process referred to as carbon capture and storage. If effective ways of doing this are developed, this would allow humanity to significantly reduce the atmospheric and oceanic CO_2 levels and thus a prevent a dangerous rise in global temperature.

Inorganic Chlorine Compounds

Chlorine Oxides

Chlorine Oxides are compounds of chlorine and oxygen: Cl_2O, ClO_2, Cl_2O_6, Cl_2O_7, and Cl_2O_8. Chlorine monoxide, Cl_2O (an ahydride of hypochlorous acid), is a yellow-brown gas, with a pungent odor similar to the odor of chlorine. It has a melting point of –121°C and a boiling point of 2 °C. A vigorous oxidizing agent, it is an unstable compound, which slowly decomposes under normal conditions and may explode spontaneously in concentrated and condensed form: $2Cl_2O = 2Cl_2 + O_2$. It is readily soluble in water (about 200 volumes of Cl_2O dissolve in one volume of water) and carbon tetrachloride. Upon dissolving in water, it yields hypochlorous acid, HClO. Chlorine monoxide may be produced by the reaction of chlorine with freshly precipitated dry mercuric oxide: $2HgO + 2Cl_2 = HgCl_2 + Cl_2O$. In industry, aqueous solutions of chlorine monoxide are produced by the chlorination of carbonates of alkali or alkaline-earth metals in water. Chlorine monoxide is used in the production of calcium hypochlorite.

Chlorine dioxide, ClO_2 (a mixed anhydride of chlorous acid and chloric acid), is a yellow-orange gas, with an unpleasant odor and a melting point of –59 °C and a boiling point of 10 °C. It is a strong oxidizing agent, especially in an acidic medium, and is explosive in concentrated form. Chlorine dioxide decomposes gradually in the light, a process accompanied by explosion above 50°C. It is readily soluble in water, sulfuric acid, acetic acid, and carbon tetrachloride. Upon reaction with alkaline solutions, it forms chlorites and chlorates. Chlorine dioxide is produced by the partial reduction of chlorates by SO_2, oxalic acid, or hydrochloric acid and, in small amounts, by the action of chlorine on sodium chlorite:

$$2NaClO_2 + Cl_2 = 2NaCl + 2ClO_2$$

It is commonly used in the form of an aqueous solution or in gaseous form in a mixture with air to bleach and sterilize wood pulp, fabrics, and flour. The reaction of ClO_2 with aqueous solutions of NaOH in the presence of a reducing agent is the basis of the industrial production of chlorites.

Chlorine trioxide, Cl_2O_6 (a mixed anhydride of perchloric acid and chloric acid), is a fuming, dark red liquid, with a melting point of 3.5 °C and a boiling point of 203 °C. It explodes upon contact with readily oxidizable compounds. In the crystalline state, it has the structure of chloryl perchlorate, $ClO_2^+ClO_4^-$. It reacts vigorously with water. Chlorine trioxide may be produced by the oxidation of chlorine dioxide with ozone or by the reaction of chlorates with fluorine: $2KClO_3 + F_2 = 2KF + Cl_2O_6$. It is always present in the products of the thermal decompostion of perchloric acid. It has no practical applications.

Chlorine heptoxide, Cl_2O_7 (an anhydride of perchloric acid), is a colorless liquid, with a melting point of −93°C and a boiling point of 83 °C. It decomposes slowly upon storage, with the production of colored decomposition products that are lower oxides of chlorine. It is spontaneously explosive, especially in the presence of its decomposition products. Chlorine heptoxide dissolves in carbon tetrachloride at room temperature and reacts with water to form perchloric acid. It explodes upon contact with iodine. Chlorine heptoxide may be obtained by the dehydration of perchloric acid in the presence of phosphorus pentoxide or oleum or by the low-temperature electrolysis of concentrated perchloric acid on a platinum anode. It has no practical applications.

Cl_2O_8, an intermediate in the electrolysis of perchloric acid and perchlorate salts, has not been isolated as a separate compound.

In addition to the oxygen compounds of chlorine mentioned above, Cl_2O_4 is also mentioned in the literature. This compound has the structure of chlorine perchlorate, $ClOClO_3$, and may be obtained by the reaction of chlorine fluorosulfonate and $CsClO_4$.

Inorganic Nitrogen Compounds

Nitrogen Hydrides

Hydride of nitrogen is basic due to lone pair of electrons on nitrogen atom. The common hydrides of nitrogen are hydrazine,diazene,ammonia out of which only ammonia is thermally stable.

Ammonia

Ammonia is the most important compound of nitrogen. It can be manufactured by Haber's process. In this process, a mixture of N_2 and H_2 in the ratio of 1 : 3 is passed over heated *Fe* at 650 −800*K* as catalyst and *Mo* as promotor, $N_2 + 3H_2 \rightleftharpoons 2NH_3$, $\Delta H = -93.6\ mol^{-1}$ This is a reversible exothermic reaction.

Ammonia is prepared in the laboratory by heating ammonium salt (NH_4Cl) with a strong alkali like *NaOH* $NH_4Cl + NaOH \rightarrow NH_3 + H_2O + NaCl$ Ammonia can be dried by passing over quick lime (*CaO*). However, it cannot be dried with dehydrating agents such as conc. H_2SO_4, $P_2O_5$and anhydrous $CaCl_2$ because ammonia reacts with these compounds.

NH_3 is a colourless gas with a characteristic pungent smell called ammonical smell. It is highly soluble in water and its solution is basic in nature, $NH_3 + H_2O \rightleftharpoons NH_4^+ + OH^-$ $NH_3$is expected to have a tetrahedral geometry, but the lone pair distorts its geometry and the molecule has pyramidal geometry with N – H bond length of 101.7 *pm* and a bond angle of 107.5°. Liquid ammonia is widely used as a refrigerant due to its high heat of vaporization.

Hydrazine ($NH_2 - NH_2$)

Hydrazine is prepared commercially by boiling aqueous ammonia or urea with sodium hypochloride in the presence of glue or gelatin.

$$2NH_3 + NaOCl \rightarrow N_2H_4 + NaCl + H_2O$$

The resulting solution is concentrated and anhydrous hydrazine may be obtained by further distillation over barium oxide. Alternatively, the hydrazine present in the resulting solution is precipitated as sparingly soluble crystalline hydrazine sulphate on treatment with sulphuric acid, $NH_2NH_2 + H_2SO_4 \rightarrow N_2H_4.H_2SO_4$.

The precipitate is removed and treated with an alkali when hydrazine hydroxide H2N.NH3OH is obtained. This is distilled under reduced pressure, over barium oxide to liberate free hydrazine.

$$H_2N.NH_3OH + BaO \xrightarrow{\text{distil}} NH_2NH_2 + Ba(OH)_2$$

Physical properties: Anhydrous hydrazine is a colourless fuming liquid (m.p. 2 °C and b.p. 114 °C) soluble in water in all proportions. It is also soluble in alcohol. It is strongly hygroscopic.

Chemical properties: It behaves as a diacid base, Thus with hydrochlorides it forms hydrazine monochloride $H_2N.NH_3Cl$ and hydrazine dichloride $ClH_3N.NH_3Cl$.

- Hydrazine burns in air with the evolution of heat.

$$N_2H_4 + O_2 \rightarrow N_2 + 2H_2O \quad \Delta H = -622 \text{ kJ}$$

The alkyl derivatives of hydrazine are used as rocket fuels.

- It reduces Fehling's solution to red cuprous oxide and iodates to iodides.

$$4Cu^{2+} + NH_2NH_4 \rightarrow 4Cu^{+} + 4H^{+} + N_2 ;\ 2IO_3^{-} + 3NH_2NH_2 \rightarrow 2I^{-} + 6H_2O + 3N_2$$

Uses: Hydrazine is used as a rocket fuel. It is also used as a reagent in organic chemistry.

Nitrogen Oxoacids

In the gas phase nitrous acid can be made by the following reaction:

$$NO + NO_2 + H_2O \rightleftarrows 2HNO_2$$

The gas phase structure as determined by IR spectroscopy is shown in Figure, in which the nitrogen is planar with sp² hybridization.

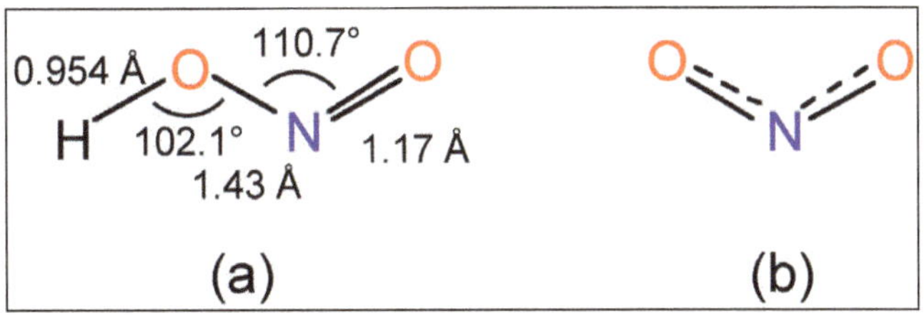

In basic aqueous solution the same reaction results in the formation of the nitrite ion, which can be precipitated as the Ba2+ salt. After separation the addition to sulfuric acid yields a solution of nitrous acid. However, it is not possible to concentrate by heating since decomposition occurs.

$$NO + NO_2 + 2OH \rightleftarrows 2NO_2 + H_2O$$

$$3H_2O \rightarrow H_3O^+ + NO_3^- + 2NO$$

Nitrous acid is a fairly weak acid in water (pK = 5.22); however, many salts are known of the nitrite ion, NO_2^-. Complexes of the nitrite ion can be monodentate with bonding via nitrogen (nitro) or oxygen (nitrito). Both isomers can be isolated in the case of an inert metal, i.e., substitutionally inert d6 octahedral complexes. Bidentate and bridging modes of coordination are also known for the nitrite anion.

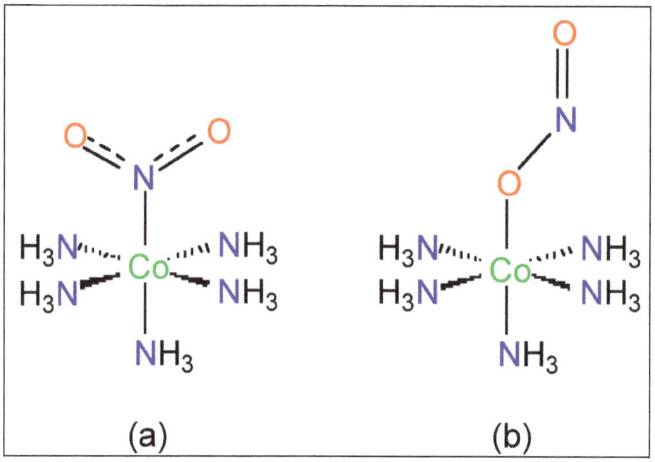

Structures of (a) orange nitro and (b) pink nitrito isomers of $[Co^{III}(NH_3)_5(NO_2)]^{2+}$.

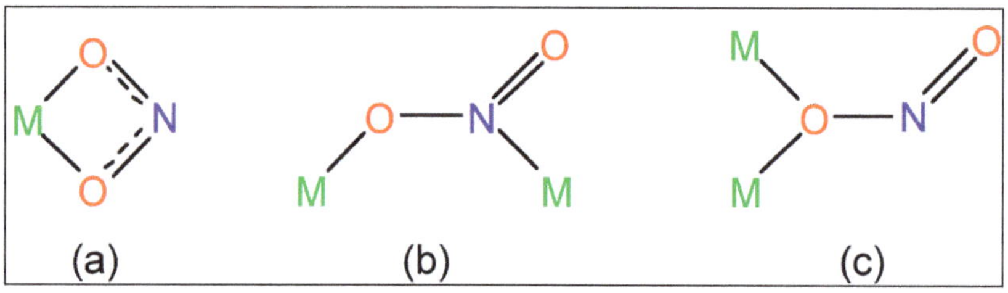

Bidentate (a) and bridging (b and c) modes of coordination of the nitrite anion.

Nitrite can act as either an oxidizing agent, or a reducing agent.

$$NO_2^- + 2H^+ + Fe^{2+} \rightarrow NO + H_2O + Fe^{3+}$$

$$NO_2^- + H_2O + 2Ce^{4+} \rightarrow NO_3^- + 2H^+ + 2Ce^{3+}$$

The most important use of nitrous acid is in the diazotization reactions in which nitrous acid is generated by acidifying nitrite solution.

Nitric Acid

Nitric acid (HNO_3), also known as aqua fortis and spirit of nitre, is made by the dissolution of nitrogen dioxide in water. Nitric acid can be concentrated by distillation from concentrated sulfuric acid.

$$3NO + H_2O \rightarrow 2HNO_3 + NO$$

The structure of HNO_3 in the gas phase is planar at the sp^2 nitrogen.

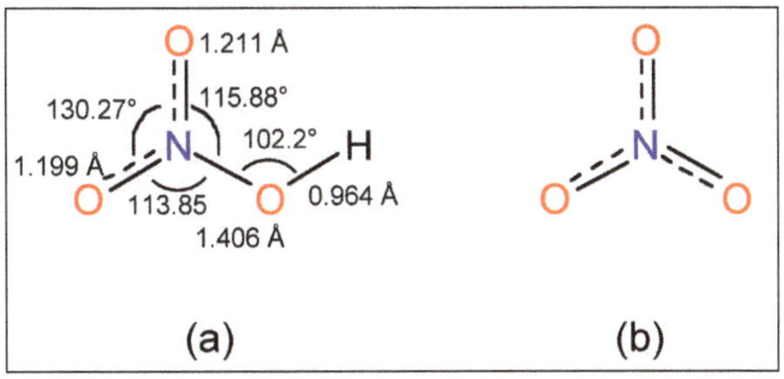

The structures of (a) nitric acid and (b) the nitrate anion.

Pure 100% nitric acid is a very corrosive liquid that is strongly acidic and protonates and dissolves organic species. In the liquid phase it is slightly dissociated. It is also a powerful oxidizing agent, converting non-metal elements to either the oxide or oxoacid. In contrast with metals it forms either salts or complexes in which the metal is in its highest oxidation state. It is unstable and decomposes upon heating or photolysis:

$$3HNO \rightleftarrows H_3O^+ + NO_2^+ + 2NO_3^-$$

The pure acid has the highest self-ionization of pure liquid acids. However, the loss of water results, such that the overall reaction can be described by.

$$2HNO_3 \rightleftarrows H_2NO_3^+ + NO_3^+$$

$$H_2NO_3^+ \rightleftarrows H_2O + NO_2^+$$

$$2HNO_3 \rightleftarrows NO_3^- + NO_2^+ + H_2O$$

The common concentration of nitric acid is 70%. While the pure acid is colorless, samples often take on a yellow color due to the photochemical decomposition of nitric acid to give brown NO_2.

$$4HNO_3 \rightarrow 4NO_2 + 2H_2O + O_2$$

The term the "acid test" is derived from the medieval practice of debasing of gold and silver currencies (often by the Monarchs who issued them) by debasing with copper. With ducats (Milan), livres (France), florins (Florence), maravedies (Spain) and bezants (Constantinople) in widespread use, and each with a nominal gold or silver content, it was important for a merchant to be able verify the worth of any particular coin. If a drop of dilute nitric acid was placed onto a silver coin adulterated with copper, it turned green, due to the formation of copper(II) nitrate. Conversely, if a gold coin reacted in any way with the nitric acid it was not pure. In both cases the coins failed the "acid test".

Aqua regia (so called because it dissolves gold) is a mixture of 70% nitric acid and hydrochloric acid in a 1:3 ratio. Aqua regia is a very powerful oxidizing agent (it contains Cl2) and stablizes some metals as their chloro complexes. If HF is added in place of HCl, tantalum may be dissolved with the formation of [TaF2]-.

Metal	Chloride salt formed
Au	$[AuCl_4]^-$
Pt	$[PtCl_6]^{2-}$

Fuming nitric acid (100% nitric) is exceedingly corrosive and should not be used. In 100% sulfuric acid, nitric acid acts as a base and gets protonated, and acts as a powerful nitrating agent.

$$HNO_3 + H_2SO_4 \rightarrow ON(OH)_2^+ + HSO_4^-$$

$$ON(OH)_2^+ HSO_4^- + H_2SO_4 \rightarrow H_3O^+ + NO_2^+ + 2HSO_4^-$$

In water dilute nitric acid is fully ionized.

$$HNO_3 + H_2O \rightarrow H_3O^+ + NO_3^-$$

The nitrate ion is planar (Figure b) and forms many salts and complexes. The nitrate anion is most commonly a monodentate ligand, but can also be a bidentate ligand (Figure a) or a bridging ligand (Figure b and c).

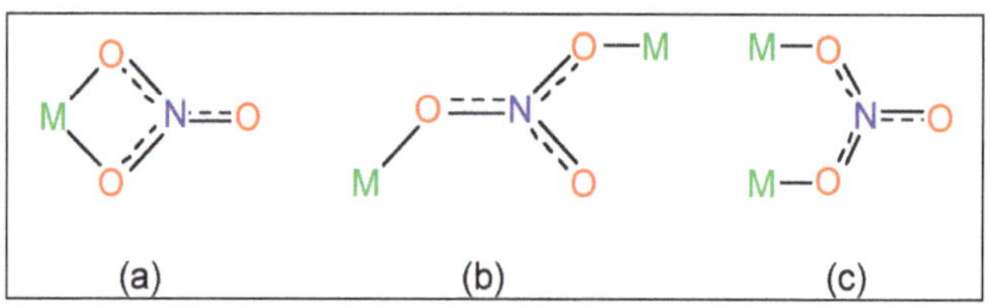

Coordination modes of the nitrate anion.

Inorganic Phosphorus Compounds

Phosphorus Oxoacids

Oxoacids are basically acids that contain the element oxygen. As such, Phosphorus is known to form a number of oxoacids, for example: H_3PO_4, H_3PO_3, etc. In oxoacids of phosphorus, it is tetrahedrally surrounded by other atoms. Generally, all these acids are known to form at least one P=O bond and one P–OH bond.

P–P or P–H bonds are also found in addition to P=O bonds and P–OH bonds in oxoacids of phosphorus where the oxidation state of phosphorus is less than +5. These acids are generally seen too disproportionate to both lower and higher oxidation states. For instance, when phosphorous acid is heated, it gives phosphine and phosphoric acid.

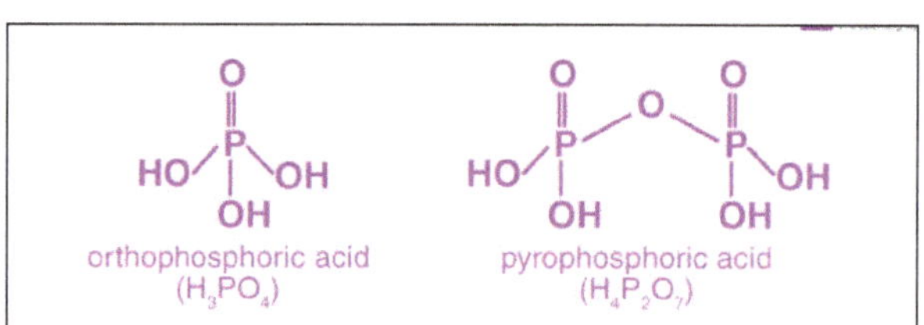

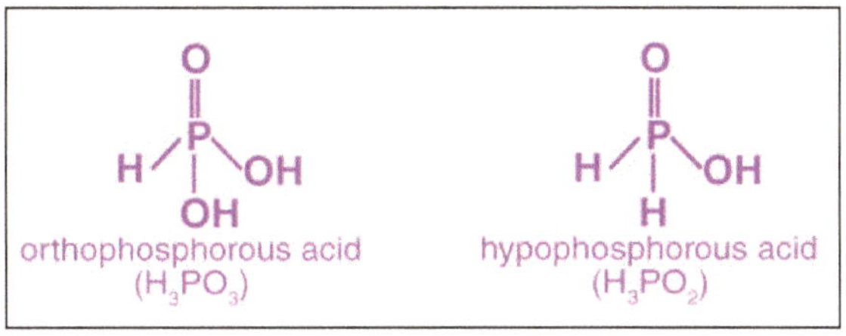

$$4H_3PO_3 \rightarrow 3H_3PO_4 + PH_3$$

The P-H bonds in oxoacids cannot go through ionization to give H+ ions whereas the H atoms which are attached with oxygen in P-OH form are ionizable. Hence, we can say that only the H atoms attached with oxygen cause basicity. As a result phosphorous acid, H_3PO_3 is dibasic due to the presence of two P-OH bonds whereas phosphoric acid, H_3PO_4 is tribasic due to the presence of three P-OH bonds. Oxoacids of phosphorus having P-H bonds have strong reducing properties. For instance: hypophosphorous acid containing two P-H bonds acts as a good reducing agent.

$$4AgNO_3 + 2H_2O + H_3PO_2 \rightarrow 4Ag + 4HNO_3 + H_3PO_4$$

Few Popular Oxoacids of Phosphorus

Phosphorus acid, H_3PO_3: Phosphorous acid is a diprotic acid that is, it ionizes two protons. It is better described with the structural formula HPO $(OH)_2$ Phosphorous acid is prepared by hydrolysis of phosphorus trichloride with acid or steam.

$$PCl_3 + 3H_2O \rightarrow HPO(OH)_2 + 3HCl$$

Phosphoric acid, H_3PO_4: Phosphoric acid is a triprotic acid that is, it ionizes three protons. It is a non-toxic acid, when pure and is a solid at room temperature and pressure. Phosphoric acid is prepared by adding sulphuric acid to tricalcium phosphate rock:

$$Ca_5(PO_4)_3X + 5H_2SO_4 + 10H_2O \rightarrow 3H_3PO_4 + 5CaSO_4.2H_2O + HX$$

- X can be F, Cl, Br and OH

Meta Phosphoric $(HPO_3)n$

This acid is formed by warming orthophosphoric acid at around 850 K. Metaphosphoric acid exists as a cyclic trimer, polymer or cyclic tetramer but not as a monomer.

$$H_3PO_4 \rightarrow HPO_3 + H_2O$$

Hypophosphoric Acid ($H_4P_2O_6$)

Hypophosphoric acid is formed by conducting a controlled oxidation of red phosphorus with sodium chlorite. When the disodium salt of the acid is formed it then moves via cation exchanger that ultimately gives hypophosphoric acid. The acid is of tetrabasic nature.

$$2P + 2NaClO_2 + 2H_2O \text{ ———> } Na_2H_2P_2O_6 + 2HCl$$

$$Na_2H_2P_2O_6 + 2H \text{ ——resin——> } H_4P_2O_6 + 2Na - \text{resin}$$

Pyrophosphoric Acid ($H_4P_2O_7$)

When orthophosphonic acid is heated at about 250 °C it forms tetrabasic acid.

$$2H_3PO_4 \text{ ———> } H_4P_2O_7 + H_2O$$

Orthophosphoric Acid (H_3PO_4)

This acid is formed when P_4O_{10} is treated with bubbled water. Orthophosphoric Acid is basically a tribasic acid.

$$P_4O_{10} + 6H_2O \text{ ————> } 4H_3PO_4$$

Phosphorus Pentoxide

Phosphorous pentoxide is an inorganic chemical compound, composed of four phosphorus atoms and ten oxygen atoms. It is generally found as a dimer of P_2O_5 that's why it is known as phosphorous pentoxide. Its chemical name is tetraphosphorous decaoxide. It has several other names such as phosphoric anhydride, phosphorous (V) oxide, diphosphorus pentoxide etc.

It is a white crystalline solid which is used as a powerful dehydrating agent. It is a strong desiccant. It is an anhydride of phosphoric acid. Its molecular formula is P_4O_{10} but it is found as a dimer of P_2O_5.

Formula of Phosphorous Pentoxide

S. No.	Formula of Phosphorous Pentoxide	
1.	Empirical Formula	P_2O_5
2.	Molecular Formula	P_4O_{10}
3.	Structural Formula	Image will be uploaded soon

Structure of Phosphorous Pentoxide

It shows polymorphism. It means it is found in more than one crystalline form. Its metastable form is the most is 160 pm while its P=O bond length is 143 pm. O-P-O bond angle is 102° and P-O-P bond angle is 123°.

Preparation of Phosphorous Pentoxide

It is prepared by oxidation of tetraphosphorous. P_4 is burned in presence of oxygen and it gives P_4O_{10}. Reaction is given below:

$$P_4 + 5O_2 \rightarrow P_4O_{10}$$

It is used as a source to provide pure concentrated phosphoric acid.

Properties of Phosphorous Pentoxide

Properties of phosphorous pentoxide are listed below:

- It is a white crystalline solid.
- Its molar mass is 283.9 g/mol.
- It can absorb moisture easily. It is a very deliquescent compound.
- It is an odorless compound.
- Its density is 2.39 g.cm^{-3}.
- Its melting point is 340 °C.
- Its boiling point is 360 °C.
- It may undergo explosive decomposition when subjected to heat.
- It is not flammable.
- It is an acidic compound as it reacts with water and forms phosphoric acid.
- Its short or long exposure may cause residual injuries hydrolysis.

  $P_4O_{10} + 6H_2O \rightarrow 4H_3PO_4$ (–177 kJ)
- It reacts with primary amide and converts it into nitrile. General reaction is given below – $P_4O_{10} + RC(O)NH_2 \rightarrow P_4O_9(OH)_2 + RCN$.
- It reacts with carboxylic acid and forms corresponding anhydride. General equation of its reaction with carboxylic acid is given below:

  $P_4O_{10} + RCOOH \rightarrow P_4O_9(OH)_2 + [RC(O)]_2O$
- It can easily corrode metals, so must be kept away from them.

Applications of Phosphorous Pentoxide

Due to its unique properties, it is used in many fields. few of its applications are listed below:

- It is used as desiccant. It is an excellent absorber of moisture. In its granular form, it is used in desiccators.
- It is used as a potent dehydrating agent. but due to its tendency to form a viscous coating during the reaction, makes its use limited.
- It is used in synthesis of organic compounds.
- Its solution in dimethyl sulfoxide is used for oxidation of alcohols.
- Its desiccating powder can convert many mineral acids into their anhydrides.
- It is used in manufacturing of optical glass and heat insulating glass.

- It acts as a drying agent for many substances especially for those with which it does not react.
- It is used in medicines.
- It is used in pesticides as well.

Hazards from Phosphorous Pentoxide

It is dangerous due to its exothermic hydrolysis. It vigorously reacts with water and releases a large amount of heat in a short time span. It reacts with water containing substances such as wood, leaves and cotton etc. in the same way. It is harmful due to corrosive nature against metals. It may cause itching and burning sensation when exposed to eyes and skin. Its exposure may harm mucous membrane and respiratory tract as well even at very low concentration such as 1.5 – 2 mg.m^{-3}.

Phosphorous Pentoxide

Phosphorous Pentoxide	
Chemical formula	P_4O_{10}
IUPAC Name	Tetraphosphorous decaoxide
Molar mass	283.9 g/mol
Melting point	340 °C
Boiling point	360 °C
Density	2.39 g.cm^{-3}
Structure	Hexagonal
State at STP	Solid
Solubility in water	Soluble. Shows exothermic Hydrolysis.
Solubility in organic solvents	Insoluble in acetic acid, acetone, benzene and Chloroform.
Solubility in acids and alkalis	Soluble in sulfuric acid and reacts with some alkalis
Preparation	By burning tetraphosphorous in air or presence of oxygen – $P_4 + 5O_2 \rightarrow P_4O_{10}$
Main properties	Explosive decomposition, exothermic hydrolysis
Main use	As a desiccant and Dehydrating agent.
Disadvantage	Exothermic hydrolysis of it, at large scale, may cause many Hazards. It is corrosive to Metals.

Inorganic Silicon Compounds

Silicon Carbide

Silicon carbide is an exceedingly hard, synthetically produced crystalline compound of silicon and carbon. Its chemical formula is SiC. Since the late 19th century silicon carbide has been an important material for sandpapers, grinding wheels, and cutting tools. More recently, it has found application in refractory linings and heating elements for industrial furnaces, in wear-resistant parts for pumps and rocket engines, and in semiconducting substrates for light-emitting diodes.

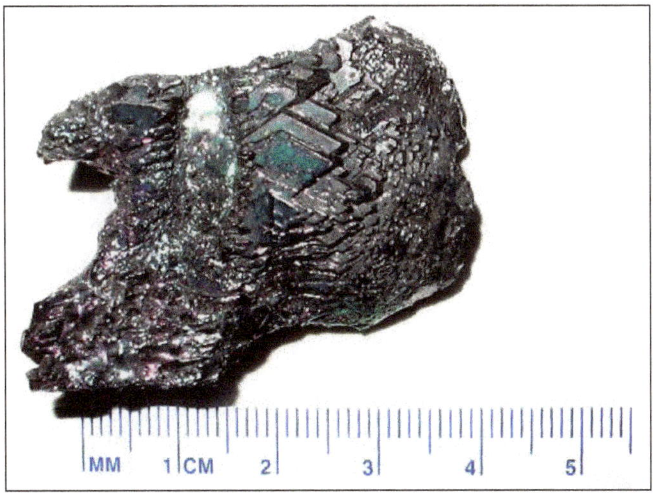

Silicon carbide.

Silicon carbide was discovered by the American inventor Edward G. Acheson in 1891. While attempting to produce artificial diamonds, Acheson heated a mixture of clay and powdered coke in an iron bowl, with the bowl and an ordinary carbon arc-light serving as the electrodes. He found bright green crystals attached to the carbon electrode and thought that he had prepared some new compound of carbon and alumina from the clay. He called the new compound Carborundum because the natural mineral form of alumina is called corundum. Finding that the crystals approximated the hardness of diamond and immediately realizing the significance of his discovery, Acheson applied for a U.S. patent. His early product initially was offered for the polishing of gems and sold at a price comparable with natural diamond dust. The new compound, which was obtainable from cheap raw materials and in good yields, soon became an important industrial abrasive.

About the same time Acheson made his discovery, Henri Moissan in France produced a similar compound from a mixture of quartz and carbon; but in a publication of 1903, Moissan ascribed the original discovery to Acheson. Some natural silicon carbide was found in Arizona in the Canyon Diablo meteorite and bears the mineralogical name moissanite.

Modern Manufacture

The modern method of manufacturing silicon carbide for the abrasives, metallurgical, and refractories industries is basically the same as that developed by Acheson. A mixture of pure silica sand and carbon in the form of finely ground coke is built up around a carbon conductor within a brick electrical resistance-type furnace. Electric current is passed through the conductor, bringing about a chemical reaction in which the carbon in the coke and silicon in the sand combine to form SiC and carbon monoxide gas. A furnace run can last several days, during which temperatures vary from 2,200° to 2,700° C (4,000° to 4,900° F) in the core to about 1,400° C (2,500° F) at the outer edge. The energy consumption exceeds 100,000 kilowatt-hours per run. At the completion of the run, the product consists of a core of green to black SiC crystals loosely knitted together, surrounded by partially or entirely unconverted raw material. The lump aggregate is crushed, ground, and screened into various sizes appropriate to the end use.

For special applications, silicon carbide is produced by a number of advanced processes. Reaction-bonded silicon carbide is produced by mixing SiC powder with powdered carbon and a plasticizer, forming the mixture into the desired shape, burning off the plasticizer, and then infusing the fired object with gaseous or molten silicon, which reacts with the carbon to form additional SiC. Wear-resistant layers of SiC can be formed by chemical vapour deposition, a process in which volatile compounds containing carbon and silicon are reacted at high temperatures in the presence of hydrogen. For advanced electronic applications, large single crystals of SiC can be grown from vapour; the boule can then be sliced into wafers much like silicon for fabrication into solid-state devices. For reinforcing metals or other ceramics, SiC fibres can be formed in a number of ways, including chemical vapour deposition and the firing of silicon-containing polymer fibres.

Properties and Applications

Until the invention of boron carbide in 1929, silicon carbide was the hardest synthetic material known. It has a Mohs hardness rating of 9, approaching that of diamond. In addition to hardness, silicon carbide crystals have fracture characteristics that make them extremely useful in grinding wheels and in abrasive paper and cloth products. Its high thermal conductivity, together with its high-temperature strength, low thermal expansion, and resistance to chemical reaction, makes silicon carbide valuable in the manufacture of high-temperature bricks and other refractories. It is also classed as a semiconductor, having an electrical conductivity between that of metals and insulating materials. This property, in combination with its thermal properties, makes SiC a promising substitute for traditional semiconductors such as silicon in high-temperature applications.

References

- Inorganic-compounds, anatomy-and-chemistry-basics, anatomy-and-physiology, study-guides: cliffsnotes.com, Retrieved 10, March 2020
- Inorganic-compound: biologyonline.com, Retrieved 19, May 2020
- Carbide, Some-carbides-and-their-uses: newworldencyclopedia.org, Retrieved 16, January 2020
- Silicon-carbide, science: britannica.com, Retrieved 11, April 2020
- Phosphorus-pentoxide, chemistry: vedantu.com, Retrieved 26, March 2020

3

Principles of Bioinorganic Chemistry

Bioinorganic chemistry deals with the role of inorganic metals and non-metals in biological systems. It includes the study of behaviors and effects of the various metalloproteins on biological systems. Biological processes like photosynthesis, respiration, metal ion transport, enzymatic actions, etc. all fall under its umbrella. This chapter has been carefully written to provide the reader with a thorough understanding of bioinorganic chemistry and its related processes.

Bioinorganic or biological inorganic chemistry is the discipline dealing with the interaction between inorganic substances and molecules of biological interest. It is a rather wide field, because it addresses the role, uptake, and fate of elements essential for life, the response of living organisms to toxic inorganic substances, the function of metal-based drugs, the synthetic production of functional models, the production of MRI contrast agents in medical applications, the development of theoretical models.

The importance of the field stems from the fact that life originated and developed on the earth's crust, i.e., within an inorganic environment. Biological inorganic chemistry has always been part of chemistry and evolved as did the research tools of chemistry. The discovery of iron in the blood and of copper and zinc as essential elements for many physiological reactions may be referred to as fundamental steps in the field. The beginning of the present flourishing may be dated back to the late 1950s to early 1960s, when the x-ray structures of myoglobin and hemoglobin were solved by the illustrious chemists J. C. Kendrew and M. Perutz, respectively. In the field, also organized by several learned societies, took place. In the 1970s, a significant share of inorganic chemists entered the field, which has been steadily growing since then, as has the flood of publications related to the field. The birth of the Journal of Inorganic Biochemistry in 1971 was quite timely. Eventually, the Society of Biological Inorganic Chemistry was founded in 1995, followed after 1 year by the Journal of Biological Inorganic Chemistry. The mid-1990s were also the time for the birth of genome-based research, which raised the new challenge of matching genomic.

Today's Mission of Bioinorganic Chemistry

Every historical period is characterized by major scientific challenges. For example, at the beginning of the 20th century, the challenge understood atomic structure. The challenge of the beginning of this century is the understanding of the molecular bases of living processes. This challenge takes advantage of the availability of genome sequences. In 1995, the first sequence of an entire genome that of Haemophilus influenzae, was reported. At present, genome sequences for 16 archaea, 87 bacteria, and 8 eukaryota are available. In addition, the genome sequences for nine other eukaryota are being completed, six of which are plants. And the amount of data is increasing like an avalanche.

From the genome sequences, it is possible to deduce the primary sequences of essentially all of the proteins that a living organism can produce. The 3D structures of these proteins are being collected at the genome scale within programs of structural genomics financed in North America, Japan, and Europe. These programs are developed around the concept of high-throughput structure determination, which per se does not pay attention to issues such as incorporation of cofactors containing metals. This unprecedented availability of data opens new perspectives for bioinorganic chemists. Particularly, it is well established that a number of metal ions are essential to life. Other metal ions are instead poisonous to living organisms, even when present in the environment at very low concentrations. Sometimes, the same element may be beneficial or noxious depending on speciation. The investigation of the linkages between inorganic elements (essential and noxious) and the information obtainable from genome sequences, as well as of the mechanisms that warrant homeostasis in cell compartments, represent a further challenge for bioinorganic chemists in the so-called postgenomic era.

The Identification of Metal-Binding Proteins in Gene Banks

A number of bioinformatic web servers and databases have been created and updated (or, in some cases, discontinued) before and after the advent of genome sequences with the aim of providing the scientific community with tools for searching gene banks, for the analysis of protein sequences, and for the prediction of a variety of protein properties. Surprisingly enough, very few of these resources are dedicated to the analysis of the interaction between metal ions and proteins. This situation was already pointed out in 2000. Genome browsing in the field of bioinorganic chemistry cannot be limited to the analysis of amino acid identity or similarity in sequences [i.e., as is the case for one of the most popular programs for sequence comparison and genome browsing, blast]. We may take advantage of known consensus sequences, i.e., by taking into account the nature and spacing of amino acids acting as metal ligands and the nature of amino acids present in the metal-binding region. When the consensus sequence is well known, it can thus be used as a further input for genome browsing, together with the primary sequence of one or more proteins evolutionarily related to the one(s) of interest, by using the program phi-blast. This poses the problem of having at one's disposal databases of consensus sequences for metal binding. An available database addressing this problem is PROMISE, which collects structural and functional information (including bibliography) on metalloproteins, with emphasis on the properties of the metal site. The proteins have been categorized on the basis of the metal cofactor bound. Unfortunately, this service has been discontinued, so the database is no longer updated. A second relevant database publicly available is the Metalloprotein Database, maintained at The Scripps Research Institute. This database provides quantitative information on geometrical parameters of metal-binding sites in metalloprotein structures deposited in the Protein Data Bank (PDB), as well as statistical information about the recurrence of consensus sequences (called patterns) within the PDB. An explicitly stated goal of the database is that of becoming a comprehensive source of useful information for metalloprotein design.

The two databases mentioned taken together provide the largest amount of information on metal-binding sites and consensus sequences available from the web. It appears that specific validated bioinformatic tools for the search of metal-binding sequences within gene data banks, possibly building on the information that the above databases contain, are still lacking. The development of such tools will most likely be crucial to extend and fully exploit the success of genome sequencing projects in the field of bioinorganic chemistry. Such tools would not only allow researchers to identify uncharacterized metalloproteins in genome sequences for subsequent experimental studies but would also enable broader applications of other bioinformatic methods, such as molecular modeling. Molecular modeling indeed has been proving extremely powerful in linking structural to functional information, also in the case of metalloproteins; however, available approaches do show some (potential) limitations in the identification and treatment of metal-binding sites, which often can be circumvented only through careful human examination.

The Problem of Metal Trafficking in Cells

Being able to browse genomes and gene banks and to identify metalloprotein sequences with a good degree of reliability would be a precious advancement for the field of bioinorganic chemistry but probably would not be sufficient to answer the key question: how do processes involving metal ions occur in cells? To answer this question, i.e., to understand the interaction between cells and metal ions (either essential or noxious) within the context of the global functioning mechanism of the cell, requires synergistic efforts across a number of different disciplines, ranging from genetics to structural biology, from analytical chemistry to biophysics, from enzymology to drug discovery, and beyond. A bioinorganic chemist of the 21st century should be knowledgeable of all of this.

Schematically displays what is currently known about the trafficking pathway of copper in yeast. This is probably the most intensely studied and thus better understood, "global" scheme of interaction between one metal and a cell. Cu (II) is reduced to Cu(I) at the external surface of the membrane and is then pumped in the cell by Ctr. At this point, in one pathway, Ctr delivers the Cu (I) ion to the Atx1 chaperone, which delivers it to Ccc2 to subsequently reach Fet3, the final target; in a second pathway, CCS (the copper chaperone for superoxide dismutase) gets the Cu (I) ion from Ctr to deliver it to the final target, superoxide dismutase; in the third pathway, Cox17 shuttles Cu(I) from Ctr to Sco1 in the mitochondrion, where it reaches cytochrome c oxidase through Sco2. Regulation of this system is comparatively less clear. It is also very important to note that the trafficking mechanism for a given metal may vary somewhat when comparing bacteria and eukaryota. In addition, an extreme variability is expected and found among bacteria, given their fast evolution processes. Of course, some aspects of these mechanisms can be instead quite well conserved even in different organisms, in which case bioinformatic approaches relying on multiple sequence alignments, phylogenetic trees, and molecular modeling can be extremely informative on the functional features of entire pathways. For instance, phylogenetic studies clearly point out that copper chaperones and copper ATPases evolved in parallel but independently from a single ancestor, common to proteins from eukaryota and prokaryota. Interestingly, it appears that even in operons coexpressing a copper chaperone and an ATPase, the two proteins have generally evolved independently, suggesting that their coregulation is the result of specific genetic pressure.

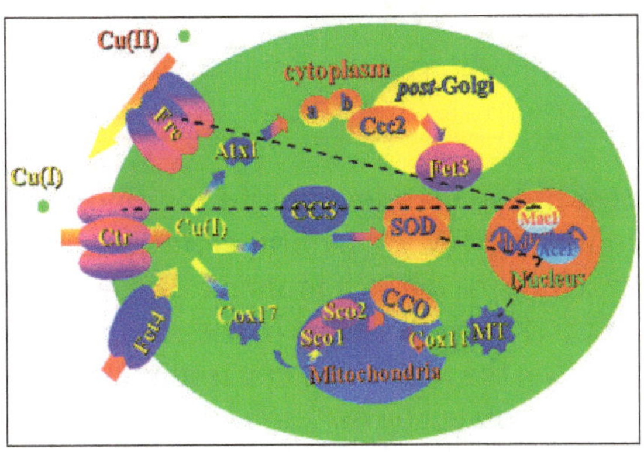

Display of the proposed copper trafficking pathways in yeast. Information on the directionality of copper transport, as well as on the identity of the partners involved in each transfer of the copper ion, is obtained mainly from experimental data but in some instances can be based on speculation or computational predictions. Regulation of the system is maintained at different points: Mac1 regulates the transcription of the genes encoding Fre and Ctr, and Ace1 regulates the transcription of the genes encoding superoxide dismutase and metallothionein (MT).

Copper trafficking in cells constitutes an example that may be quite typical of how the field of bioinorganic chemistry may evolve to link genetic information and analytical and spectroscopic methods to unravel the cellular processes at the basis of the role of metal ions in biology. The topic of copper trafficking was initiated by the identification of the chaperone function of the soluble Cu(I) receptor Atx1 and of its 3D structure.

The existence of complex molecular machinery in cells controlling uptake, transport, and delivery of copper ions in cells. Genetic studies have been at the basis of the identification of Atx1 as a key factor in metal homeostasis, before the identification of its chaperone function. In the subsequent years, a huge amount of data has accumulated on copper trafficking in a variety of organisms, as well as in the various organelles within single eukaryotic cells. Structural biologists have made outstanding contributions in detailing, within the general picture of copper homeostasis, the atomic and molecular determinants of the interaction between the individual proteins and the metal ion. The investigation at the atomic level of protein–protein interactions responsible for the key step of transfer of the metal ion along with the delivery has proved more difficult, possibly because of the transient nature of these interactions. Here, the bulk of high-resolution structural information on the physiologically relevant interactions has been produced by NMR spectroscopy, a technique particularly suitable for the study of this kind of interaction in solution. However, through a clever use of site-directed mutagenesis or thanks to repeated trials in a variety of conditions, high-resolution information can be gained also through x-ray crystallography. Figure shows the example of the interaction between two partners in the yeast cell: Atx1 and Ccc2. Molecular modeling and experimental studies indicate that the same mode of interaction is adopted by the homolog proteins in bacteria. However, the directionality of Cu (I) transfer may be species-dependent.

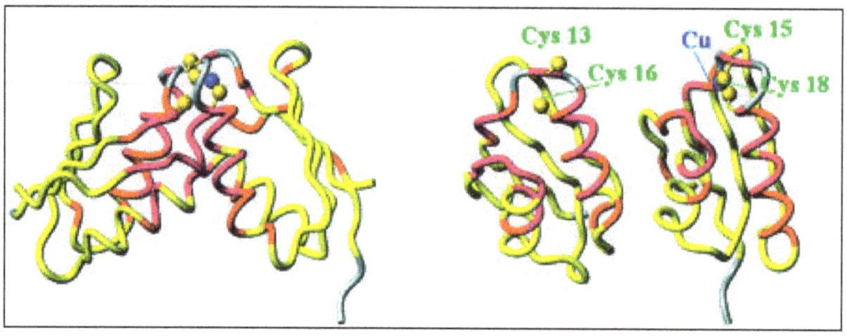

Display of the complex formed by the first domain of Ccc2 (molecule, Left) and Atx1 (molecule, Right) in solution, as determined by NMR chemical-shift mapping studies, based on the independently solved solution structures of the two proteins. The configuration of the adduct is supported by the dimerization of Cu (I) Atx1 observed in the crystal structure. Atx1 delivers Cu (I) to Ccc2 for subsequent delivery within the post-Golgi apparatus. (Left) The proposed structure of the complex. (Right) The two proteins rotated by 90° along the vertical axis, to allow the reader to observe the interaction regions. The figure is color-coded according to the variations in chemical shift of the backbone amide protons experimentally observed: regions in pink feature chemical-shift variations (in absolute value) >0.1 ppm, those in red feature variations between 0.04 and 0.1 ppm, and those in yellow feature variations <0.04 ppm. No data could be obtained for regions in gray. The Sγ atoms of the coordinating cysteines and the Cu (I) ion are also shown (gold and blue balls, respectively).

A tight regulation is indeed crucial to maintain the delicate balance of metal partitioning among the various cellular organelles, as well as between the cell and the environment. If we consider again the case of copper homeostasis, here the largest amount of information is probably available for Enterococcus hirae. The cop operon in E. hirae consists of four genes (copY, copZ, copA, and copB) and regulates copper uptake, availability, and export in this bacterium. The CopY protein regulates expression of all genes in the operon in dependence of the level of intracellular copper. CopZ is a metallochaperone homologous to Atx1, whose structure has been solved, which delivers Cu(I) to CopY, triggering its detachment from the promoter region of the cop operon and thus resulting in expression of the latter. In vitro studies coupled to site-directed mutagenesis have provided detailed information about the metal-binding properties in the two proteins, as well as on some details of their mode of interaction, even in the absence of a detailed structural model for CopY. It is interesting to note that CopA in E. hirae is responsible for copper uptake under copper-limiting conditions, whereas the homolog protein in Bacillus subtilis is most likely responsible for copper efflux from the cell.

Finally, it is worth noting that there is yet another very interesting aspect of the topic of metal homeostasis: resistance to toxic metals. Some well documented examples are those of resistance to arsenic, mercury, and, once again, copper. This last metal is used against pathogens affecting some plants. A strain of Pseudomonas syringae was found to have acquired resistance against copper, the genes responsible for resistance being localized in a copper-inducible operon on a 35-kb plasmid, which contains four genes (copA, -B, -C, and -D). For one of these, CopC (note that the single operon contained in the mentioned plasmid is called cop but does not bear any relationship to the aforementioned cop operon of E. hirae), the solution structure has been recently solved. The role of CopC in vivo is most likely that of receiving or donating copper ions from CopA, possibly in cooperation with CopD. It has been proposed that copper binding takes place through an unprecedented consensus sequence; furthermore, it appears that the mechanism of interaction between the protein and the metal ion presents peculiar features, shedding new light on the mechanism of copper resistance in P. syringae.

We have focused on the interaction between cells and copper, both as an essential metal (e.g., in the yeast cell) and as a noxious metal (e.g., for P. syringae). The reason for this choice is that copper is the essential metal for which the largest amount of information is available, ranging from genetic analyses to high-resolution structural information on both individual proteins involved in copper homeostasis and on protein–protein complexes mediating copper transfer. In addition, detailed insights are available on the mechanisms of regulation. Following on the example of copper, several studies addressing some aspects of the regulation and homeostasis of other essential metals. For instance, the tight regulation of the amount of free intracellular zinc has been demonstrated, a zinc-specific export pump has been identified and structurally characterized, and a few details about the zinc-binding site in the zinc uptake regulator are now known. The Ni chaperone (UreE) for Ni-urease has been structurally characterized, and is flourishing about iron, manganese, etc. However, we are still far from being able to draw a picture analogous to that of for the homeostasis of any metal other than copper, itself representing only a part of the overall mechanism of copper homeostasis. This is an area where bioinorganic chemists will have to be proactive in the coming years. It is also important to mention that metal homeostasis and its regulation are not "isolated" cellular processes but are tightly connected with all of the metabolic pathways in which metal-binding proteins are involved. (Just to give an example with reference to, copper homeostasis is linked to respiration via the enzyme cytochrome c oxidase, which necessitates copper to perform its function, or to iron metabolism via the Fet3 oxidase.) Also, it is often the case that the homeostasis of different metals is interlinked, so that in conditions where there is a shortage of a given metal, a cell can still maintain the functions that would require availability of this metal by producing proteins that use different metal ions [e.g., Monoraphidium braunii can use either plastocyanin or cytochrome c6 as an electron carrier in photosynthesis depending on the availability of copper].

The hundreds of final users of the metal ions, i.e., the enzymes (and some other metalloproteins) using the metal to perform their catalytic function. Such processes may also control illnesses. At present, there may be a tendency to consider them passive actors, anxiously waiting for their metal to be delivered. It can be anticipated, however, that this picture is not entirely true, and that these systems may actually be active in signaling their need for metal ions, thus playing a more relevant role in the broader picture of the cellular processes. Finally, metal ions play a prominent role in protein folding/unfolding, as well as neurodegenerative diseases.

Essential and Trace Metals

Of the 109 known elements, 30 elements are believed to be essential for the survival of living organisms. Nineteen of the 30 are trace elements, out of which 12 are transition metals. We can account for the biological activity of about one-half of the essential trace elements which function in metalloenzymes, including Fe, Zn, Cu, Mn, Mo, Co, Ni and Se.

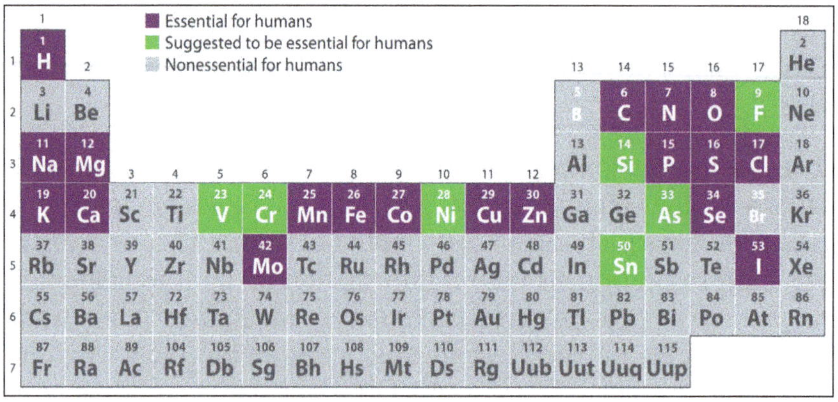

Essential elements shown in the periodic table.

Knowledge of the biological function of the trace elements has lagged far behind. The treatment of anemia with iron and the association of iodine deficiency with goiter marked these as the only two trace elements recognized as essential for animals. In an early 20th century discovery of the essentiality of copper, zinc, cobalt, manganese, and molybdenum in animals were the main thrust.

These efforts have resulted in evidence supporting the essentiality of selenium, chromium, tin, vanadium, fluorine, silicon, nickel, lead, cadmium, arsenic, and lithium.

Criteria for Essential Elements

The criteria for an element to play an essential role in our body are as follows:

- A physiological deficiency appears when the element is removed from a purified diet.
- The deficiency can be relieved by the addition of one specific element.
- A specific biochemical function is associated with a particular element.

The simplest definition of an essential element is that, it is an element required for the maintenance of life; its absence results in death or a severe malfunction of the organism. This rigorous criterion cannot always be satisfied and this has led to a broader definition of essentiality. An element is considered essential when a deficient intake produces an impairment of function and when restoration of physiological levels of that element relieves the impaired function or prevents impairment. The organism can neither grow nor complete its life cycle without the element in question. The element should have a direct influence on the organism and be involved in its metabolism. The effect of the essential element cannot be wholly replaced by any other element.

The essential trace elements provide a classic example of required nutrients. An organism passes through several stages as the concentration of essential nutrient progresses from deficiency to excess. In absolute deficiency, death may occur.

Determination of Trace Elements

However, trace elements are not only blessing but also curse by generating environmental and health-related concerns when exceeding certain concentration levels. In the context of environmentally sensitive trace elements, it is often not easy to draw a clear line between the desired concentration range of a trace element and the range in which it already exerts toxicity. Aquatic environments, soil, organisms, food, fodder, energy carriers like coal, and airborne particles are targets, which are potentially contaminated with trace elements; this, in turn, provokes the need for modern, reliable, and fast tools for determination of trace element in diverse matrices. Although many problems caused by trace elements are already well managed in the meanwhile, constant vigilance in environmental exposure, application, and nutritional supply of trace elements still displays the conditio sine qua non, and calls for intensified research in trace element determination. For example, during mining, beneficiation, and combustion of coal as well as during metal mining and follow-up treatments, a total of about 26 trace elements, among them toxic heavy metals like mercury, needs to be considered and quantified. In the biological field, many plants are described to heavily accumulate trace elements from soil (metal "hyperaccumulator" plants); to understand these uptake mechanism and how the plants develop the required resistance factors needed to withstand the high concentration of trace elements, which often are highly toxic heavy metals, needs highly precise detection devices. The determination of type and concentration of trace elements in nutrition, body tissue and liquids, coal, water, soil, and so on is regarded as the first and most important step to follow the mechanisms controlling the dispersal and accumulation of trace elements. Element speciation in different media (water, soil, food, plants tissue, coal, biological matter, food and fodder, minerals, etc.)

is pivotal to assess an element's toxicity, bioavailability, environmental mobility, and biogeochemical performance. Classical methods for elemental analysis, such as gravimetric, titrimetric, calorimetric, and so on, do not sufficiently address the precision requirements when analyzing elements in trace concentrations. Therefore, new analytical techniques have been developed, which greatly simplified the quantitation of many trace elements and considerably extended their detection range. In this context, the development of reproducible and accurate techniques for trace element analysis in different media using spectroscopic instrumentation is continuously advanced. Here, inductively coupled plasma mass spectroscopy (ICP-MS) analysis of trace elements is of increasing importance; among the different ICP-MS techniques, modern developments like dynamic reaction cell inductively coupled plasma mass spectrometry (DRC-ICP-MS) was successfully used to simultaneously determine 17 trace elements in blood samples. Similar techniques encompass sector field inductively coupled plasma mass spectrometry (SF-ICP-MS), which was successfully used for determination of aluminum, beryllium, cadmium, cobalt, chromium, mercury, manganese, nickel, lead, and vanadium in urine, serum, blood, and cerebrospinal fluid. Moreover, laser ablation inductively coupled plasma mass spectrometry (LA-ICP-MS) constitutes a quantitative microbeam technique for rapid and accurate determinations of trace element concentrations in the sub-ppm range in various target materials. A classical method, atom absorption spectroscopy (AAS), is still widely used to quantify various trace elements, for example, the well-known graphite furnace atomic absorption spectrometry, which was, inter alia, successfully used for trace element determination in fish samples. Prior to AAS measurement, biological samples to be investigated for trace elements need to be solubilized by different ashing techniques; here, a multitude of methods exists, with the use of nitric acid for "wet ashing" often being the method of choice. Decades ago, X-ray fluorescence analysis was recognized as a viable tool to determine trace elements in geological samples; in this context, a study reported by Leoni and Saitta in 1976 illustrated the viability of this technique to determine a total of 29 metals in rock and mineral samples in an astonishingly broad concentration range between 1 and 5000 ppm. This AAS technique was later advanced to make it more custom-oriented and better applicable to different types of samples. Moreover, radiochemical trace element analysis is also widely used, mainly using radiochemical neutron activation analysis. By resorting to this method, trace elements were accurately determined in various samples, which are as diverse as human tissue, food products, and also in terrestrial, lunar, and meteoritic rock samples.

Essential Ultra Trace Elements

The essential ultratrace elements are required for growth and survival of cells and organisms. They normally occur and function in cells at extremely low concentrations, usually far less than 1μM and as low as 10^{-8} to 10^{-9} M.

The Essential Ultratrace Metals

Of the essential ultratrace metals only four (Mn, Mo, Co, Ni) have been clealy identified as forming metallo-enzymes. Manganese metalloenzymes are present for several important enzymes superoxide dismutase (mitochondrial), arginase, pyruvate carboxylase, and glycosyl transferase. In molybdo-metalloenzymes, the metal exists as a Mo cofactor, a complex of the metal with a novel organic molecule, called molybdopterin. The role of cobalt is the best understood of any of the essential ultratrace metals. Both vitamin B_{12} and coenzyme B_{12}, have a cobalt ion complexed in their equatorial positions by four nitogens of a macrocyclic ligand called corrin.

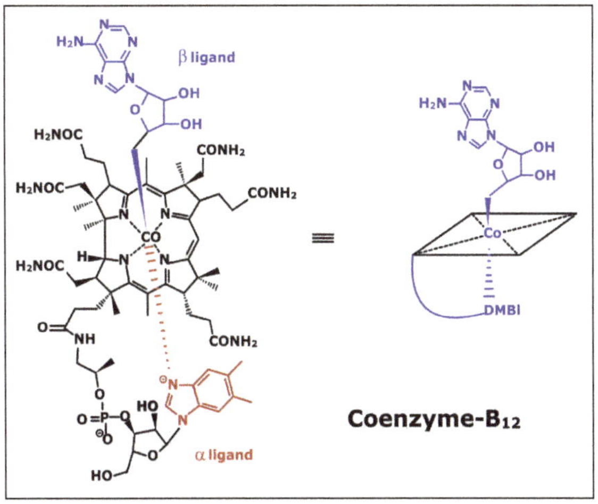

Co ion in coenzyme B_{12}.

The Essential Ultratrace Non Metals

The essential non-metals comprise a much more heterogenous group than do the transition metals, whose properties are all closely related. The two essential halogens, F and I are highly specific and very different. Fluoride has a remarkable anti – dental caries effect. This may be related to its ability to replace OH, thereby stabilizing the structural matrix of bones and teeth. Fluoride also can inhibit strongly certain key enzymes, enolase, pyrophosphatase. Among the non-metals the most important is selenium. Despite its high toxicity, selenium has been shown to be a component of several enzymes involved in essential oxidation-reduction reactions.

Toxicity and Deficiency

Most of the essential elements are also toxic when taken in excess. Too much copper causes necrotic hepatitis and hemolytic anemia. Selenium is the most toxic of the essential element, and intakes greater than 4 μg g-1 is serious. Too much manganese causes psychic and neurological disorders. The dual action of trace elements, in relation to toxicity and deficiency is represented by a dose-response curve (Figure). The cut-off points, between deficiency, normal health and toxicity, vary from nutrient and nutrient, and from person to person. Both insufficient and an excess of iodine produces thyroid problems. Deficiency of Cu and Se may contribute to kwashiorkor. Deficiency in Co, a component of vitamin B_{12}, can lead to anaemia.

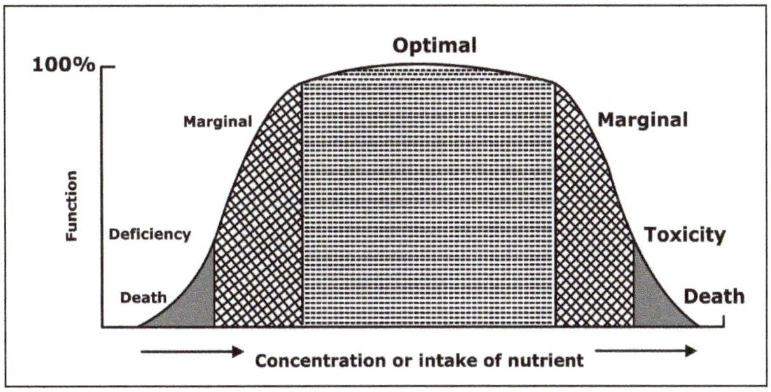

Dose response range of an essential element.

The nature of the chemical form of an element influences the dose response relationship. For example, the cations Sn_{2+} and Ni_{2+} are not particularly toxic, the chemical species $Sn(CH_3)_3{}_+$ and Ni(CO)4 are dangerously toxic. In these cases the toxicity is due to the whole species, not just the trace element. In the case of $Ni(CO)_4$ the toxicity arises from the dissociation of the CO groups. Chromium (VI) is more toxic than chromium (III) which is due to its strong oxidising properties.

Toxic Effects of Trace Elements

Lead

Lead is so widespread in our society; it is probably the most serious toxic metal. Around 70% to 90% of the lead assimilated goes into the bones, the next major accumulator is the liver, then the kidneys. Divalent lead readily replaces Ca_{2+} in bone, and either becomes firmly fixed or reversibly fixed. Lead in the latter form may be released into the blood stream and a person who has been purged of lead may have a reoccurrence of lead poisoning at a later date to release from bone tissue.

Symptoms of lead poisoning are many and not always the same for every person. The principal danger areas are, the hemopoietic system, leading to anaemia, the nervous system, leading to irreversible brain damage, and the renal system. Young children are particularly susceptible, the damage showing up in behavioural and educational abnormalities, and may be mental retardation.

Arsenic

Arsenic has similar toxic properties to Pb, Hg and Cd, as regards bonding to sulphur and inhibiting enzyme action such as pyruvate dehyrogenase. Arsenic, which is found mainly in the liver, kidneys, lungs and intestinal walls, is readily absorbed if water soluble. Since P and as are in the same periodic group, and since P is a major essential element, it is not surprising that arsenic interferes with phosphorous metabolism. Thus arsenic is toxic by replacement by phosphorous in ATP and in coagulation of proteins.

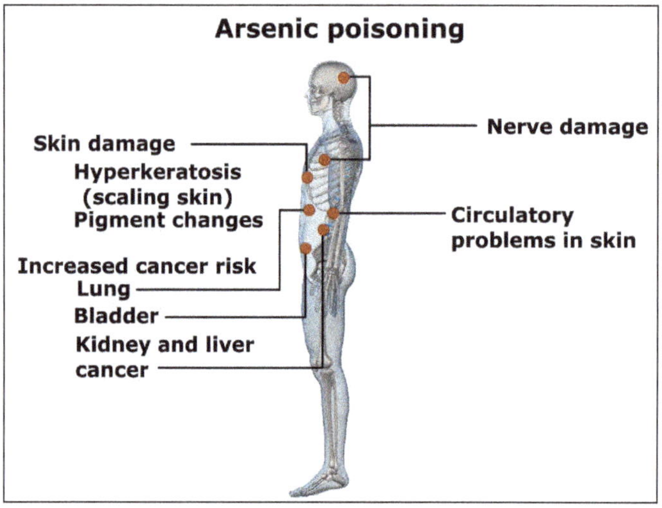

Arsenic poisoning.

Cadmium

Cadmium in lower levels has an adverse effect on human beings. Tobacco contains around 1 μg g-1 of Cd, so up to 2 to 4 μg of metal are inhaled from smoking one packet. Approx. 25 to 50% of inhaled cadmium is absorbed into the body, while from food the absorption is about 6%. A lethal dose of Cd is > 350 mg. Cadmium accumulates in the liver and kidneys produces irreversible kidney damage. Respiratory and pulmonary damage is reported to occur from the breathing of cadmium vapour or particulates. Cadmium cannot cross the placental membrane and the mammary gland is an effective barrier so it does not affect the central nervous system. Therefore new-born babies are quite free of any cadmium. When excessive amounts of cadmium are ingested, it replaces zinc at key enzymatic sites, bringing about metabolic disorders.

Mercury

Mercury is the most toxic heavy metal, and many serious incidences have resulted from mercury poisoning (Figure). The toxicity of mercury is related to the chemical form. Liquid mercury appears to have little effect, but mercury vapour is readily absorbed into the bloodstream producing brain damage.

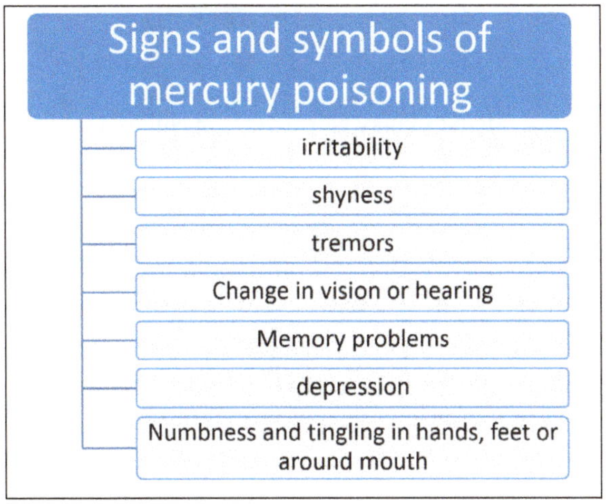

Symptoms of Mercury poisoning.

Mercury as a toxic metal came to the limelight after the incidence of "Minamata disease" in 1953-60 in Japan. Nearly 111 cases of mercury poisoning were reported among persons who had eaten mercury contaminated fish from Minamata Bay. Among them, about 45 people died. Genetic defects had been observed in 20 babies whose mothers had eaten seafood from the bay.

Elemental mercury has been fairly inert and non-toxic. If it is swallowed, it gets excreted without causing serious damage. Hg vapour, when inhaled, enters the brain through the blood stream, giving rise to severe damage of the central nervous system. Alkyl mercurals are even more dangerous.

Copper

Copper is essential for all forms of life, problems arise when it is deficient or in excess. In some respects intake of essential elements is more critical than for toxic elements. The daily human consumption of Cu is around 2-5 mg, while the required amount for a child is 1 to 1.6 mg per day, and for adult 2mg per day while less than 0.3 mg per day produces deficiency. Copper is mainly distributed in the liver, kidneys and intestines.

Copper deficiency means reduced protein and enzyme activity. Signs of this are: anaemia, loss of hair pigment, reduced growth and reduced arterial elasticity. Menkes kinky hair disease in children (Figure) is due to a deficiency of cytochrome oxidase, and is fatal by the age 3-5 years. Wilson's disease occurs with excess Cu where the biosynthesis of ceruloplasmin is suppressed, it is an inherited disease.

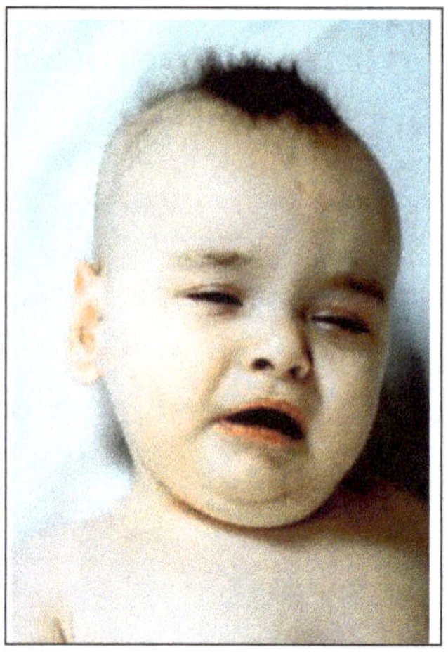

Menkes kinky hair disease in children.

Zinc

The element zinc occurs in over 80 proteins and enzymes, mainly peptidases and anhydrases. The metallothioneins are zinc storage proteins which are useful in combating Cd and Pb poisoning. The common problem with zinc is having a deficiency. A daily consumption of < 5 mg will produce deficiency symptoms, such as dwarfism, dermatistis, loss of taste, immature gonads and delayed wound healing.

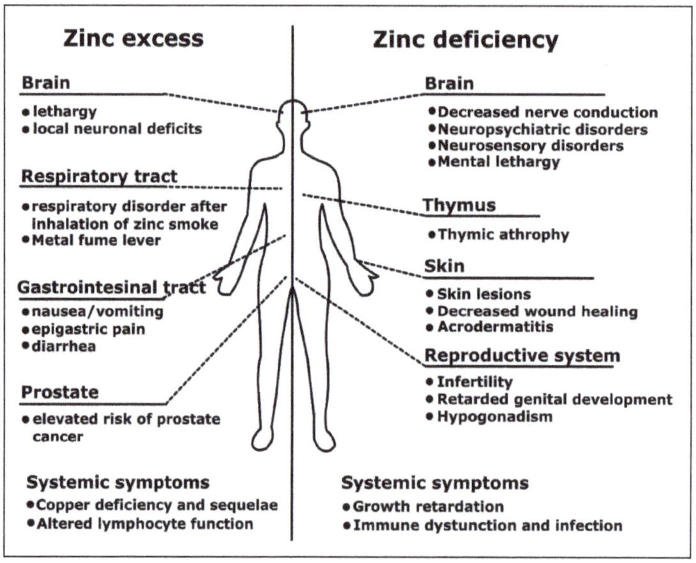

Effects of deficiency and excess of zinc.

Biomineralization

Biomineralization is the process by which living organisms produce minerals. Biomineralization processes often lead to the hardening or stiffening of the mineralized materials. This includes the formation of silicates in algae and diatoms, carbonates in invertebrates, and calcium phosphates and carbonates in the hard tissues of vertebrates.

Examples of biominerals.

The production of morphologically complex minerals according to a genetically determined blueprint mainly results from the necessity for robust support and defense structures. In principle, there is no natural preference for inorganic or organic support materials in endo- or exoskeletons; for instance, the relatively rapidly assembled chitin (polysaccharide) structures of the invertebrates and the skeletons of sharks consist largely of organic-chemical material. In fact, most biomineral constructions feature an organic/inorganic composite texture; the bones of vertebrates thus consist of the calcium 'mineral' hydroxyapatite and an organic matrix. The advantage of the inorganic component is its hardness and pressure-resistance, which allows for the existence of larger land living creatures; the organic matrix consisting of collagen fibers, glycoproteins and mucopolysaccharides guarantees elasticity as well as tensile, bending and breaking strength. Because of such advantages much of the modern material sciences are focused on the development of composite materials, in particular fiber-enhanced materials, and microstructural physical methods of analysis such as high – resolution electron microscopy are now being used in both areas of natural and synthetic composite systems. The most important biominerals, some in different polymorphic. Forms are summarized in table with respect to occurrence and main functions.

Table: Most important biominerals.

Chemical composition	Mineral form	Solubility exponent pK_{sp} at pH 7	Occurrence and function
Calcium carbonate			
$CaCO_3$	Calcite	8.42	Exoskeletons (eg egg shells, corals, mollusk shells), spicules, gravity sensor.
	Aragonite	8.22	
	Vaterite	7.6	
	Amorphous	7.4	

Calcium phosphates			
$Ca_{10}(OH)_2(PO_4)_6$	Hydorxyapatite	13	Endoskeleton (vertebrate bones and teeth).
$Ca_{10}F_2(PO_4)_6$	Fluoroapatite	14	
Calcium oxalate			
$CaC_2O_4.nH_2O$	Whewellite, wed-delite	8.6	Calcium storage and passive defense of plants, calculi of excretory tracts.
Metal sulfates			
$CaSO_4{\cdot}_2H_2O$	Gypsum	4.2	Gravity sensors
$SrSO_4$	Celestite	6.5	Exoskeletons
$BaSO_4$	Baryte	10.0	Gravity sensors
Amorphous silica			
$SiO_2.nH_2O$	Amorphous		Valves of diatoms and radiolarians, defense functions in plants.
Iron oxide			
Fe_3O_4	Magnetite		Magnetic sensors, teeth of chitons.
α,γ – Fe(O)OH	Goethite		Teeth of chitons

The biominerals can occur as pure or mixed phases, in amorphous or (micro)-crytalline form, or as composites with polymeric organic 'matrix' materials such as proteins, lipids or polysaccharides. They can be formed intracellularly, at the cellsurface (epicellularly) or in the extracellular space. Overall, the following biominerals are predominant:

- The calcium carbonate phases aragonite and calcite which often contain some Mg^{2+}.
- The calcium phosphates, mostly as hydroxyapatite.
- Amorphous silica.
- The iron oxide/hydroxide ferrihydrite and magnetite.

All other biominerals occur either as minor components or in only a few species. In addition to the already mentioned mechanical support function the biominerals also assume a storage function because of the large amounts usually present. On the other hand, the formation of biominerals can have a deposition character or even a detoxification function, particularly in microorganisms. This implies the existence of active regulatory and transport systems which control (de)mineralization and regeneration.

Magnetite, for instance, is a widely distributed mineral in the earth's crust which is formed geologically at high temperatures and pressures. In contrast, 'magnetotactic' bacteria are able to synthesize this mineral in biologically useful form under physiological conditions.

Important functions of the biominerals consist in:

- Their use as essential constituents of mechanically robust instruments and weapons, eg. of teeth for killing and processing of food.

- The formation of sensor components eg. of specifically heavy crystallites in gravity sensitive organs or of magnetic microcrystallites in magnetotactic bacteria.
- The passive mechanical protection of animals (e.g. the shells of mollusks) and plants (e.g. silica containing spikes) against predators or climatic effects.

Nucleation and Crystal Growth

Nucleation and crystal growth are processes that occur in supersaturated media and that have to be carefully controlled in a directed mineralization process (Figure). These requirements can be met in an organism through highly regulated active transport mechanisms and through specific modulation of the surface reactivity. The transport mechanisms may include transmembrane ion flux, ion (de)complexation, enzymatically catalyzed gas exchange (CO_2, O_2 and H_2S), local changes in redox potential (FeII/III) or pH, and variations in the ionic strength of the medium. All these factors can create and sustain a supersaturated solution in a biological compartment. Nucleation, on the other hand, is linked to the kinetics of (heterogenous) surface reactions; processes such as cluster formation, anisotropic crystal growth and phase transformartions are also determined by the properties of the surface. In the biological realm there are also a number of surface structures that specifically prevent undesired nucleation; for instance, fishes in polar waters can thus protect themselves from ice formation in their bodily fluids at temperatures slightly below 0 °C.

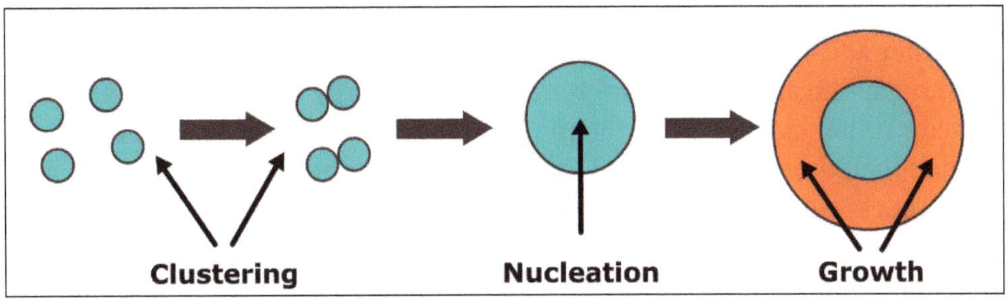

Nucleation and crystal growth.

The growth of a crystal or of an amorphous solid from the nucleus which is formed in an activation step (crystal nucleation energy) can proceed directly from the surrounding solution or by the continuous supplementation of the necessary ions or molecules. On the other hand, the diffusion of particles can be drastically altered through significant changes in the viscosity of the medium. e.g. through gel formation of a biological matrix – a mechanism which is presumably responsible for the deposition of amorphous silica in plants.

One of the most important and most fascinating aspects of biominerals is the fact that their shapes do not have to coincide at all with the regular crystallographic forms of the inorganic material. Far more significant for the final morphology are the spatial limitations through biopolymers, membranes or vesicles during the formation of biominerals, even if the completed structures are finally assembled outside such confinements. Of course, modifications of crystal growth may be induced nonbiologically; in fact, rather simple chemicals in the growth medium can influence the shape of the mineral formed. For example, the spindle shape of calcite crystals in gravity sensors can be reproduced by crystallization in the presence of 5 mM malonic acid, and monomolecular layers of stearic acid induce the crystallization of $CaCO_3$ as disc-shaped vaterite crystals rather than as rhombic calcite.

Mechanism

Crystallization from a supersaturated solution of calcium phosphates. Calcium phosphate solutions are complex and the ion species present depend upon the pH and ionic strength. The phosphate species and their equilibrium constants, at 25 °C in water, are:

$$H_3PO_4 \leftrightarrows H_2PO_4^{(-)} + H^+, K_1 = 7.11\times10^{-3},$$
$$H_2PO_4^{(-)} \leftrightarrows HPO_4^{(2-)+}H^+, K_2 = 6.34\times10^{-8},$$
$$HPO_4^{(2-)} \leftrightarrows PO_4^{(3-)} + H^+, K_3 = 4.30\times10^{-13}.$$

In a solution mixture containing total Ca = 1.5 × 10^{-3}M, total phosphate = 0.9 × 10^{-3}M at pH 7.0 and 25 C in 0.15 M KNO_3 these equilibria lead to the concentrations;

H_3PO_4 = 3.59 × 10^{-9} M; $H_2PO_4^-$ = 3.41 × 10^{-4} M; HPO_4^{2-} = 5.137 × 10^{-4} M; and PO_4^{3-} = 9.37 × 10^{-9} M.

Thus, around physiological pH and ionic strength 0.15, the predominant ion species are an essentially equimolar mixture of $H_2PO_4^-$ and HPO_4^{2-} both of which interact avidly with Ca^{2+} ion. According to the classical nucleation theory crystal formation from a supersaturated solution of the constituent ions requires local fluctuations in concentrations of the interacting ions. At some critical concentration of those ions their favorable interactions stabilize the ion cluster so that the surface free energy gain by dissociation of the ions in the cluster surface is less than the free energy gain by further addition of ions to the dense cluster or crystalline phase. That is, the process is driven thermodynamically by the difference in free energy between crystalline and solution phases, but is controlled kinetically by the magnitude of the energy barrier of critical size cluster formation, which includes the free energy of formation of a new surface. In simple 1–1 salts the crystallization process can proceed as a homogeneous reaction. The reaction free energy pathway is thus:

Supersat.Soln. Gss ⇆ Ion clusters, Gc ⇆ Critical cluster, Gn∗ → Stable crystal, Gcrt

The generalized Gibbs equation for the free energy for crystal nucleation, ΔG(n), is given in equation (Illustrated in Fig) in which n is the number of particles in the cluster and μ is the chemical potential per particle.

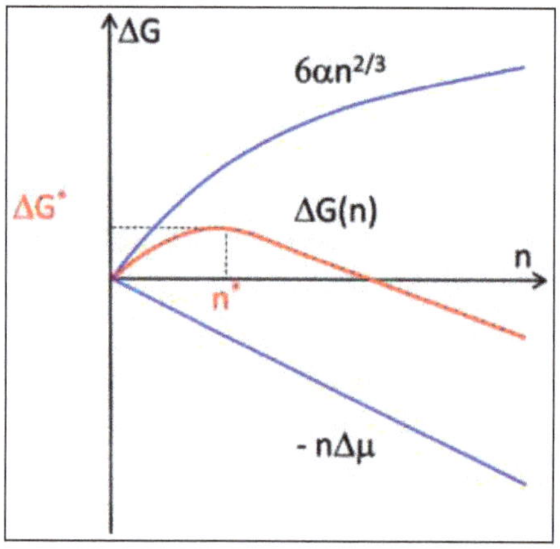

The generalized Gibbs formulation of crystal nucleation, from Equation. The upper blue curve is

the unfavorable gain in free energy because of creation of the new surface interface, the lower blue curve is the favorable decrease in free energy because the stability of the crystal relative to the initial solution free energy. G* is the critical free energy at cluster size n*, the cluster size at which the probability of release of particles from the surface of the cluster is balanced with the probability cluster growth. When n > n* the cluster will have a greater probability of further growth. From Vekilov with permission.

$$\Delta G(n) = -n\Delta\mu + 6\alpha n^{2/3}$$

Thus, the aggregation of the particles from the supersaturated solution to a more stable crystalline phase becomes more favorable as more particles are added to the aggregate. This is countered in the second term which expresses the work required to create the new interface between the supersaturated solution and the crystal aggregate. This second term is problematic since it requires knowledge of the shape of the aggregate, but this is represented and generalized by scaling the surface area to volume to $n^{2/3}$, taking α as an adjustable factor expressing the averaged interfacial energy per particle in the cluster. We can define the critical cluster size, n*, as the point at which the absolute value of $n^*\Delta\mu$ equals $6\alpha n^{2/3}$ and hence at $\Delta G_n^* = G_c - G_{SS}$ the probabilities of cluster growth and cluster dissolution are balanced. To become more specific one can assume a particular shape and correct the second term more precisely. In the classical form of this generalized Gibbs equation the shape of the aggregate is assumed to be a sphere and imposes specific geometric boundary conditions. Schmelzer et al. examined the thermodynamic, surface and kinetic parameters for homogeneous nucleation in detail and concluded that the classical Gibbs approach considerably overestimated the work of critical cluster formation but also demonstrated that the generalized Gibbs equation, with the adjustable parameter α, was still applicable. Some specific terms need definition before undertaking further mechanistic interpretation of the nucleation process. These are provided in Table and utilized in the ensuing discussion.

Table: Essential Definitions of Nucleation Intermediate States.

• Aggregate: A collection or cluster of ions, molecules and/or particles.
• Prenucleaction cluster: A cluster of ions, molecules and/or particles, stable enough to exist (resisting decay) but have neither the size nor structure sufficient enough to grow into a crystal on its own.
• Posner cluster: A prenucleation cluster that is an assembly of calcium and phosphate ions with a ratio of 1.5:1, with a size 0.9–1.0nm.
• Dense Liquid: A prenucleation cluster that exists as a separate amorphous liquid-like precursor phase distinct from either the supersaturated solution or crystal phases. The dense liquid has a defined interface, and is more dense than the surrounding solution, but less dense than the crystal form. The dense liquid is a hydrated aggregate and can be composed of ions, molecules, and/or prenucleation clusters. The water present is bound as hydration layers. The dense liquid is larger than a crystal nucleus. Crystal nuclei can form within the dense liquid.
• Critical Cluster: A cluster of ions, molecules and/or particles of sufficient size (n = n*) and structure to have an equal probability of growth or dissolution.
• Nucleation cluster: A stable cluster of ions, molecules and/or particles of sufficient size (n ≥ n*+1) and structure to continue to grow rather than decay. Has a greater probability of growth over decay. This represents the nuclei of a crystal phase.

The next interpretative advance in understanding crystal nucleation was by ten Wolde and Frenkel who introduced the concept that the crystallization could more readily take place within the prenucleation cluster rather than directly from the supersaturated mother liquor. Since the initial cluster is presumed to be disordered the initial prenucleation clusters can be considered as "dense liquid" (DL) phases and it has been suggested that crystallization may take place within that phase. Vekilov has provided a visualized description of the formation of the dense phase, which we have further elaborated upon as shown in Figure. The red dashed line (ΔG_1^*) in. depicts the formation of transient, unstable, unstructured aggregates which readily disassociate when n < n*. When fluctuations take place that lead to clusters larger than the critical value crystals may form (green dash-dot line). The diagonal arrow represents the path of cluster formation directly from solution and corresponds to the free energy landscape in. Whereas the heavy green dash-dot line in Fig follows the free energy landscapes in figure. depicts the formation of the DL cluster, and the subsequent growth of the crystal within the DL aggregate. Two possible two step mechanisms describing the free energy reaction coordinate are also shown. In both upper and lower reaction coordinate schemes of the free energy barrier to critical cluster n1* formation is the same, ΔG_1^*, and further aggregation continues to form a DL phase. If the DL is more ordered and marginally more stable than the initial supersaturated solution the lower reaction coordinate applies, then the DL intermediate will be long-lived. Regardless of the stability of the DL compared to the supersaturated solution, the development of crystallinity within the DL will be governed by the value of $\Delta G_2^* < \Delta G_1^*$.

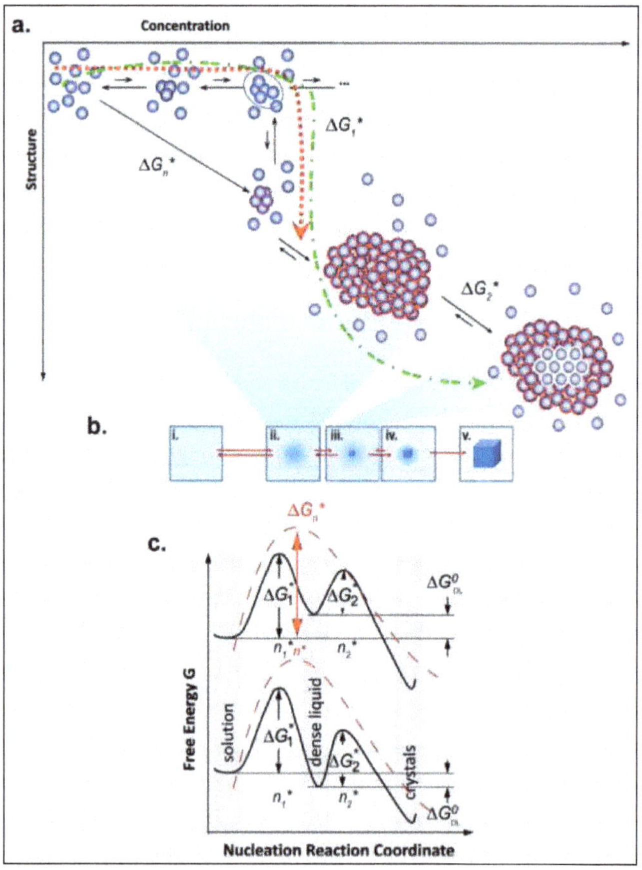

Schematic illustration of the two-step mechanism of nucleation of crystals, modified from work presented by Vekilov. In this two-step mechanism a dense liquid cluster forms, and a crystal nucleus may form inside the cluster. (a) Microscopic viewpoint shows crystal formation in the plane of two order parameters (Concentration and Structure) inspired by the work of ten Wolde and Frenkel ; (b) macroscopic viewpoint of events along the energy pathways (solid lines) depicted in (c). (i) The supersaturated solution. (ii) The formation of the dense liquid depicted in most of (a). (iii) The formation of the critical cluster for the crystal at n2* within the dense liquid, also shown in (a). The creation of the nucleation cluster and subsequent growth of the crystal. (v) The fully formed crystal. (c) The pathway for the change in free-energy ΔG, along two possible versions of the two-step nucleation mechanism. If dense liquid is unstable and ΔG_0 DL > GSS(ΔGoDL standard free energy of formation of dense liquid phase), the upper curve applies; if dense liquid is stabilized with the introduction of a foreign interface, ΔG_0 DL < 0, the lower curve applies. $\Delta G_1{}^*$ is the barrier for formation of a cluster of dense liquid, $\Delta G_2{}^*$ is the barriegr for a formation of a crystalline nucleus inside the dense liquid.

To move from these generalizations to the mechanism of crystal nucleation from a prenucleation, DL precursor, we need to consider the experimental verification of the idea of the prenucleation cluster in solution as part of a homogeneous nucleation process. The best direct demonstration of this came from studies of the calcium carbonate system. Wolf et al. suspended a droplet of a calcium carbonate solution between a piezoelectric vibrator that generated an acoustic wave and a concentrically adjusted sonic reflector. The droplet could be held in place without any extraneous constraints other than the droplet air-water interface. The concentration of the droplet increased as the solvent evaporated and the state of the solute was followed by in situ measurement of the wide angle x-ray scattering determination of crystallinity over time. In parallel transmission electron microscopy (TEM) of droplets collected at suitable time points, along with scanning electron microscopy (SEM) and cryogenic SEM, data led to the conclusion that "in this completely contact-free system a homogeneous formation of $CaCO_3$ proceeded via an amorphous liquid-like state, unambiguously without artifacts." These data demonstrated the development of nanoscale liquid-like cluster formation in pure calcium carbonate solution, free of any organic polyions or proteins, showing unequivocally that metastable liquid-like cluster formation generated by density fluctuations in the inorganic ions in the bulk phase reflected the direct interactions of the inorganic ions involved. The intervention of organic polyions or other heterogeneous particles or surfaces could perhaps alter kinetics of liquid-like prenucleation complex formation, but clearly the heterogeneous factors are not required for crystal nucleation. Mahamid et al. And Beniash et al. proposed that amorphous calcium phosphate precursor phases were involved, respectively, in mineralizing zebra fish fin bone and tooth enamel formation. Much earlier Posner had found that that biological apatites had an amorphous component, and proposed that a cluster of Ca and PO4 ions with composition $Ca_9(PO_4)_6$ and a diameter of 0.9 to 1.0 nm was the stable prenucleation cluster.

Recently, Dey et al. Have shown that stable prenucleation clusters of calcium phosphate of the size predicted as Posner clusters, can be observed in a simulated body fluid (SBF) at 37 °C. The clusters began to aggregate and after 24 h in the presence of a monolayer of arachidic acid, the clusters aggregated and densified on the monolayer. Initially, selected area electron diffraction (SAED) showed the clustered calcium phosphates to be amorphous as ACP, but ultimately the SAED indicated that crystallinity developed with an orientation determined by the monolayer surface. Thus, the clustering is a multistep process that begins as a homogeneous reaction. Dey et al. proposed that "the presence of the nucleating surface induces structural and compositional changes that enable the denser packing of the clusters and their subsequent fusion to form ACP" and ultimately apatitic crystals. However, crystallization could also occur in the absence of heterogeneous nucleators.

Crystallization Enhancement by Heterogeneous Surfaces and Polymeric Additives

Nudelman et al. directly examined the potential role of collagen fibrils as a templating surface for the deposition of amorphous clusters or ACP directly on the collagen to nucleate apatitie crystal formation. They examined the in vitro mineralization of reconstituted collagen fibrils isolated from horse tendon. This is a collagen which does not normally, in vivo, undergo mineralization and thus should be free of mineralization related non-collagenous SIBLING proteins. In the absence of any additives other than SBF, deposition of needles of apatite was random along the surfaces of in vitro formed fibrils. When polyaspartic acid (polyAsp), used as a model for the SIBLING proteins, was added to the mineralizing solution, calcium phosphate mineral was found on the surface of the fibrils after 24h and the mineral deposits were loosely associated with the collagen gap and overlap zones. Cryo energy dispersive X-ray spectroscopy and low-dose selected-area electron diffraction (LDSAED) showed the presence of ACP. Within 48 hours, the ACP was partly converted to apatite crystals, and these developed in 72 h to elongated hydroxyapatite (HAp) crystals, but still within a bed of ACP, comparable to typical bone apatite. Remarkably, however, if the polyAsp was first absorbed on the collagen, before the collagen was exposed to the mineralizing buffer, no mineralization took place except on the surface. The in vitro mineralization with apatite deposition within the fibrils was found only when the polyAsp and SBF solution were mixed before placement on the collagen fibrils. This directly implicated the participation of the SIBLING analog polyAsp in the formation of cAp prenucleation clusters, or some other aggregate, and their interaction with the collagen.

The formation of the prenucleation cluster and its subsequent crystal growth in the presence of anionic polymer has been treated as a Polymer-Induced Liquid-Precursor (PILP) process, and a very complete discussion of the phenomenology of the PILP process has been given recently by Gower, particularly focusing on the in vitro production of biomimetic minerals. Although the nature of the PILP process is clear, the mechanism of enhancement of the nucleation by the participation of the inducing polymer, for example the polyAsp in the heterogeneous nucleation process described above, has not been described in equal depth. How this is accomplished has great importance in the actual or real biological systems in which biomineralization takes place. The questions posed above as to where the prenucleation clusters form, the specificity of placement of the mineral phase in the collagen matrix, the regulation of crystal growth within the fibrous matrix, are general to all systems, but in each case of mineral deposition we know that bone, dentin and even mineralizing tendon utilize different members of the SIBLING family proteins to promote and/or inhibit mineral deposition and produce bones and teeth with near equivalent physical properties, but with different crystal sizes and arrangements tuned to the required mechanical and dynamic properties of the particular tissue involved.

Nudelman et al. have argued that the placement of the prenucleation clusters on the collagen and their role in moving the mineral into the collagen fibrils as well as leading to mineralization along the fibril surfaces, is related to the observation that the presence of the SIBLINGS and acidic non-collagenous proteins slows rather than accelerates the precipitation of the calcium phosphate prenucleation clusters. It has been observed many times that the SIBLING proteins may act to accelerate or inhibit mineralization depending on the factors of concentration and temperature. From this point of view, the SIBLING proteins may all be considered as mineralization inhibitors. The duality of activity of these proteins depending on the conditions and context illustrates their role as regulators of crystal nucleation/growth kinetics.

SIBLING Proteins and their Interaction with Collagen Fibrils

The SIBLINGs are among the class of proteins now designated as Intrinsically Disordered Proteins (IDP), that is, in solution they are flexible polypeptides with no long range regular structures although they can have limited sequences exhibiting, for example, local α or β conformations, and further structural order may be induced by interaction with other proteins. Electron micrographs of 0.1 μg/mL of DPP in 10 mM Ca^{2+} at pH 7.5 showed the DPP as roughly globular structures of about 16–18 nm in diameter. When the same preparations were added to EM grids which had turkey tendon collagen fibers deposited on them, the DPP decorated the surface of the fibrils with similar sized DPP structures, distributed along the fibril with perhaps 75% of the collagen bound DPPs associated with the fibril "e"-bands. Thus it was suggested that this e-band position, corresponding to the collagen fibril gap region, was a favored site for DPP binding. Further EM studies of the collagen-DPP interaction products in ammonium formate buffers without Ca2+ ions and using soluble monomeric collagen and purified DPP at different mixing ratios showed that there was a preferred DPP binding site near the amino-terminal region of the collagen monomers. This specific binding was differentially detectable only at very high collagen/DPP ratios. As the concentration of DPP was increased relative to the soluble collagen, additional DPP particles were seen bound non-specifically along each single collagen molecule. These collagen monomers +DPP readily aggregated into asymmetric fibrillar structures. The intensity of binding at a molar ratio of 1 to 1 rather than at electrostatic equivalence suggested that a non-electrostatic binding was involved at the primary specific interaction site, emphasizing the probable specificity in DPP sequence responsible for its primary collagen localization. Though polyAsp can do the in vitro job of nucleating cAp, the evidence is strong that, although the sequence of DPP contains approximately 40% Asp residues, the phosphorylation of the Ser residues of DPP (50% of the amino acids) is crucial for its function in mediating mineralization and, most recently, it has been shown that the introduction of a DPP cDNA into a non-mineralizing fibroblast cell leads to expression and secretion of phosphorylated DPP and the induction of HAp mineralization in the fibroblast cultures.

Delivery of Ca and PO_4 Ions to the Mineralization Front

The in vivo situation is much more complex than the in vitro model experiments suggest, and relatively few biomineralization discussions include the problem of the Ca and PO_4 delivery to the mineralization front, although such discussions have appeared since the pioneering studies of Bonnucci, Anderson, Lehninger and Greenwalt et al. regarding the role of the mitochondria in taking in large amounts of Ca and phosphate ions, accumulating them and delivering calcium phosphate in membrane bound vesicles (Matrix Vesicles, MV) to the extracellular spaces of mineralizing cartilage and bone. The mechanisms by which vesicles may be broken open, and their mineral contents transferred and localized to the collagen are still a subject of current study and debate. There is, indeed, abundant evidence that the mineral containing intracellular vesicles do exist, and their contents in different situations contain amorphous calcium phosphate, or apatitic needle-like crystals. The questions which remain unresolved deal with the mechanisms by which the mineral is transferred from the mitochondria to the mineralizing matrix, and then reordered, in essence dissolved and recrystallized, on the collagen fibril matrix on the fibril surfaces, or within the fibrils themselves.

Transiting from the In Vitro to the In Vivo Situation

All of the preceeding discussion has examimed the mineral ions and other components in idealized and controlled mineralization conditions in vitro. It is worthwhile to consider these same factors in the in vivo situation and determine how that might modulate the various component interactions. Particularly first from the perspective of the cytoplasm and ECM themselves, and the crowding of the macromolecules within them. Different types of cells vary in macromolecular content, over the range of 17 to 35% by weight, and the cell water content is divided between two phases, about half as the macromolecular/protein-bound water of hydration and the other half "bulk" or free water. Other components of the cell, such as the inorganic cations and anions in aqueous solution in the cell also have layers of bound water of hydration that are more ordered than in the free water. Thus, one must specifically ask about the role of the hydration layers in the aggregation of the ions into clusters, and their continued densification during crystallization, as well as the question of how the ion clusters are localized to the collagen matrix. The unusual properties of water are directly related to the fact that at ambient temperatures free (bulk) water molecules form extensive hydrogen bonded networks, albeit with energies of only a few kcal/mole (H-bond between two waters ~4.3 kcal/moL, van der Waals interaction ~1 kcal/mol, ionic bond 4–7 kcal/mol, a C-C bond ~84.0 kcal/mol). In liquid water the time scale of randomization of the networks by the rapid formation and dissociation of hydrogen bonds is on the order of approximately 2.6 ps. Fayer has shown that the water orientational relaxation requires H-bond rearrangements, and that these relaxations are slowed by the presence of interfaces of all sorts, from hydrophobic surface films, to organic macromolecules (regardless of being charged or uncharged), and in the structure of the hydration layers around inorganic cations and anions. By studying the state of the water molecules in nano-confined systems, created as the water held within spherical reverse micelles, or water held between the layers of water confined between lipid monolayers, it is clear that the first shell of bound water has substantially different properties than that in the bulk water, and much longer relaxation times, on the order of 18 ps, although liquid water clearly has no long range stability.

The first hydration shell of 6 H_2O molecules coordinated to a Ca^{2+} ion plus the next shell of 12 H_2O molecules is depicted in Figure for the most stable conformation of $(Ca(H_2o)_6)^{2+}(H_2o)_{12}$). The oxygen atoms of the first shell water molecules are at 2.35 Å from the central Ca atom, whereas the second shell O atoms are at 2.79 Å from the first shell O forming a significantly less stable H-bonded network. The relaxation time for the second shell water is 2 to 3 times longer than the bulk water relaxation time, while the relaxation time for the first shell water can be upwards of 50 times longer than the bulk water. In addition to the stability difference, the water layers shield the electrostatic potential of the Ca ion. The water of the first hydration shell of the phosphate ion is similarly more tightly held than the second shell water with a retention time about 2 times longer than the bulk water, Figure, but the first shell arrangement is more complicated. Two types of water are evident, a strongly bonded P=Op- H-Ow of a water molecule in a linear arrangement, equivalent for each phosphate Op and set of interstially located waters that appear to bridge between two Op resulting in weaker bonds to the water hydrogens with a length of 2.0–2.2 Å. In total the first shell of bound water is comprised of 12 water molecules, although molecular dynamics predicted that on average, 13 water molecules would be included. The polarity of the H-bonded bound water around phosphate is opposite to that of the hydration layer around the Ca ion. It is interesting that in the simulations, all 4 phosphates O's appear to be equivalent in bonding to water. Similar water binding and polarity organization would be found in the arrangement of first layer water in organic phosphates on phosphoserine and on carboxyl groups on Asp or Glu in polypeptides. Thus phosphorylated polypeptides and proteins such as phosphophoryn, with sequences such as $(DS^*S^*)_n$ or $(DS^*DS^*)_m$ (S* = phosphorylated serine) should, even for IDP, exhibit structured ionic patches of hydrated surface with similar strongly bound first shell water. At the nano-scale where the initial nucleation clusters are forming all water molecules are not equivalent and the water should be treated as a direct reactant, as important as the Ca or phosphate ions or protein side chains.

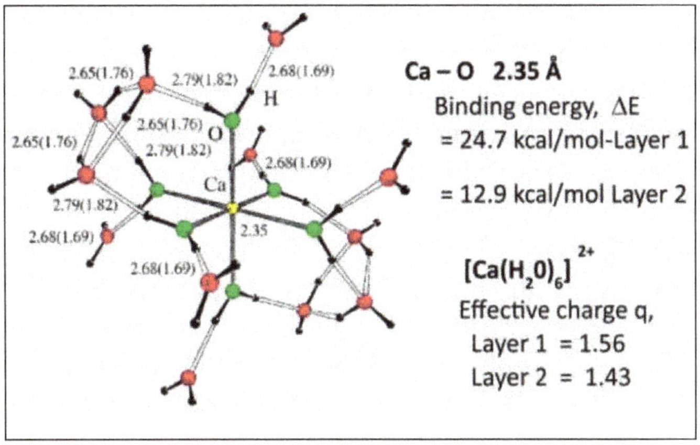

Structured water around a Ca^{2+} ion. The central Ca ion is indicated in yellow, the 6 oxygen atoms of the first water shell, coordinated directly to the Ca are in red. The oxygen atoms of the 12 second shell waters are in green. The polarity of the water molecules is quite well fixed and reduces the effective charge on the central Ca ion in so far as long range electrostatic interactions between ions is considered. Pavlov et al., with permission.

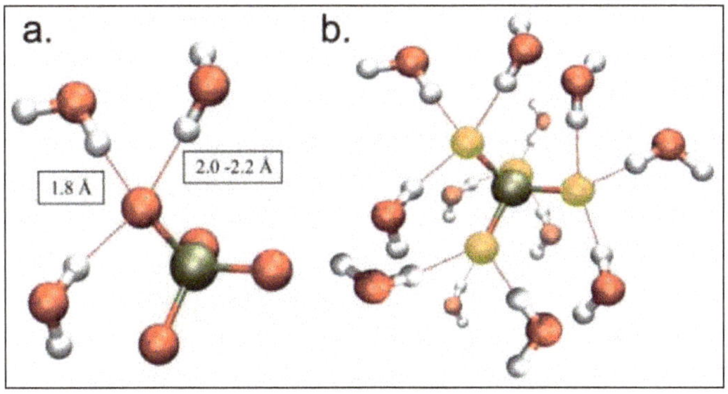

Structured water around a PO43− ion. (a) The central P atom is indicated in gold, 3 waters of the first shell are shown coordinated directly to the phosphate molecule. (b) The central P atom is indicated in gold with the oxygen atoms on the phosphate indicated in yellow. The 12 waters of the first shell are illustrated. (a) From Pribil et al. with permission. (b) Is our construction of the complete first shell, as suggested, with the understanding that the orientations of the non-axial waters will be affected by the waters of the second shell.

The Nucleation Mechanism

As depicted schematically in Figures and the related discussions, the Ca and phosphate ions in the supersaturated aqueous solution are now seen as each carrying closely bonded water, differentiated from the bulk water in stability. We propose that, as the electrostatic interactions attract the oppositely charged ions (such as Ca^{2+} and PO_4^{3-}), the interaction is modulated by the restrictions on the mobility of the bound first layer water. Thus, the initial aggregates remain in a solvated state without long range order, but now in the DL state with the free, 2nd shell water released. The enhanced stability of the DL water allows time for the final steps of densification to establish order and crystallinity from the nucleation clusters. From this perspective it is easy to see why and how the introduction of a stable but compatible polymeric polyion interface also having a restricted water layer, can be so important in modulating and /or guiding the final location and rate of formation of the final stable crystals.

There are many unanswered questions that need to be examined in the light of the mechanism proposed above. The specificity of the nucleation must reside in the nature of the interfaces presented to the physiological, supersaturated calcium phosphate solution. Obviously, heterogeneous nucleation can be enhanced by the presence of interfaces which allow the aggregation and densification of the liquid-like prenucleation clusters. The polymeric polyions, as exemplified by the SIBLING proteins of the ECM that are both interactive with the Cations and localize their presence to the matrix fibrils need to be examined for their dual roles of localization of the crystals and control of the rate of deposition within the structural matrix. The effect of the presence of other ions, such as Mg^{2-}, HCO_3^-, CO_3^{2-}, should be studied with respect to the rate of formation of the prenucleation clusters, and their effect on the stepwise densification to the DL state. The question of the role of the collagen fibrils and the mechanism of entry of the crystal nuclei or the DL phase is an important target for further studies. Finally, the basic step of the delivery of the calcium and phosphate ions into the extracellular space needs to be reexamined. If the matrix vesicle mechanism is important for delivery of the contents of partitioned vesicles into the ECM, the question of how the mineral is transferred from a MV to the inner regions of the bone and dentin collagenous matrices needs careful analysis.

Examples of Biominerals

Calcium Phosphate in the Bones of Vertebrates

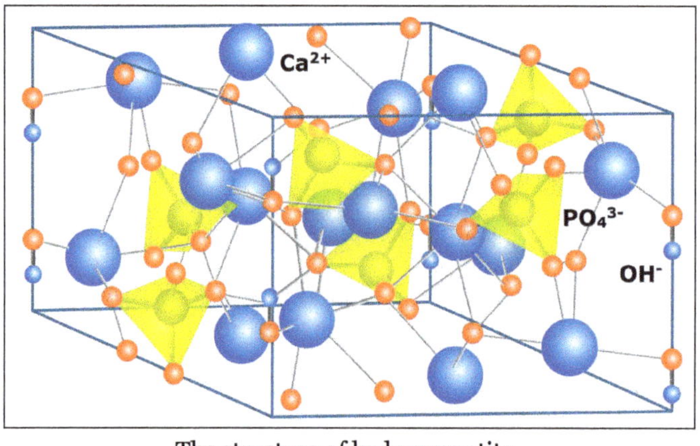

The structure of hydroxyapatite.

The dry support structure of vertebrate long bones consists of elastic fibrous proteins (approximately 30%, mainly collagen) and the inorganic components which are imbedded in 'cementing' glycoprotein: calcium phosphate, microcrystallized mainly as hydroxyapatite (55%), small amounts of calcium carbonate, silica, magnesium carbonate, other metal ions and citratre (15%). Collagens are fibrous proteins with a molecular mass of about 300 kDa; three polypeptide chains are wound together in a fibrilles as a superhelix. Apatite is also a well-known as a nonbiological mineral; it has assumed great importance as fertilizer and as the basic material for phosphorus chemistry. The inorganic material of bones, the bone meal, is also being used as fertilizer while the main organic component can be converted to collagen glue. The structure of hydroxyapatite is given in figure.

Calcium Carbonate

In egg and molluscan shells, the $CaCO_3$ crystals grow in a preformed arrangement of proteins and polysaccharides. The pure organic matrix can be obtained from these shells after dissolution of $CaCO_3$ through chelating agents such as EDTA. The matrix is composed of a water soluble protein and oligosaccharide component and of water – insoluble hydrophobic proteins, also with significant polysaccharide content. The soluble proteins and sulfate- containing oligosaccharides are strongly acidic and therefore well suited to bind Ca2+. Significant amounts of carboanhydrase are also present; they are obviously necessary to produce HCO_3 - supersaturated solutions, thus allowing for a rapid growth of shells.

The limestone deposits in corals reefs which display a highly species- specific architecture are thus a direct consequence of the symbiosis of polyps with photo-synthetically active algae. CO_2 uptake and $CaCO_3$ precipitation by marine organisms may depend on temperature, salinity, buffer capacity of the medium, and on the original pH value. $CaCO_3$ deposition is particularly favored in compartmentalized cells with small volumes since diffusion effects are negligible in such cases.

Amorphous Silica

While silicon in form of the silicates comes second relative to all other elements in the earth's crust with regard to quantity, it plays only a marginal role in the biosphere. This can be partially attributed to the low solubility of 'silicic acid', H_4SiO_4, and its oligomeric condensation product 'amorphous silica'; in water between pH 1 and 9 this solubility is about 100 – 140 ppm. In the presence of cations such as calcium, aluminium or iron, the solubility decreases markedly, and in sea water, the solubility is only about 5ppm. In the biosphere, dissolved amorphous silica is resorbed by organisms and then polymerized or connected to other solid structures. Amorphous silica as biomineral is featured mainly in unicellular organisms, in siliceous (glass) sponges and in several plants, where it occurs in the cell membranes of grains, grass or horsetail in the form of 'phytolothes' with a passive deterrent function. Accumulated in deposits of diatomaceous earth, the siliceous remains of diatoms illustrate that these are a rather old group of diverse unicellular organisms. The morphology of biogenic silica is determined by membrane proteins where nucleation can take place through condensation between silicic acid or its derivatives and the hydroxyl groups of organic matrix.

Iron Biominerals

The biomineralization of iron oxide is a relatively well documented field of bioinorganic chemistry. In addition to the iron (III) hydroxide condensation products, biogenic iron oxide also exists in the form of magnetite (Fe_3O_4) (Figure). The FeIII,II – containing mineral was detected in fairly ubiquitous magnetotactic bacteria, in chitons, mollusks, pidgeons and even in humans.

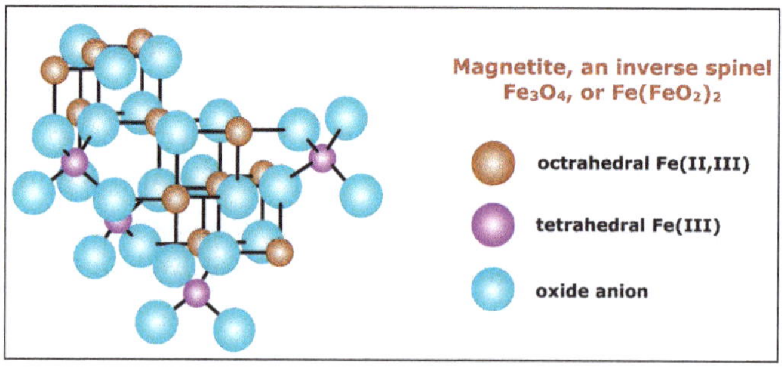

Strontium and Barium Sulfates

The unicellular plankton organism Acantharia (Figure) which belongs to the radiolarians features exoskeletons consisting of exactly 20 strontium sulfate single crystals which can assume very complex shapes. The origin of these forms is connected to the morphology of the cell since nonbiogenic strontium sulfate (celestite) crystallizes as flat rhombs when left undisturbed. The higher symmetry and simplicity of this special mineral have made the Acantharia a preferred object for studies of biomineralization.

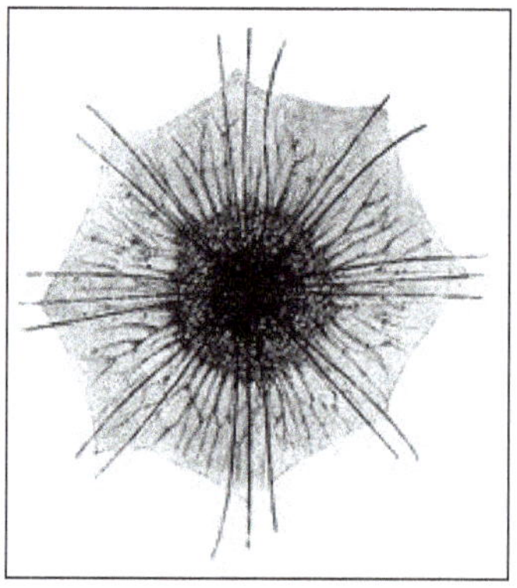

Twenty – spicule Acantharia species.

Certain unicellular algae (Desmidiacea) contain vesicles with barium sulfate in their crystalline form as baryte. The cells of Closterium, for example, are bent in a crescent-like shape with two barite crystal containing vesicles at opposing ends of the cell which suggests a function as inertia sensors. The crystals show the normal rhombic form of baryte i.e. morphology of the cell does not influence the crystal shape in contrast to the Acantharia.

Metalloproteins

Several types of proteins are available in nature, each of which is associated with different components and hence belong to various categories. Proteins containing only amino acid residues and no other chemical constituents are termed as simple proteins, examples include enzymes ribonuclease A and chymotrypsinogen. Whereas, proteins that contains both chemical components and amino acids are called as conjugated proteins. The non–amino acid part of a conjugated protein is commonly called its prosthetic group. Further classification of conjugated proteins is on the basis of the chemical nature of their prosthetic groups. Therefore, proteins containing lipids are termed as lipoproteins, those containing sugar groups are termed as glycoproteins, and the proteins containing metal ions are called as metalloproteins (Figure). A number of proteins have more than one prosthetic group. Usually the prosthetic group plays a substantial part in the protein's biological function.

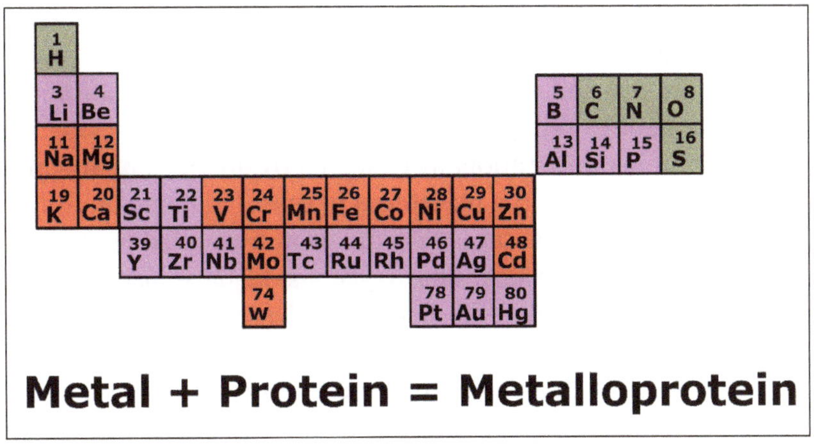

Metalloproteins.

Eversince the first X-ray crystal structure of a protein, sperm whale myoglobin, indicated the presence of an iron atom, metalloproteins drew immense attention of biochemists and chemists. Metalloproteins are responsible for catalysing important biological processes, like as photosynthesis, respiration, water oxidation, molecular oxygen reduction and nitrogen fixation and are account approximately half of all proteinsin nature. This is validated by the active-site structures of the enzymes involved, which are often reminiscent of minerals. Although metal centres form the basis of reactions, the protein matrix regulates reactivity.

Function of Metalloproteins

Metalloproteins are a class of biologically important macromolecules and account for nearly half of all proteins in biology. They are responsible for performing some of the most difficult yet important functions, including photosynthesis, respiration, water oxidation, oxygen transport, electron transfer, oxygenation and nitrogen fixation. These diverse functions of metalloproteins have been supposed to depend on the ligands from the amino acid, coordination structures, and protein structures in the immediate vicinity of metal ions. Huge amount of efforts have been devoted toward understanding the structure and function of metalloproteins. It is estimated that approximately half of all proteins contain a metal and about one quarter to one third of all proteins are supposed to have metals in order to carry out their functions. It is due to the presence of the metal ions that allows metalloenzymes to perform functions which cannot be performed by the functional groups found in amino acids. Metalloproteins deal with almost all aspects related to the intracellular and extracellular metal-binding proteins, including their structures, properties and functions. The biological roles of metal cations and metal-binding proteins are limitless. Hence, metalloproteins play many different roles in cells, such as storage and transport of proteins, enzymes and signal transduction proteins.

Design of Functional Metalloproteins

Metalloprotein Design in De Novo Designed Scaffolds

The purest and most challenging form of metalloprotein design is de novo design or design "from scratch". De novo design of a metalloprotein involves constructing a polypeptide sequence not directly related any natural protein, to precisely fold into a defined three-dimensional structure that binds a metal ion[4]. Most work in this area has focused on introducing metal-binding sites into designed α-helical bundles, which are among the first de novo designed proteins. These α-helical bundles are a common scaffold for a number of heme proteins. Therefore, not surprisingly, heme centers have been among the first and most common metal centers incorporated into de novo α-helical bundles5, from mono heme binding to four heme maquettes mimicking long-range electron transfer proteins. These proteins were rationally designed by direct comparison to native proteins, such as cytochrome bc1, for clues as to the placement of essential heme binding and structure-stabilizing residues. In addition to electron transfer8 and proton-coupled exchange functions, a functional aniline hydroxylase10, peroxidase11, and oxygenase12 have also been engineered into de novo proteins. An exciting development in this field is the design of a di-heme containing four-helix bundle with oxygen transport properties, whose O_2 affinities and exchange timescales match those of natural globins13. Remarkably, this designed protein binds O_2 even tighter than CO. A key to the success is reduction of helical-interface mobility to exclude water from the heme binding site, thereby reducing heme oxidation and increasing O_2 binding stability.

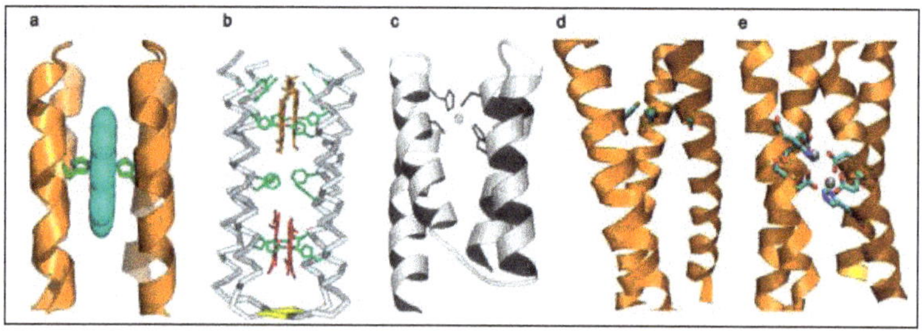

Designed metalloproteins using de novo designed scaffolds a, Computer model of a bis-His ligated mono-heme α-helical bundle. Adapted from ref6 courtesy of William DeGrado. b, Computer model of a bis-His ligated multi-heme four α-helical bundle. Reproduced from ref. Computer model of Zn(II) bound His2Cys2 motif of a Zn-finger mimic. Reproduced from ref. d, X-ray crystal structure of As(III) bound three-stranded coiled-coil (PDB 2JGO)16. e, X-ray crystal structure of di-Zn(II) Due Ferro 1; PDB 1EC5).

In addition to heme, other metal ions/cofactors have been engineered into de novo designed helical bundles by introducing metal-binding ligands at specific locations to mimic those in native proteins. Examples include the His_2Cys_2 ligand set found in Zn-finger proteins and the His3 set in carbonic anhydrase The x-ray crystal structure of an arsenic bound three-stranded coiled-coil protein, resembling the putative tris Cys metal-binding environment of the natural arsenic binding protein, ArsR, has been reported, affording the first biologically relevant small molecule mimic of an AsS3 coordination. Metal-binding ligands Glu and His have also been introduced by retrostructural analysis, whereby the active site of a protein can be separated into idealized secondary structural components (e.g., an α-helix and a β-hairpin), and mathematically described as a root-mean-square deviation from the idealized structure. This strategy led to the construction of dinuclear metalloproteins in α-helical bundles that can bind two Zn(II), Co(II), or Fe(II), with the Zn(II) crystal structure shown to be structurally analogous to the diiron site of ribonuclease reductase and ferritin. One of these diiron proteins was reported to catalyze the two electron oxidation of 4-aminophenol, similar to the reaction catalyzed by diiron phenol oxidases.

Taking a cue from naturally occurring Cys rich coordination sites in proteins, the introduction of Cys ligands led to the development of α-helical coiled-coil metalloproteins capable of binding Cu(II), Zn(II), Cd(II), Hg(II), and As(III). One exciting aspect of de novo design is the demonstration that these de novo proteins can stabilize unusual metal coordination states, such as three-coordinate Hg(II) by a tri-helical bundle over the bis-coordination often preferred by Hg(II). In other cases, metal-binding can actually direct the folding and assembly of an α-helical bundle from either a random coil, less folded, or misfolded state, as was the case of Cd(II) and Hg(II) binding26 to Cys containing peptides predisposed to form helices. Interestingly, geometry-selective binding of two different Cd(II) in one designed protein has been achieved and confirmed by the correlation of 113Cd NMR and 111mCd Perturbed Angular Correlation spectroscopy.

In stark contrast to the design of α-helical proteins, de novo design of β-structure proteins is still in its infancy, because there were few model systems with which to study β-sheets in isolation from other protein structures. The de novo design of a redox active rubredoxin mimic, RM1, is a rare example of a structural and functional metallo-β-sheet protein29. RM1 was shown to bind iron and reversibly cycle between the Fe(II/III) oxidation states, mimicking native rubredoxin. The RM1 β-hairpin structure was designed computationally by constraining the positions of the peptide that were strategic in mimicking and/or stabilizing the β-structure and subsequent use of another computer program, SCADS to select the most probable amino acids for the remaining positions.

Metalloprotein Design in Native Protein Scaffolds

While designing metalloproteins using de novo designed scaffolds offers the prospect of complete control over a protein's structure and metal-binding properties, our current knowledge of protein folding limits the number of de novo designed scaffolds to only a few types, such as the α-helical bundles. It has been observed that β-sheet containing proteins tend to support more rigid and more pre-organized metal-binding sites than α-helical proteins, which are inherently more flexible31. Unlike de novo designed scaffolds, the Protein Data Bank contains ~1000 natural protein scaffold types most of which maintain the same fold and similar stability even after numerous mutations. Therefore, metalloprotein design using native protein scaffolds provides many more scaffold choices that are more tolerant to mutations, allowing for the incorporation of metal-binding sites into proteins with less concern over decreased protein stability. When chosen carefully, native proteins are also relatively easier to crystallize than de novo designed scaffolds, making 3D characterization possible, an objective that is critical to success. Perhaps the strongest argument for designing metalloproteins using native scaffolds is that Nature employs the same approach since the same scaffold is often seen in numerous proteins with diverse metal-binding motifs and functions, suggesting that these native scaffolds are robust and modifiable. For example, the Greek key β-barrel fold has been shown to be used by ~600 different types of proteins with diverse functions, ranging from oxidase, reductase, amylase, and dismutase activities, for instance. Unearthing the secret of how Nature is able to use the same scaffold to design a variety of metal-binding sites is an important goal of metalloprotein design.

Biochemical techniques, such as site-directed mutagenesis, have been used extensively to study the function of metalloproteins. The loss of function accompanied by certain mutations (usually to highly conserved residues) allows identification of residues essential for function. Although serving a different purpose, the same mutagenesis techniques can be used for metalloprotein design to impart new function into a protein scaffold by the introduction of residues that bind metal ions. This approach can be in the form of redesigning existing metal-binding sites to introduce novel functionality, introducing mononuclear metal-binding sites to proteins that do not natively bind metal ions, or by introducing homonuclear or heteronuclear metal-binding sites into proteins. The designs are often aided by empirical approaches through knowledge/experience, rational design using computer programs, or combinatorial selections.

The redesign of an existing metal-binding site to introduce novel function or metal specificity is the simplest form of metalloprotein design, as this approach relies on the structural differences between template and target proteins, and is amenable to empirical designs through knowledge/experience. Despite its simple form, this approach is still quite powerful in elucidating structural features important for the change, or gain of function by redesign. Heme proteins have been extensively redesigned in this way. They are one of the most diverse classes of proteins, with functions ranging from electron transfer, small molecule (e.g., oxygen and NO) transport, sensing, and activation. By systematically changing these characteristics of heme proteins, researchers have been able to convert one type of heme protein into another, as well as introduce novel function or substrate specificity into a heme protein.

The selectivity of designed metalloproteins for specific metals has also been utilized for the purpose of metal-sensing applications. Ratiometric Zn(II) sensors have been created by designing Zn(II) binding sites into green fluorescent proteins and by taking advantage of the resulting fluorescent signal variations due to Zn(II)-dependent conformational changes. The metal-binding site of NikR, a DNA binding protein, has been redesigned to bind UO^{22+} rather than its native Ni^{2+} cofactor, resulting in an artificial protein that binds DNA only in the presence of UO^{22+}. In addition to sensors, designed metal-binding sites have allowed modulation of protein-protein interactions, which are at the heart of many diverse biological functions. Because of complicated non-covalent interactions, it has been difficult to control the interactions well. These designed metal-binding sites allowed specific control of these interactions.

More complex metal-binding site redesign may require rational design using computational biology, such as the simultaneous incorporation and adjustment of functional elements to convert the GlyII enzyme from a glyoxalase, which hydrolyzes thioester bonds, to a β-lactamase. In order to impart the new functionality, the metal-binding site had to be significantly redesigned in order to change the metal-binding geometry, as well as the substrate binding pocket.

While redesign of metal-binding sites can offer insight into the different structural features between the template and target proteins, those structural features common to the metal-binding sites of both proteins will remain hidden. Introducing metal-binding sites into a protein location where no native metal-binding site is found, can overcome this limitation, although this approach requires a higher level of complexity. A common strategy is to base the design on structural homology between the template protein which contains no metal ions and the target metalloprotein, to introduce metal-binding sites at the corresponding positions. Using this strategy, new Zn(II) binding sites have been introduced into charybdotoxin42 and retinal binding protein43 to mimic carbonic anhydrase (CA).

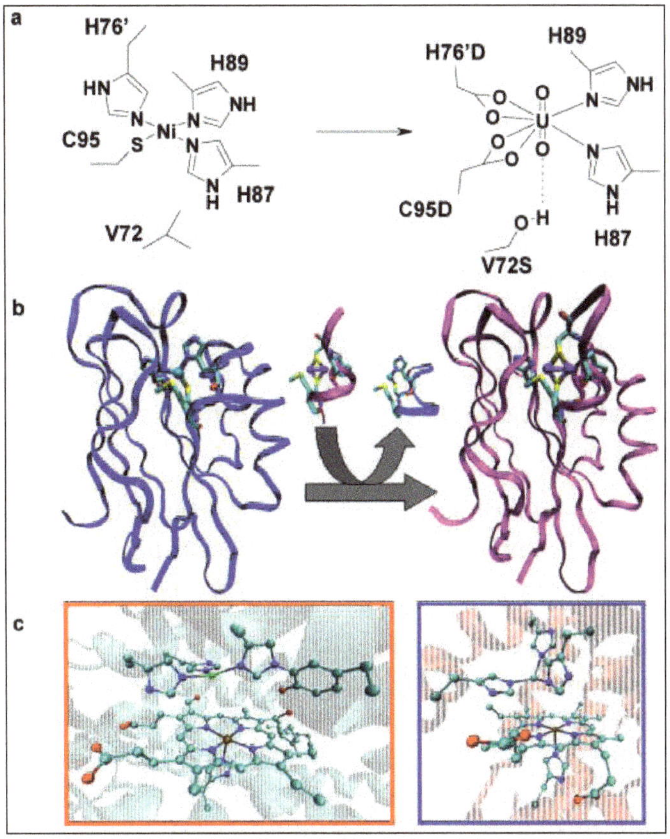

Designed metalloproteins using native scaffolds.

a. Native nickel binding site of NikR (left) and the re-engineered UO_2^{2+} binding site of the mutated NikR (right). b. Loop directed mutagenesis of blue copper azurin to yield the dinuclear, purple CuA azurin construct. c. Catalytic heme-copper center in CcO (left) and the designed heme-cooper model in sperm whale Mb (right).

When there is no structural homology, computer aided design using programs such as Metal-Search and Dezymer, has become indispensable. A Ca(II), Pb(II), and [Fe(Cys)] rubredoxin center have all been designed into scaffold proteins using computer aided design. In the case of an artificial rubredoxin designed into the non-metalloprotein thioredoxin, it was found to undergo several rounds of air oxidation and reduction by β-mercaptoethanol, similar to native rubredoxins.

An even higher level of difficulty in metalloenzyme design is the introduction of dinuclear metal-binding sites or metal clusters into proteins. An example is the binuclear CuA center, consisting of two copper ions bridged by two cysteine thiolates, with each copper ion also coordinated to a histidine nitrogen. New CuA centers were designed into the cupredoxins azurin (Az) and amicyanin using a technique called loop directed mutagenesis (LDM), in which the copper binding loop of cupredoxins is replaced by the corresponding loop in CcO that share similar structural homology. LDM has also been used to introduce lanthanide binding sites into a helix-turn-helix motif by drafting a calcium binding EF-hand loop, and such a designed protein has been shown to promote DNA and phosphate hydrolysis.

Heteronuclear bimetallic centers are also found in many natural proteins and they perform important biological functions such as oxygen reduction, lignin degradation and nitric oxide reduction. Designing metalloproteins to mimic the enzymes that carry out these complex reactions is even more challenging than designing homonuclear metal-binding sites because of the selective placement and binding of two or more different metal centers. An example is the heme-CuB center in CcO where three histidines bind to a Cu(II) ion that is ~4.5 Å away from the heme iron. Computer-aided design was used to introduce the histidines above the heme of myoglobin (Mb) at positions corresponding to CcO. Copper binding was confirmed by multiple spectroscopic techniques. Further studies showed the importance of the CuB site in O_2 binding and reduction, as well as the significance of a proton network54. This designed protein also displays interesting functional properties, such as nitric oxide reduction, similar to certain members of CcOs. A Mn(II) binding site has also been designed into cytochrome c peroxidase, resulting in a heme-Mn(II) protein with Mn(II) oxidation activity, as in native manganese peroxidase (MnP), a protein involved in lignin biodegradation, which is a critical step in biomass conversions from plant materials to biofuels.

In addition to discrete metal-binding sites, metal nanoparticles have been introduced into proteins, including Pt nanoparticles in ferritin and a small heat shock protein. The former catalyzes size-selective olefin hydrogenation while the latter catalyzes hydrogen production.

Even though empirical approaches through knowledge/experience and rational design using computer programs have been quite successful in designing metalloproteins, it is still difficult to design certain metalloproteins, especially functional ones, due to the large number of potential interactions within the active site of a metalloenzyme. In this case, combinatorial design and directed evolution of novel metalloenzymes is another powerful choice. For example, directed evolution has been extensively applied to antibodies to develop novel metalloenzymes. Utilizing phage display, the $^{7}G^{12}$ antibody protein was evolved to bind heme, with which the oxidation of tyramine could be catalyzed60. A similar technique has recently been expanded to find antibodies for inorganic ligands, such as BINAP and BINOL, with the goal to develop completely novel metalloenzymes. A striking example of this approach is the use of directed evolution to obtain a new cyt P_{450} with the ability to selectively hydroxylate propane, a function that has not yet been achieved by small molecule catalysts.

Designing Novel Metalloproteins Containing Unnatural Amino Acids or Non-Native Metal Cofactors

Metalloprotein design using de novo designed or native protein scaffolds has succeeded in providing valuable insights into the fundamental rules that govern protein structure and function. Replicating native protein structure and function, although quite challenging, is not the only goal of protein designers; they want build proteins with properties beyond those found in Nature. In recent years, the repertoire of functional groups such as amino acids or cofactors available to protein designers has been expanded beyond those available in biological systems.

A myriad of techniques have been developed for the incorporation of unnatural amino acids into a protein scaffold, each with its own advantages and limitations. Total synthesis of proteins using solid phase peptide synthesis enables incorporation of any synthetic amino acid for which successful protection/deprotection techniques can be applied, but cost and size constraints limits the use of total synthesis to smaller proteins (~60–100 amino acids). Native chemical ligation (NCL), involving covalent attachment of two synthetic peptides, has expanded this size regime65, but cost is still an issue, at least for large proteins. Expressed protein ligation (EPL), the coupling of a bacterially expressed peptide fragment with a synthetic peptide fragment containing unnatural amino acids, has reduced cost and extended the protein size even further, but it often requires a cysteine residue at the point of ligation to maximize ligation efficiency66, although methods have been developed to use other residues at the point of ligation, with less efficiency. Methods not restricted by protein size include auxotroph generation, chemical residue modification, and cavity complementation69, but these methods have limited versatility in terms of the variety of unnatural amino acids that can be incorporated. A promising and versatile technique is the use of custom evolved tRNA synthetase molecules for in vitro and in vivo translation of new residues through recognition of the amber stop codon. A seminal contribution from this approach was the design of a DNA cleaving protein through site-specific genetic incorporation of the DNA cleaving agent (2,2′-bipyridin-5-yl)alanine into a DNA binding protein. Upon addition of Cu(II) or Fe(II) and reducing agent to the protein, a double stranded DNA substrate was cleaved at its consensus sequence. The variety of unnatural amino acids this technique can incorporate is higher than those of auxotroph generation, chemical residue modification, or cavity complementation, but lower than those of total synthesis, NCL or EPL.

The expansion of the amino acid regime has made significant contributions toward determining systematically the role of each amino acid in tuning the activity of the metal center toward a given reactivity, and in increasing the breadth of chemical transformations performed by proteins. For instance, replacement of heme binding His residues in de novo designed four-α-helical bundles with 4-β-(pyridyl)-L-alanine and 1-methyl-L-His has resulted in an increase in the reduction potential of the metal site for the pyridyl system and a preference for five-coordinate low-spin heme in the 1-methyl-L-His system. In another system, introduction of bulky penicillamine ligands in place of cysteines led to the creation of an open metal-binding site in a ferredoxin model peptide. This methodology resulted in the isolation of the first water-stable CdS3 complex through subtle manipulation of steric crowding around the metal center. Furthermore substitution of L-amino acids for their D-enantiomers resulted in fine control between 3- and 4-coordinate Cd(II) complexes in a de novo designed metallopeptide. Finally, the role of a conserved Tyr residue near the active site of rubredoxin was probed via a set of unnatural Tyr analogs in which the para $^{-}$OH group was replaced with $^{-}$H (native Phe), $^{-}$F, $^{-}NO_2$, and $^{-}$CN, spanning a range of electron withdrawing strengths. Interestingly, a linear correlation was found between the reduction potential of the nearby heme and the Hammett σp value, but no such trend is observed with dipole moment or residue size.

A powerful advantage to the introduction of unnatural amino acids is the ability to probe the role of the amide backbone, since site-directed mutagenesis using natural amino acids cannot modify the protein backbone. Both the carbonyl and nitrogen functionalities of the protein backbone have been implicated in critical and specific hydrogen bonding interactions, either directly to the metal, or to the metal ligands. For example, the Fe_4S_4 cluster of high potential iron proteins (HiPIPs) has such an amide interaction, which was found to be responsible for stabilizing the reduced form through attenuation of a metal bound sulfur. Replacement of this amide linkage with an ester linkage indeed resulted in a lowering of the reduction potential by ~100 mV.

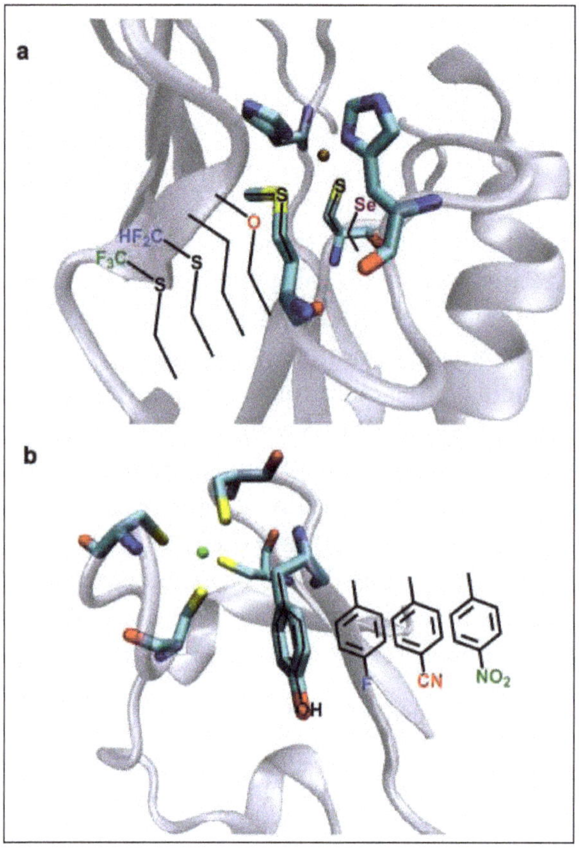

Site specific incorporation of unnatural amino acids into a protein scaffold for tuning of metal properties Crystal structures of a, rubredoxin (PDB: 1CAA) and b, azurin (PDB: 4AZU) showing variant residue location for unnatural amino acid incorporation.

Through the use of EPL, the metal bound Cys and Met residues in the electron transfer protein azurin were replaced with isostructurally similar unnatural amino acids. Replacing Cys with selenocysteine resulted in drastically different spectral features, but did little to affect the reduction potential. In contrast, replacing methionine with selenomethione and other isostructural variants resulted in spectral features similar to the wild type protein, but with drastically different reduction potentials. The use of unnatural amino acids helped to identify the hydrophobicity of the residues at this position as the major factor in tuning the reduction potential of the protein.

The repertoire of non-native functionalities has also been expanded through the incorporation of non-native metal containing cofactors into the protein scaffold. Biological systems incorporate metal cofactors such as heme through both non-covalent and covalent means. Among the covalent approach, Nature uses both single and dual point anchoring strategies to position the prosthetic group within the protein scaffold. These strategies have all been utilized for incorporation of non-native metal cofactors, with the non-covalent approach being the most extensively used strategy. For example, replacement of the native heme cofactors of myoglobin and horseradish peroxidase (HRP) with the less symmetrical structural isomer, porphycene results in enhanced peroxidase activity, highlighted by a twelve-fold increase in oxidation of thioanisole when compared to native HRP. In addition to native protein scaffolds, de novo designed α-helical bundles have also been engineered to bind non-biological heme cofactors preferentially over biological hemes.

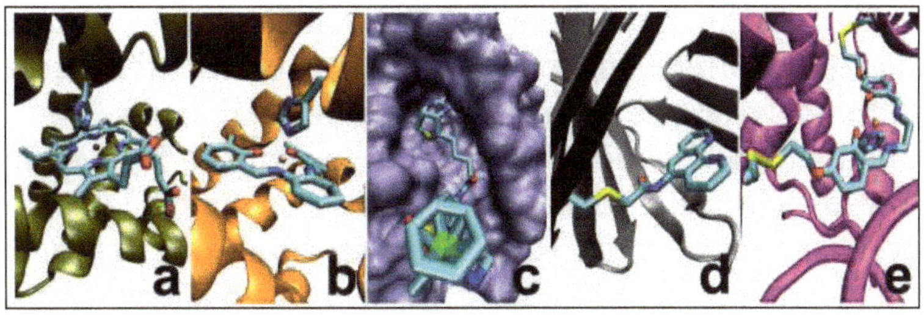

Strategies for non-native cofactor incorporation into a protein scaffold Native cofactor substitution of heme b in myoglobin for a, Fe-porphycene and b, 3,3-Cr salophen, exploiting the structural and dative-bonding similarities between the native cofactor and the surrogate. Affinity tagging of the catalyst with a protein-selective linker such as the biotin-streptavidin couple, c, allows for versatile attachment of non-native cofactors. Covalent strategies include a single attachment strategy as in d, showing the adipocyte lipid binding protein (ALBP) linked through Cys to a phenanthrolene complex; and dual covalent attachment as in e, showing dual anchoring of a Mn-salen complex into the myoglobin scaffold. Description of streptavidin complex obtained from ref91. Crystal structure of Fe porphycene, Cr (3,3′-Me2-salophen)-Mb complex, biotin-streptavidin complex, and phenanthroline-ALBP were obtained from the Protein Databank (PDB codes: 2D6C, 1J3F, 2QCB, and 1A18, respectively). Computer model of Mb (L72C/Y103C) generated from an overlay with native heme b structure.

In addition to heme analogs, other planar moieties similar to heme have also been substituted into heme proteins. For example, strong binding of Mn, Fe, and Cr Schiff base complexes into a myoglobin scaffold was observed after careful design of the protein to relieve steric repulsion in the active site. Such rational design has resulted in novel asymmetric catalysts capable of oxidation rate enhancement and chiral induction.

To incorporate non-planar metal-containing cofactors into proteins where there is no native binding site is much more challenging. One way to meet such a challenge is through derivatization of the catalyst with a moiety exhibiting natural bio-affinity for a protein target, though, for example, the use of the high-affinity biotin-streptavidin couple ($K_a = 1 \times 10^{15}\ M^{-1}$). In this way, metal cofactors have been incorporated into the streptavidin scaffold through covalent attachment to a biotin moiety resulting in artificial biocatalysts with new activities including hydrogenation of alkenes, transfer-hydrogenation of ketones, and C-C bond formation through allylic alkylation, and up to 99% ee.

While the use of biotin and avidin/streptavidin to incorporate non-native metal cofactors is very powerful, not all cofactors are amenable to such a strategy, as binding affinity and location may not be readily controlled. Covalent attachment is an effective and versatile approach to attach a cofactor to the protein at a defined location through selective conjugation techniques involving engineered or native Cys or Lys residues. This strategy has been utilized to attach DNA cleaving agents such as -phenanthroline-copper and iron-EDTA to DNA binding proteins for the generation of artificial nucleases, 92, and 1,10-phenanthroline-copper for enantioselective hydrolysis.

While the single-point covalent approach makes it possible to attach a non-native cofactor at a defined location, it does not necessarily fix the conformation of the cofactor inside the protein. For example, single covalent attachment of a 4-(2-methanesulfonylthio-ethoxy)salicylidene-1,2-ethanediamino-manganese(III) (Mn-Salen) cofactor to apo myoglobin through a single cysteine-selective linker resulted in only 12% ee toward sulfoxidation of thioanisole95, suggesting that the Mn-Salen conformation may be too flexible inside Mb to facilitate high enantioselectivity. Addition of a second linker to the cofactor and selective cavity positioning led to a significant increase in catalytic efficiency (over 7-fold increase) and selectivity (from 12% to 51%). Furthermore, it has been shown that such incorporation of Mn-Salen into a myoglobin scaffold enhances significantly the chemoselectivity in oxidation of thioanisole with H_2O_2, virtually eliminating over oxidation to the sulfone product through electrostatic exclusion of the sulfoxide product from the site.

Measures of Success, Lessons Learned and Future Challenges

Tremendous progress has been made in this exciting field of synthetic biology. Success is measured not only by how close the designed protein is to the native protein, but also by the insight gained from the process. For example, the crystal structure of a designed CuA site is almost identical to that of native CuA, including both primary and secondary coordination spheres, indicating that metalloprotein design can lead to structures closely resembling those in Nature. More importantly, two metal-binding sites, native blue copper and the designed CuA center, were placed in the same azurin scaffold, allowing the electron transfer properties of two different metal centers to be compared in the same protein, a task that is difficult, if not impossible to do by studying native proteins. Such a comparison provided deeper insight that the CuA center is a more efficient ET center than blue copper.

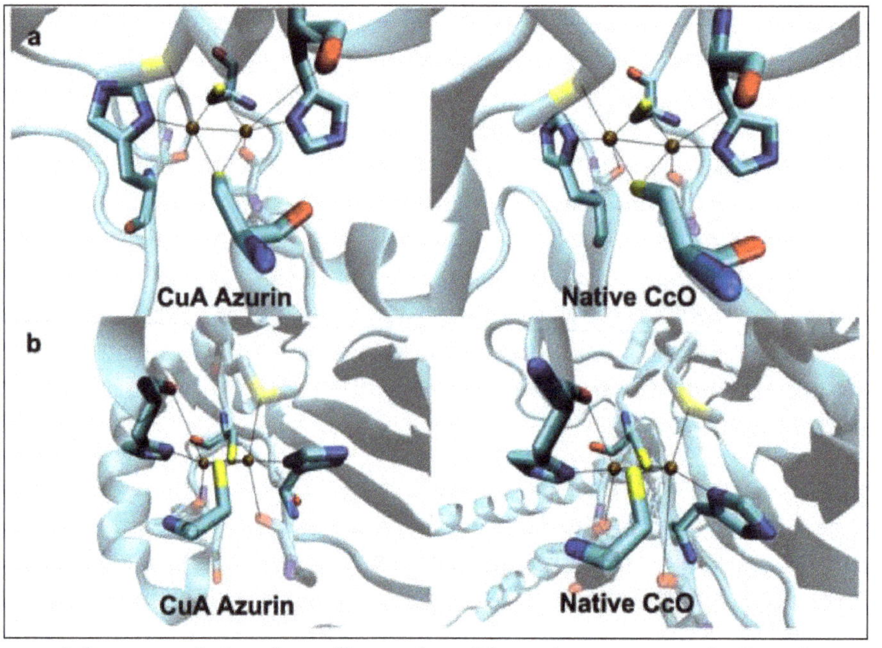

Close match between a designed metalloprotein and its native target protein Crystal structures of a designed CuA site in azurin (PDB code 1CC3)49 and a CuA site in native CcO (PDB code: 1AR1) as viewed from above the $Cu_2(Scys)_2$ plane (a), and along the $Cu_2(Scys)_2$ plane.

Given the initial successes, it is a good time to summarize the lessons learned and to recognize future challenges. Essential to success is the recognition that structural biology must play a major role in the design process. Some early metalloprotein designs have generated much excitement in the field, only to be shown by x-ray crystallography to contain a different metal-binding site than originally designed. While spectroscopy is invaluable in providing support for the design, more concrete evidence from 3D structural characterization is essential. In addition to confirming the initial design, it provides detailed information on what has or has not been designed correctly, thus providing insights on the next level of design.

Furthermore, primary coordination sphere design is not enough in many cases, to confer structure and/or function. Consideration of the environment surrounding the primary coordination sphere is critical to successful design. One primary example is water inside the protein, where the associated hydrogen bonding interactions play essential roles in many metalloproteins, including Cyt P_{450}, heme oxygenase (HO), and CcO_{33}. For example, even though a CuB center was designed into myoglobin that structurally mimics the heme-copper center in CcO_{53}, its function mimics HO rather than CcO when O2 is used as an oxidant. When H_2O_2 is used as an oxidant, a ferryl species, a putative intermediate in CcO activity, was observed. These results suggest that the extra protons introduced through H_2O_2 are important for function. The introduction of an additional hydrogen bonding donor through the replacement of the b-type heme in Mb with an o-type heme, resembling the heme found in CcO, allowed suppression of HO activity and promotion of CcO activity.

Metalloprotein design has focused on the positive design of structural features responsible for correct metal-site geometry and function. However, negative design to prevent competing metal-site geometries and unwanted functions is also critical. For example, in designing dinuclear metalloproteins in a helical bundle, there was a possibility that undesirable heterotetrameric motifs and homooligomers would form since the protein was comprised of different peptides. To circumvent this problem, a computational algorithm was included to destabilize these interactions.

The final lesson is that metalloprotein design often results in lower stability because of changes made to the protein scaffold. Sometimes, the stability is so low that even if the design is valid, not enough stable protein can be obtained to prove the design. The stability of the protein must be improved, even if function may not be improved directly. A beautiful example is demonstrated in cytochrome P_{450} where only after the protein was stabilized, could the highly destabilizing mutations introduced confer novel activity.

The majority of the successes in the metalloprotein design field are derived from the construction of a geometrically correct, sterically compatible primary coordination sphere sufficient to reproduce the structure and function of a desired metal center, such as a bis-His heme or His_2Cys_2 tetrahedral Zn(II) site. However, many metal-binding sites in proteins, such as the type 1 copper and CuA centers are not necessarily in their preferred metal ion geometric states. Therefore, one future challenge is to design metal sites and metal clusters with unique geometric requirements. Other future challenges include the design of metal sites at the interface of two or more proteins, which occur often in proteins but have not been designed until recently, and the design of metal-binding sites from membrane proteins, which has also been rare. An even higher level of challenge is the design of protein metal-binding sites requiring helpers called chaperones. Consideration of the interactions between metal ions and protein hosts is not enough to confer function, making their interactions with chaperones necessary in the design process. In addition to metal-binding site design, functional metalloproteins also require the design of a substrate-binding site for catalysis. An even higher level of complexity includes coupling the redox reaction of a designed metal-binding site to proton transfer, conformational change, or charge separation.

To meet these challenges, advances in a number of fields are required. The most important requirement is the development of better computer programs for designing metal-binding sites in proteins, including consideration of structural features beyond the primary coordination sphere. In contrast to well-defined force fields for peptide bonds and amino acid side chains, most force fields for metal-binding sites are ill-defined because there are many more variables involved in defining a metal-binding site. These variables include variations on the identity of the metal ion, different oxidation states of the same metal ions, and different ligand and geometric preferences of different metal ions. Density functional theory has been excellent in defining small metal complexes. Extending its success to metal-binding sites in proteins, together with other leading algorithms, such as QM/MM (quantum mechanics/molecular mechanics), is required for the successful rational design of metalloproteins. Despite these challenges, metalloprotein design will remain a vibrant field where exciting discoveries of new insights and novel enzymes will occur every day.

Coordination Chemistry Behind Metalloproteins

The metal ions in metalloproteins are typically coordinated to nitrogen, oxygen or sulfur centres that belong to amino acid residues of the polypeptide chain incorporated into the protein. Amongst all, some important substituents are the imidazole, thiolate and carboxylate groups in histidine, cysteinyl, and aspartate residues, respectively. Bestides this, deprotonated amides and amide carbonyl oxygen centers are the donor groups which is provided by peptide backbone. Moreover, a large number of organic cofactors also act as donors to the metal ions, such as the tetradentate N4 macrocyclic ligands incorporated inside the heme protein and inorganic ligands such as sulfide and oxide.

Four Classes of Electron Transferases

In most of the biological processes, such as photosynthesis, respiration, collagen synthesis, steroid metabolism, the immune response, drug activation, neurotransmitter metabolism, and nitrogen fixation, electron-transfer reactions metalloproteins play major roles. Out of which, both respiration and photosynthesis are extremely significant processes as they provide most of the energy that is required to sustain life.

Metalloprotein which has a single type of redox cofactor can be classified into two general groups: electron carriers and proteins involved in the transport or activation of small molecules. It is assumed that proteins that function as electron transferases typically place their prosthetic groups in a hydrophobic environment where in addition to ligand; hydrogen bonds may be provided which help to assist in stabilization of both the oxidized and the reduced forms of the cofactor. In order to minimize inner-sphere reorganization, metal-ligand bonds remain intact upon electron transfer. Many of the complex metalloenzymes containing multisite such as: cytochrome c oxidase, xanthine oxidase, and the nitrogenase FeMo protein have redox centers that function as intramolecular electron transferases.

There are four classes of electron transferases: blue copper proteins, cytochromes, flavodoxins, and iron-sulfur proteins, each of which contains many members that exhibit important structural differences.

Flavadoxins

The flavodoxins are proteins with molecular weights in the range of 8-13 kDa and are found in many species of bacteria and algae. These proteins contain an organic redox cofactor, flavin mononucleotide (FMN cofactor) that is found at one end of the protein, near the molecular surface, but only the dimethylbenzene portion of FMN is significantly exposed to the solvent. FMN can act as either a 1- or a 2-electron redox center. In solution, the semiquinone form of free FMN is unstable and disproportionate to the oxidized form called quinine and reduced form called hydroquinone (Figure). Hence, free FMN works as a two-electron reagent. On the other hand, FMN in flavodoxins, can function as a single-electron carrier. This is easily differentiated on the comparison of reduction potentials for free and protein-bound FMN.

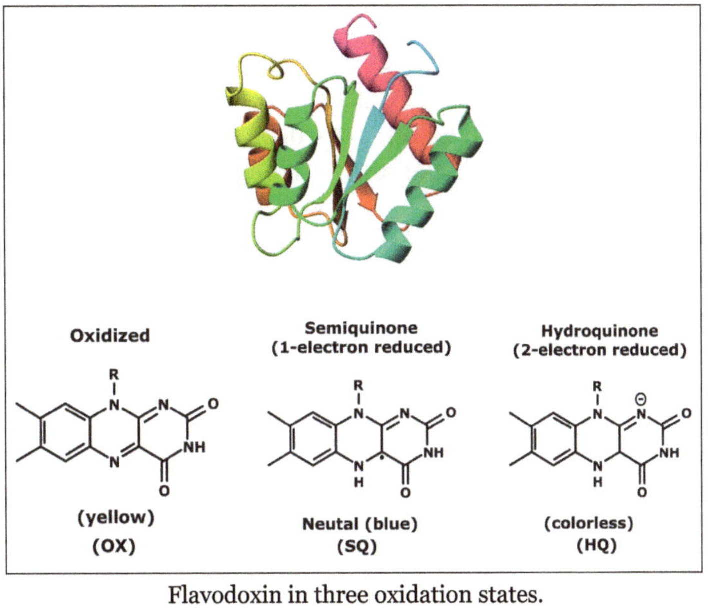

Flavodoxin in three oxidation states.

Evidently, the protein medium is considered to be responsible for this drastic alteration in oxidation- state stability. It was also concluded from the NMR study of the M. elsdenii flavodoxin quinone/ semiquinone and semiquinone/hydroquinone electron self-exchange rates, that the latter is approximately 300 times faster than the former, in keeping with the view that the physiologically relevant redox couple is semiquinone/ hydroquinone.

Cytochromes

Out of the four classes of the electron transferases, the cytochromes are the most thoroughly characterized class and were among the first to be identified in cellular extracts due to their unique optical properties and prominent intense absorption in the region of 410-430 nm known as the Soret band.

By definition, cytochromes are proteins containing one or more heme cofactors (Figure). These proteins are generally classified on the basis of heme type. There are three most commonly encountered types of heme: heme a possesses a long phytyl "tail" and is found in cytochrome c oxidase; heme b is found in b-type cytochromes and globins; heme c is covalently bound to c-type cytochromes via two thioether linkages.

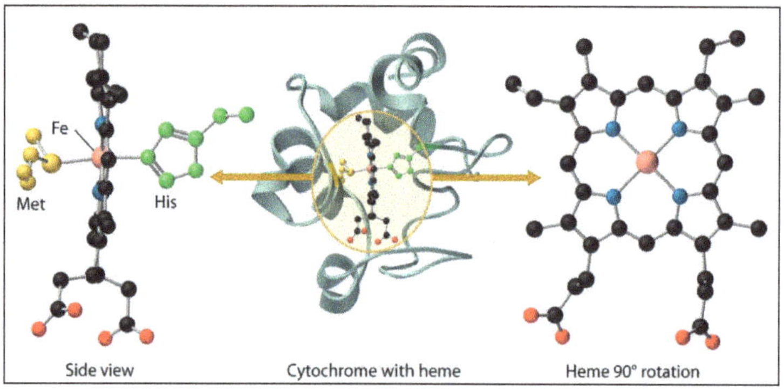

Cyctochrome c.

The real challenge lies in the nomenclature of these cytochromes. Some of them are designated according to the historical order of their discovery, e.g., cytochrome C_2 in bacterial photosynthesis. Others are designated according to the Amax of the band in the absorption spectrum of the reduced protein (e.g., cytochrome C_{55}]). Of all, cytochrome c is most dominant in nature. Ambler divided these electron carriers into three classes on structural grounds.

- Class I cytochromes c contain axial His and Met ligands, with the heme located near the Nterminus of the protein. These proteins are globular and X-ray studies of Class I cytochromes c from a variety of eukaryotes and prokaryotes clearly demonstrated an evolutionarily preserved "cytochrome fold," with the edge of the heme solvent-exposed. Cytochromes possess positive reduction potentials (200 to 320 mV).

- Class II cytochromes c (EO I ~ - 100 mV) are found in photosynthetic bacteria, where they serve an unknown function. Unlike Class I, these c-type cytochromes are high-spin where the iron is fivecoordinate with an axial His ligand. The vacant axial coordination site is buried in the interior of protein. These proteins, generally referred to as cytochromes c, are four-ex-helix bundles.

- Class III cytochromes c, also called cytochromes C3, contain four hemes, each ligated by two axial histidines. These proteins are found in a restricted class of sulfate-reducing bacteria and may be associated with the cytoplasmic membrane. The low molecular weights of cytochromes C3 (~14.7 kDa) require that the four hemes be much more exposed to the solvent than the hemes of other cytochromes, which may be in part responsible for their unusually negative (- 200 to - 350 mV) reduction potentials. These proteins possess many aromatic residues and short heme-heme distances, two properties that could be responsible for their anomalously large solid-state electrical conductivity.

Iron-sulfur Proteins

The iron-sulfur proteins play essential role as electron carriers in nearly all living organisms, and take part in plant photosynthesis, nitrogen fixation, steroid metabolism, and oxidative phosphorylation, as well as many other processes.

The simplest of all iron-sulfur proteins are the rubredoxins, which are primarily found in anaerobic bacteria. Rubredoxins are small proteins (6 kDa) and contain iron ligated to four Cys sulfurs in a distorted tetrahedral arrangement. The reduction potentials of iron-sulfur proteins are typically quite negative, indicating a stabilization of the oxidized form of the redox couple as a result of negatively charged sulfur ligands. Ferredoxins [2Fe-2S] are the other class of iron-sulfur proteins (10-20 kDa) which are found in plant chloroplasts and mammalian tissue. The structure of Spirulina platensis ferredoxin confirmed earlier suggestions, based on EPR and Mossbauer studies, that the iron atoms are present in a spin-coupled [2Fe-2S] cluster structure.

The optical spectra of all iron-sulfur proteins are very broad due to numerous overlapping chargetransfer transitions that impart red-brown-black colors to these proteins. In contrast, the EPR spectra of iron-sulfur clusters are quite distinctive, and have significant value in the redox chemistry studies of these proteins.

Four-iron clusters [4Fe-4S] are also termed as ferredoxins, found in many strains of bacteria and in most of these bacterial iron-sulfur proteins, two such clusters are present in the protein. These proteins are small with molecular weight of 6-10 kDa and have reduction potentials in the - 400 mV range. Each of these clusters contains four iron centers and four sulfides at alternate corners of a distorted cube where each iron is coordinated to three sulfides and one cysteine thiolate. The irons are strongly exchange-coupled, and the [4Fe-4S] cluster in bacterial ferredoxins is paramagnetic when reduced by one electron. The so-called "high-potential iron sulfur proteins" (HiPIPs) are found in photosynthetic bacteria, and exhibit anomalously high (~350 mV) reduction potentials. The C. vinosum HiPIP (10 kDa) structure demonstrates that HiPIPs are distinct from the [4Fe-4S] ferredoxins, and that the reduced HiPIP cluster structure is significantly distorted, as is also observed for the structure of the oxidized P. aerogenes ferredoxin. In addition, oxidized HiPIP is paramagnetic, whereas the reduced protein is EPR silent.

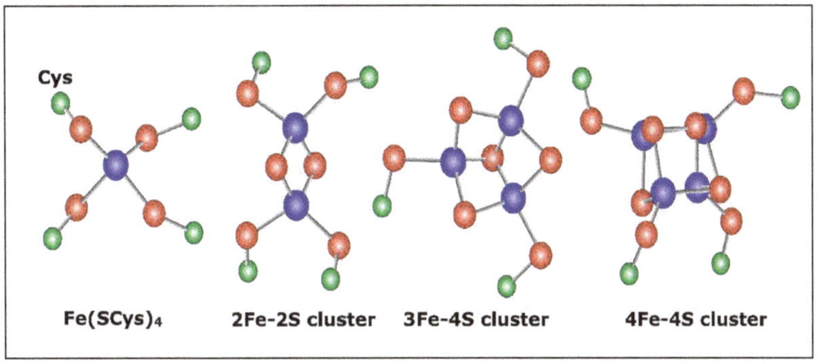

Four types of iron-sulfur clusters with one, two, three and four iron atoms respectively.

Blue-Copper Proteins

The blue copper proteins are characterized by intense S(Cys)$\rightarrow$Cu charge transfer absorption near 600 nm, an axial EPR spectrum displaying an unusually small hyperfine coupling constant, and a relatively high reduction potentia1. With few exceptions (e.g., photosynthetic organisms), their precise roles in bacterial and plant physiology remain obscure. X-ray structures of several blue copper proteins indicate that the geometry of the copper site is approximately trigonal planaras illustrated by the Alcaligenes denitrijicans azurin structure(Figure). In all these proteins, three ligands (one Cys, two His) bind tightly to the copper in a trigonal arrangement. Suppose, EO = 276 mV for A. denitrijicans azurin, whereas that of P. vulgaris plastocyanin is 360 mV. These differences in bond lengths are expected to stabilize copper in azurin to a greater extent than in plastocyanin, and result in a lower E^0 value for azurin.

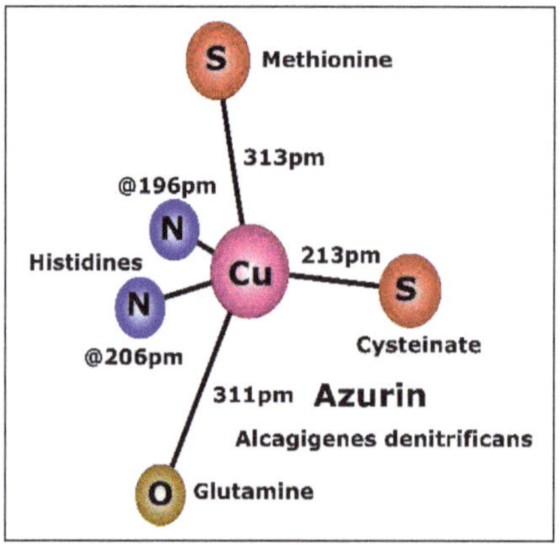

Azurin.

Bioenergetics

Bioenergetics means study of the transformation of energy in living organisms. The goal of bioenergetics is to describe how living organisms acquire and transform energy in order to perform biological work. The study of metabolic pathways is thus essential to bioenergetics.

In a living organism, chemical bonds are broken and made as part of the exchange and transformation of energy. Energy is available for work (such as mechanical work) or for other processes (such as chemical synthesis and anabolic processes in growth), when weak bonds are broken and stronger bonds are made. The production of stronger bonds allows release of usable energy.

Adenosine triphosphate (ATP) is the main "energy currency" for organisms; the goal of metabolic and catabolic processes are to synthesize ATP from available starting materials (from the environment), and to break- down ATP (into adenosine diphosphate (ADP) and inorganic phosphate) by utilizing it in biological processes.

In a cell, the ratio of ATP to ADP concentrations is known as the "energy charge" of the cell. A cell can use this energy charge to relay information about cellular needs; if there is more ATP than ADP available, the cell can use ATP to do work, but if there is more ADP than ATP available, the cell must synthesize ATP via oxidative phosphorylation.

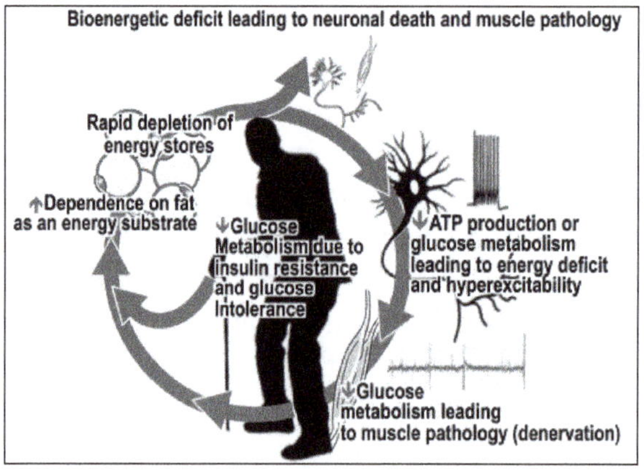

Living organisms produce ATP from energy sources via oxidative phosphorylation. The terminal phosphate bonds of ATP are relatively weak compared with the stronger bonds formed when ATP is hydrolyzed (broken down by water) to adenosine diphosphate and inorganic phosphate. Here it is the thermodynamically favorable free energy of hydrolysis that results in energy release; the phosphoanhydride bond between the terminal phosphate group and the rest of the ATP molecule does not itself contain this energy.

Types of Bioenergetics Reactions

Exergonic Reaction

Exergonic implies the release of energy from a spontaneous chemical reaction without any concomitant utilization of energy. The reactions are significant in terms of biology as these reactions have an ability to perform work and include most of the catabolic reactions in cellular respiration.

Most of these reactions involve the breaking of bonds during the formation of reaction intermediates as is evidently observed during respiratory pathways. The bonds that are created during the formation of metabolites are stronger than the cleaved bonds of the substrate. The release of free energy, G, in an exergonic reaction (at constant pressure and temperature) is denoted as:

$$\Delta G = G_{products} - G_{reactants} < 0$$

Endergonic Reactions

Endergonic in turn is the opposite of exergonic in being non-spontaneous and requires an input of free energy. Most of the anabolic reactions like photosynthesis and DNA and protein synthesis are endergonic in nature. The release of free energy, G, in an exergonic reaction (at constant pressure and temperature) is denoted as:

$$\Delta G = G_{products} - G_{reactants} > 0$$

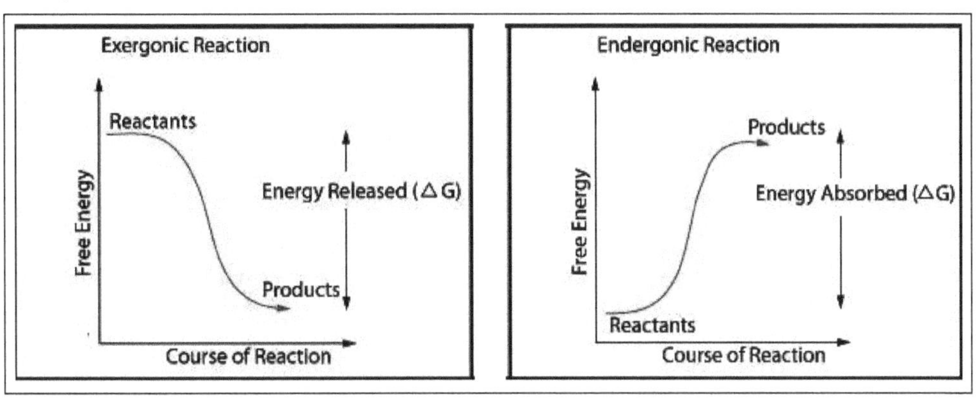

Activation Energy

Activation energy is the energy which must be available to a chemical system with potential reactants to result in a chemical reaction. Activation energy may also be defined as the minimum energy required starting a chemical reaction.

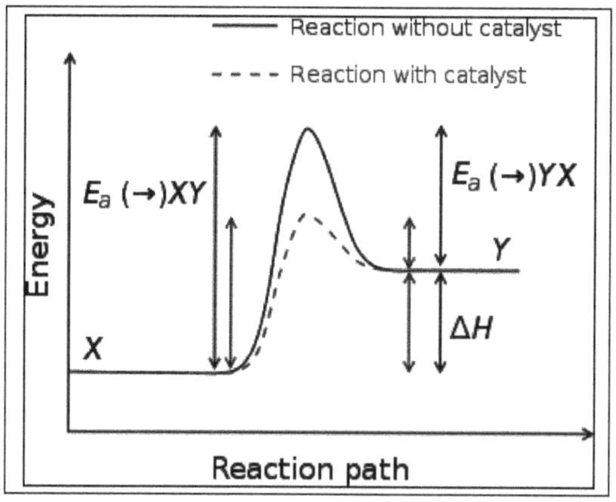

Examples of Major Bioenergetics Processes

Glycolysis is the process of breaking down glucose into pyruvate, producing net eight molecules of ATP (per 1 molecule of glucose) in the process. Pyruvate is one product of glycolysis, and can be shuttled into other metabolic pathways (gluconeogenesis, etc.) as needed by the cell. Additionally, glycolysis produces equivalents in the form of NADH (nicotinamide adenine dinucleotide), which will ultimately be used to donate electrons to the electron transport chain.

Gluconeogenesis is the opposite of glycolysis; when the cell's energy charge is low (the concentration of ADP is higher than that of ATP), the cell must synthesize glucose from carbon- containing biomolecules such as proteins, amino acids, fats, pyruvate, etc. For example, proteins can be broken down into amino acids, and these simpler carbon skeletons are used to build/ synthesize glucose.

The citric acid cycle is a process of cellular respiration in which acetyl coenzyme A, synthesized from pyruvate dehydrogenase, and is first reacted with oxaloacetate to yield citrate. The remaining eight reactions produce other carbon- containing metabolites. These metabolites are successively oxidized, and the free energy of oxidation is conserved in the form of the reduced coenzymes FADH2 and NADH. These reduced electron carriers can then be re- oxidized when they transfer electrons to the electron transport chain. Ketosis is a metabolic process whereby ketone bodies are used by the cell for energy (instead of using glucose). Cells often turn to ketosis as a source of energy when glucose levels are low; e.g. during starvation.

Oxidative phosphorylation and the electron transport chain is the process where reducing equivalents such as NADPH, $FADH_2$ and NADH can be used to donate electrons to a series of redox reactions that take place in electron transport chain complexes. These redox reactions take place in enzyme complexes situated within the mitochondrial membrane. These redox reactions transfer electrons "down" the electron transport chain, which is coupled to the proton motive force. This difference in proton concentration between the mitochondrial matrix and inner membrane space is used to drive ATP synthesis via ATP synthase.

Photosynthesis, another major bioenergetic process, is the metabolic pathway used by plants in which solar energy is used to synthesize glucose from carbon dioxide and water. This reaction takes place in the chloroplast. After glucose is synthesized, the plant cell can undergo photophosphorylation to produce ATP.

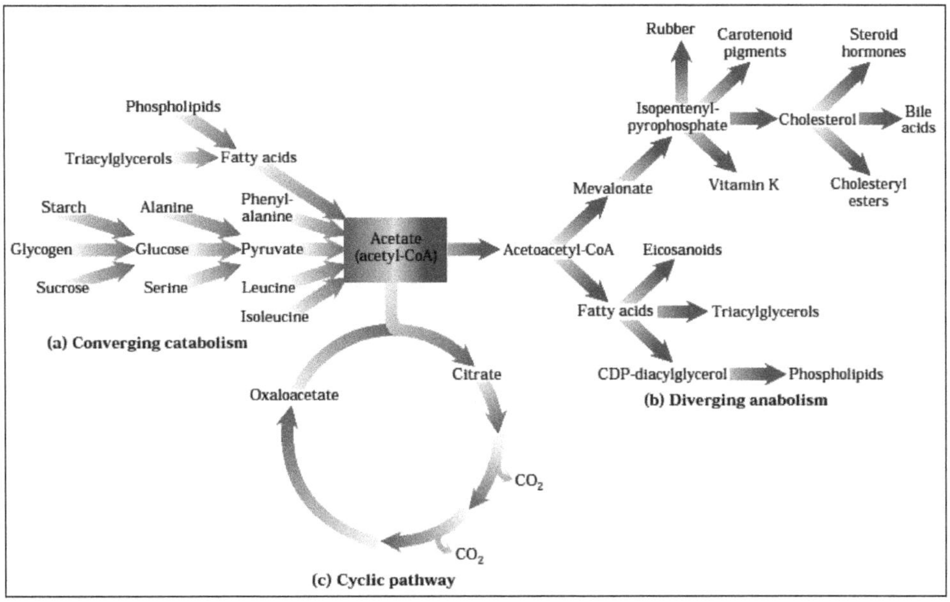

Bioenergetics Relationship between Free Energy, Enthalpy and Entropy

Every living cell and organism must perform work to stay alive, to grow and to reproduce. The ability to harvest energy from nutrients or photons of light and to channel it into biological work is the miracle of life. 1st Law of Thermodynamics: The energy of the universe remains constant. 2nd Law of Thermodynamics: All spontaneous processes increase the entropy of the universe.

The important state functions for the study of biological systems are:

1. The Gibbs free energy (G) which is equal to the total amount of energy capable of doing work during a process at constant temperature and pressure. If ΔG is negative, then the process is spontaneous and termed exergonic.
2. If ΔG is positive, then the process is nonspontaneous and termed endergonic. If ΔG is equal to zero, then the process has reached equilibrium. The Enthalpy (H) which is the heat content of the system. Enthalpy is the amount of heat energy transferred (heat absorbed or emitted) in a chemical process under constant pressure.
3. When ΔH is negative the process produces heat and is termed exothermic. When ΔH is positive the process absorbs heat and is termed endothermic. The Entropy (S) is a quantitative expression of the degree of randomness or disorder of the system. Entropy measures the amount of heat dispersed or transferred during a chemical process. When ΔS is positive then the disorder of the system has increased. When ΔS is negative then the disorder of the system has decreased.
4. The conditions of biological systems are constant temperature and pressure. Under such conditions the relationships between the change in free energy, enthalpy and entropy can be described by the expression where T is the temperature of the system in Kelvin. $\Delta G = \Delta H - T\Delta S$.

   [ΔG = Gibbs free energy; ΔH = Change in Enthalpy; T = Temperature in K; ΔS = Change in Entropy].

Three Thermodynamics Quantities			
Quantity	Symbol	Measure	Units
Enthalpy	H	Heat	Energy
Entropy	S	Disorder	Energy/K
Free energy	G	Reactivity	Energy

Energy Rich Compounds

High energy phosphates act as energy currency of cell. Three major sources of high energy phosphates taking part in energy conservation or energy capture.

Oxidative Phosphorylation

In metabolic pathway, cells use enzymes to oxidize nutrients, thereby releasing energy which is used to produce adenosine triphosphate (ATP). In most eukaryotes, this takes place inside mitochondria. Almost all aerobic organisms carry out oxidative phosphorylation. This pathway is probably so pervasive because it is a highly efficient way of releasing energy, compared to alternative fermentation processes such as anaerobic glycolysis.

The process that accounts for the high ATP yield is known as oxidative phosphorylation. In glycolysis and the citric-acid cycle generate other products besides ATP and GTP, namely NADH and FADH2. These products are molecules that are oxidized (i.e., give up electrons) spontaneously. The body uses these reducing agents (NADH and FADH2) in an oxidation-reduction reaction.

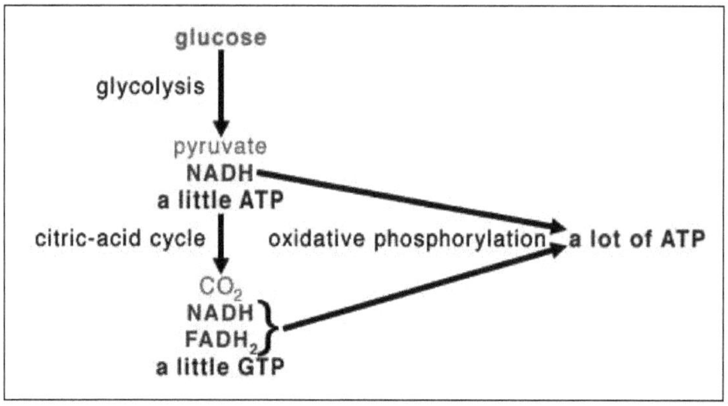

Glycolysis

- Cells use the glycolysis pathway to extract energy from sugars, mainly glucose, and store it in molecules of adenosine triphosphate (ATP) and nicotinamide adenine dinucleotide (NADH). The end product of glycolysis is pyruvate, which can be used in other metabolic pathways to yield additional energy.
- During glycolysis ATP molecules are used and formed in the following reactions (aerobic phase).

Reactions Catalyzed	ATP used	ATP formed
Stage I:		
1. Glucokinase (for phosphorylation)	1	
2. Phosphofructokinase I (for phosphorylation)	1	
Stage II:		
3. Glyceraldehyde 3-phosphate dehydrogenase (oxidation of 2 NADH in respiratory chain)		6
4. Phosphoglycerate kinase (substrate level phosphorylation)		2
Stage IV: 5. Pyruvate kinase (substrate level phosphorylation)		2
Total	2	10
Net gain	08	

- In the anaerobic phase oxidation of one glucose molecule produces 4 - 2 = 2 ATP.

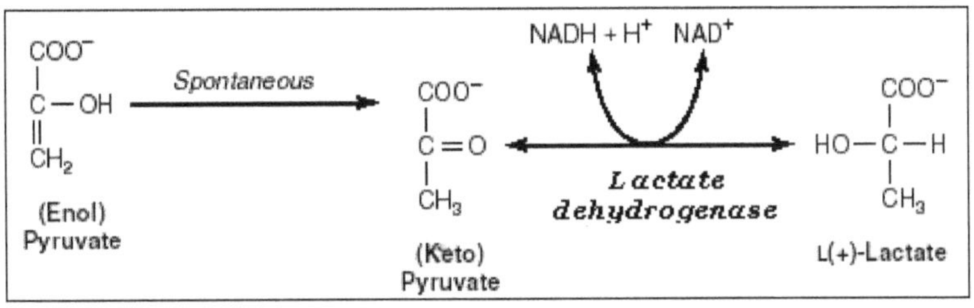

TCA Cycle

- The citric acid cycle (CAC) – also known as the tricarboxylic acid (TCA) cycle or the Krebs cycle is a series of chemical reactions used by all aerobic organisms to release stored energy through the oxidation of acetyl-CoA derived from carbohydrates, fats, and proteins into carbon dioxide and chemical energy in the form of adenosine triphosphate (ATP).
- If one molecule of the substrate is oxidized through NADH in the electron transport chain three molecules of ATP will be formed and through FADH2, two ATP molecules will be generated. As one molecule of glucose gives rise to two molecules of pyruvate by glycolysis, intermediates of citric acid cycle also result as two molecules.

	Reactions	NO. of ATP formed
1.	2 isocitrate → 2 α-ketoglutarate (2 NADH + $2H^+$) (2 x 3)	6
2.	2 α-ketoglutarate → 2 succinyl CoA (2 NADH + $2H^+$) (2 x 3)	6
3.	2 succinyl CoA → 2 succinate (2 GTP = 2 ATP)	2
4.	2 succinate → 2 Fumarate (2 $FADH_2$) (2 x 2)	4
5.	2 malate → 2 oxaloacetate (2 NADH + $2H^+$) (2 x 3)	6
	Total No. of ATP formed	24

Energy Shuttles

- NADH: An energy shuttle which delivers high energy electrons to the electron transport chain where they will eventually power the production of 2 to 3 ATP molecules. When this electron shuttle is not carrying high energy electrons, meaning it has been oxidized (lost its electrons), it is left with a positive charge and is called NAD+.
- FADH2: Another energy shuttle that carries high energy electrons to the electron transport chain, where they will ultimately drive production of 1 to 2 ATP molecules. The oxidized form of FADH2 is FAD and happens just like in NADH.
- ATP: The basic energy currency of the cell. It's a form of energy that cells can use right away.
- GTP: Similar to ATP, GTP can be easily converted to ATP in the cell.

Energy Released by Hydrolysis of Some Phosphate Compounds

Type	Example	Energy Released (kcal/mol)
acyl phosphate 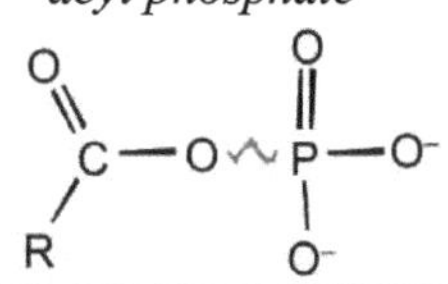	1,3-bisphosphoglycerate (BPG)	−11.8
	acetyl phosphate	−11.3
guanidine phosphates 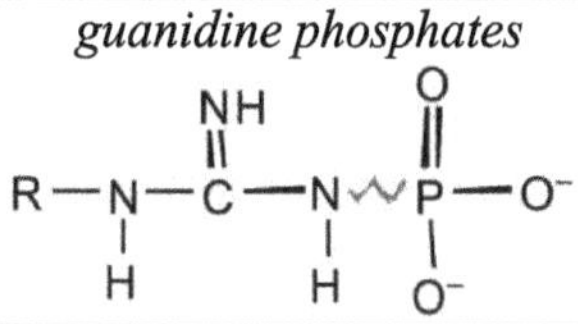	creatine phosphate	−10.3
	arginine phosphate	−9.1
pyrophosphates 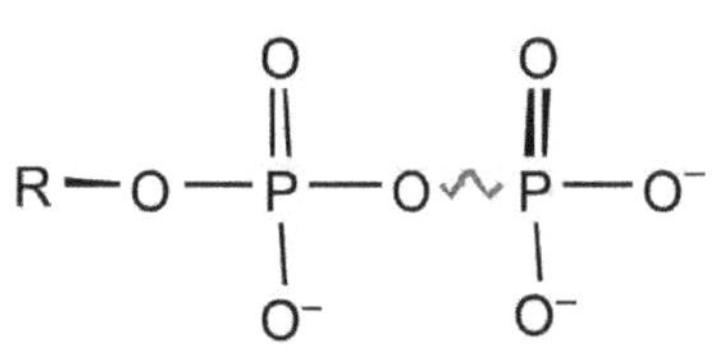	$PP_{i*} \rightarrow 2P_i$	−7.8
	$ATP \rightarrow AMP + PP_i$	−7.7
	$ATP \rightarrow ADP + P_i$	−7.5
	$ADP \rightarrow AMP + P_i$	−7.5
sugar phosphates 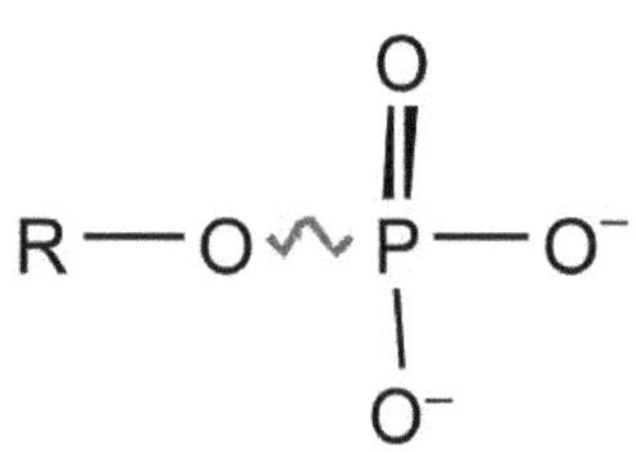	glucose 1-phosphate	−5.0
	fructose 6-phosphate	−3.8
	AMP → adenosine + P_i	−3.4
	Glucose-6-phosphate	−3.3
	Glycerol-3-phosphate	−2.2

***PP_i is the pyrophosphate ion.**

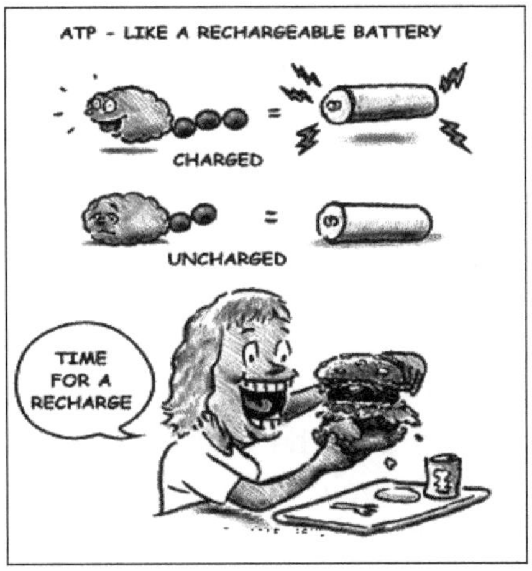

Adenosine Triphosphate (ATP)

Adenosine-5'-triphosphate (ATP) is a multifunctional nucleotide used in cells as a coenzyme. It is often called the "molecular unit of currency" of intracellular energy transfer. ATP transports chemical energy within cells for metabolism.

It is produced by photophosphorylation and cellular respiration and used by enzymes and structural proteins in many cellular processes, including biosynthetic reactions, motility, and cell division. One molecule of ATP contains three phosphate groups and it is produced by ATP synthase from inorganic phosphate and adenosine diphosphate (ADP) or adenosine monophosphate (AMP).

The structure of this molecule consists of a purine base (adenine) attached to the 1' carbon atom of a pentose sugar (ribose). Three phosphate groups are attached at the 5' carbon atom of the pentose sugar. It is the addition and removal of these phosphate groups that inter-convert ATP, ADP and AMP. When ATP is used in DNA synthesis, the ribose sugar is first converted to deoxyribose by ribonucleotide reductase.

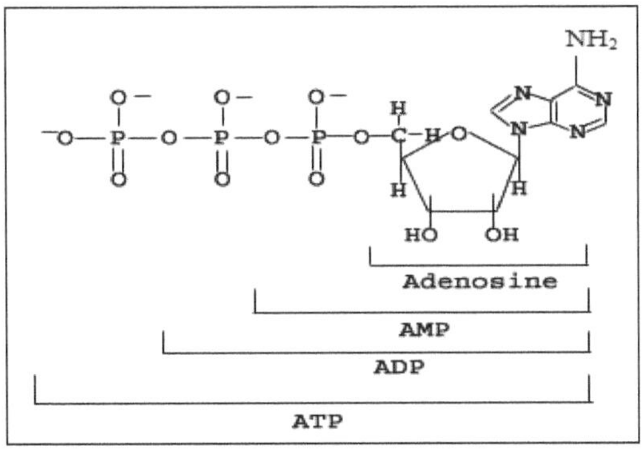

The three main functions of ATP in cellular function are:

- Transporting organic substances—such as sodium, calcium, potassium—through the cell membrane.
- Synthesizing chemical compounds, such as protein and cholesterol.
- Supplying energy for mechanical work, such as muscle contraction.

The standard amount of energy released from hydrolysis of ATP can be calculated from the changes in energy under non-natural (standard) conditions, then correcting to biological concentrations. The energy released by cleaving either a phosphate (Pi) or pyrophosphate (PPi) unit from ATP at standard state of 1 M are:

$$\text{ATP} + \text{H}_2\text{O} \rightarrow \text{ADP} + \text{P}_i\Delta G^\circ = -30.5\text{kJ/mol} (-7.3\text{kcal/mol})$$

$$\text{ATP} + \text{H}_2\text{O} \rightarrow \text{AMP} + \text{PP}_i\Delta G^\circ = -45.6\text{kJ/mol} (-10.9\text{kcal/mol})$$

These values can be used to calculate the change in energy under physiological conditions and the cellular ATP/ADP ratio (also known as the Energy Charge). This reaction is dependent on a number of factors, including overall ionic strength and the presence of alkaline earth metal ions such as Mg^{2+} and Ca^{2+}. Under typical cellular conditions, Δ G is approximately 57kJ−/ mol (14kc−al / mol) .

Cyclic Adenosine Monophosphate

It is a second messenger important in many biological processes. cAMP is derived from adenosine triphosphate (ATP) and used for intracellular signal transduction in many different organisms, conveying the cAMP-dependent pathway.

cAMP is synthesised from ATP by adenylyl cyclase located on the inner side of the plasma membrane. Adenylyl cyclase is activated by a range of signaling molecules through the activation of adenylyl cyclase stimulatory G (Gs)-protein-coupled receptors and inhibited by agonists of adenylyl cyclase inhibitory G (Gi)-protein-coupled receptors. Liver adenylyl cyclase responds more strongly to glucagon, and muscle adenylyl cyclase responds more strongly to adrenaline. cAMP decomposition into AMP is catalyzed by the enzyme phosphodiesterase.

Function: cAMP is a second messenger, used for intracellular signal transduction, such as transferring the effects of hormones like glucagon and adrenaline, which cannot pass through the cell membrane. It is involved in the activation of protein kinases and regulates the effects of adrenaline and glucagon. It also regulates the passage of Ca2+ through ion channels. cAMP and its associated kinases function in several biochemical processes, including the regulation of glycogen, sugar, and lipid metabolism by activating protein kinase.

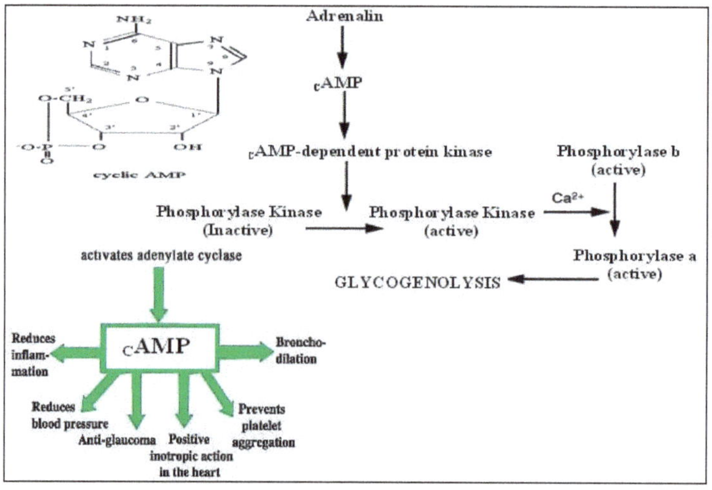

Guanosine Triphosphate (GTP)

Guanosine-5'-triphosphate (GTP) is a purine nucleoside triphosphate. It can act as a substrate for both the synthesis of RNA during the transcription process and of DNA during DNA replication. It also has the role of a source of energy or an activator of substrates in metabolic reactions, like that of ATP, but more specific. It is used as a source of energy for protein synthesis and gluconeogenesis. GTP is essential to signal transduction, in particular with G-proteins, in second-messenger mechanisms where it is converted to Guanosine diphosphate (GDP) through the action of GTPases.

Uses

Energy transfer: GTP is involved in energy transfer within the cell. For instance, a GTP molecule is generated by one of the enzymes in the citric acid cycle. This is tantamount to the generation of one molecule of ATP, since GTP is readily converted to ATP with nucleoside-diphosphate kinase (NDK).

Genetic translation: During the elongation stage of translation, GTP is used as an energy source for the binding of a new amino-bound tRNA to the A site of the ribosome.

Mitochondrial function: The translocation of proteins into the mitochondrial matrix involves the interactions of both GTP and ATP.

Synthesis of AMP and GMP from IMP

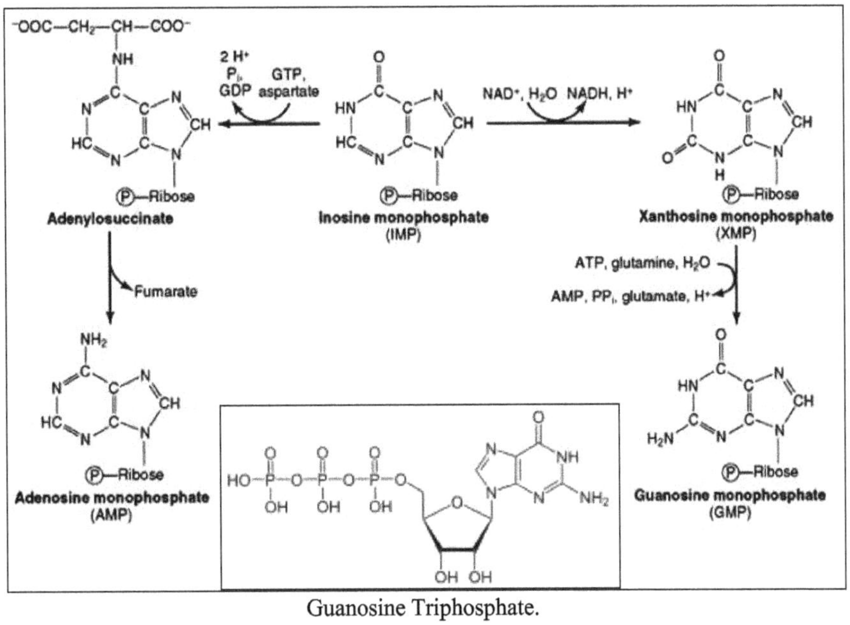

Guanosine Triphosphate.

Cyclic Guanosine Monophosphate

Cyclic guanosine monophosphate (cGMP) is a cyclic nucleotide derived from guanosine triphosphate (GTP). cGMP acts as a second messenger much like cyclic AMP. Its most likely mechanism of action is activation of intracellular protein kinases in response to the binding of membrane-impermeable peptide hormones to the external cell surface.

Synthesis: Guanylate cyclase (GC) catalyzes cGMP synthesis. This enzyme converts GTP to cGMP.

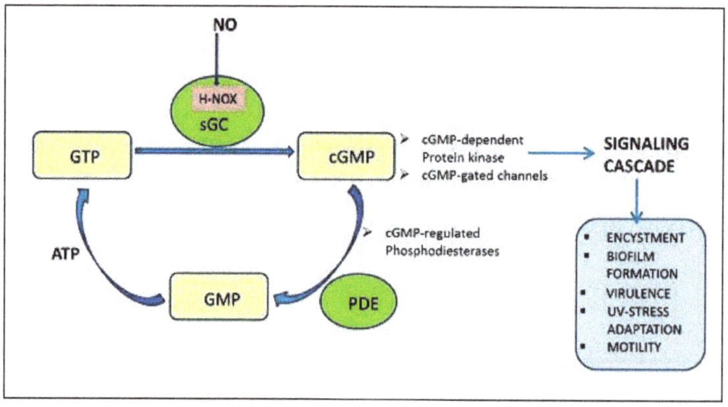

Schematic representation of synthesis, degradation, and function of cGMP. The three targets of cGMP molecules are (i) cGMP dependent protein kinases, (ii) cGMP gated ion channels and (iii) cGMP-dependent phosphodiesterases. While phosphodiesterases are involved in the degradation of cGMP to GMP, the protein kinases and activation of ion channels are subsequently involved in various bacterial signaling pathways.

Effects

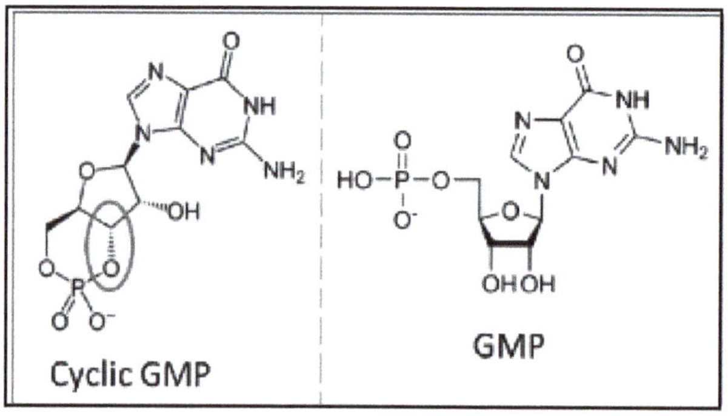

cGMP is a common regulator of ion channel conductance, glycogenolysis, and cellular apoptosis. It also relaxes smooth muscle tissues. In blood vessels, relaxation of vascular smooth muscles leads to vasodilation and increased blood flow. cGMP is a secondary messenger in phototransduction in the eye. cGMP is involved in the regulation of some protein-dependent kinases.

4

Coordination Chemistry: An Overview

Coordination compounds are the compounds that have a central atom, which is often metallic and is bounded by ligands. The study of such compounds is known as coordination chemistry. Some of the significant topics of this field includes coordination number, coordination complex, coordination sphere, trans and cis effect, Jahn-Teller effect, etc. These topics have been closely examined in this chapter.

Coordination chemistry emerged from the work of Alfred Werner, a Swiss chemist who examined different compounds composed of cobalt(III) chloride and ammonia. Upon the addition of hydrochloric acid, Werner observed that ammonia could not be completely removed. He then proposed that the ammonia must be bound more tightly to the central cobalt ion. However, when aqueous silver nitrate was added, one of the products formed was solid silver chloride. The amount of silver chloride formed was related to the number of ammonia molecules bound to the cobalt(III) chloride. For example, when silver nitrate was added to $CoCl_3 \cdot 6NH_3$, all three chlorides were converted to silver chloride. However, when silver nitrate was added to $CoCl_3 \cdot 5NH_3$, only 2 of the 3 chlorides formed silver chloride. When $CoCl_3 \cdot 4NH_3$ was treated with silver nitrate, one of the three chlorides precipitated as silver chloride.

The resulting observations suggested the formation of complex or coordination compounds. In the inner coordination sphere, which is also referred to in some texts as the first sphere, ligands are directly bound to the central metal. In the outer coordination sphere, sometimes referred to as the second sphere, other ions are attached to the complex ion. Werner was awarded the Nobel Prize in 1913 for his coordination theory. The following table is a summary of Werner's observations:

Initial compound	Resulting compounds upon adding $AgNO_3$
$CoCl_3 \cdot 6NH_3$	$\left[Co(NH_3)_6\right]^{3+}(Cl^-)_3$
$CoCl_3 \cdot 5NH_3$	$\left[Co(NH_3)_5Cl\right]^{2+}(Cl^-)_2$
$CoCl_3 \cdot 4NH_3$	$\left[Co(NH_3)_4Cl_2\right]^{+}(Cl^-)$
$CoCl_3 \cdot 3NH_3$	$\left[Co(NH_3)_3Cl_3\right]$

As the table above shows, the complex ion $\left[Co(NH_3)_6\right]^{3+}$ is countered by the three chloride ions. The multi-level binding of coordination complexes play an important role in determining the dissociation of these complexes in aqueous solution. For example,

$[Co(NH_3)_5 Cl]^{2+}(Cl^-)_2$ dissociates into 3 ions while $[Co(NH_3)_4 Cl_2]^+(Cl^-)$ dissociates into 2 ions.ot further dissociate. By applying a current through the aqueous solutions of the resulting complex compounds, Werner measured the electrical conductivity and thus the dissociation properties of the complex compounds. The results confirmed his hypothesis of the formation of complex compounds. It is important to note that the above compounds have a coordination number of 6, which is a common coordination number for many inorganic complexes. Coordination numbers for complex compounds typically range from 1 to 16.

Properties of Coordination Complexes

Some methods of verifying the presence of complex ions include studying its chemical behavior. This can be achieved by observing the compounds' color, solubility, absorption spectrum, magnetic properties, etc. The properties of complex compounds are separate from the properties of the individual atoms. By forming coordination compounds, the properties of both the metal and the ligand are altered.

Metal-ligand bonds are typically thought of Lewis acid-base interactions. The metal atom acts as an electron pair acceptor (Lewis acid), while the ligands act as electron pair donors (Lewis base). The nature of the bond between metal and ligand is stronger than intermolecular forces because they form directional bonds between the metal ion and the ligand, but are weaker than covalent bonds and ionic bonds.

Common Ligands

Monodentate ligands donate one pair of electrons to the central metal atoms. An example of these ligands are the haldide ions (F^-, Cl^-, Br^-, I^-). Polydentate ligands, also called chelates or chelating agents, donate more than one pair of electrons to the metal atom forming a stronger bond and a more stable complex. A common chelating agent is ethylenediamine (en), which, as the name suggests, contains two ammines or: NH_2 sites which can bind to two sites on the central metal. An example of a tridentate ligand is bis-diethylenetriammine. An example of such a coordination complex is bis-diethylenetriamine cobalt^{III}.

Complex compound/ion	Coordination number	Oxidation State of Metal Atom
$[Fe(CN)_6]^{4-}$	6	2+
$[Co(NH_3)_4 SO_4]^-$	5	1+
$[Pt(NH_3)_4]^{2+}$	4	2+
$[Ni(NH_2CH_2CH_2NH_2)_3]^{2+}$	6	2+

Complex ions can form many compounds by binding with other complex ions in multiple ratios. This leads to many combinations of coordination compounds. The structures of certain coordination compounds can also have isomers, which can change their interactions with other chemical agents. The binding between metal and ligands is studied in metals, tetrahedral, and octahedral structures. There are many pharmaceutical and biological applications of coordination complexes and their isomers.

Coordination Compounds

Coordination compound is any of a class of substances with chemical structures in which a central metal atom is surrounded by nonmetal atoms or groups of atoms, called ligands, joined to it by chemical bonds.

Coordination compounds include such substances as vitamin B_{12}, hemoglobin, and chlorophyll, dyes and pigments, and catalysts used in preparing organic substances.

A major application of coordination compounds is their use as catalysts, which serve to alter the rate of chemical reactions. Certain complex metal catalysts, for example, play a key role in the production of polyethylene and polypropylene. In addition, a very stable class of organometallic coordination compounds has provided impetus to the development of organometallic chemistry. Organometallic coordination compounds are sometimes characterized by "sandwich" structures, in which two molecules of an unsaturated cyclic hydrocarbon, which lacks one or more hydrogen atoms, bond on either side of a metal atom. This results in a highly stable aromatic system.

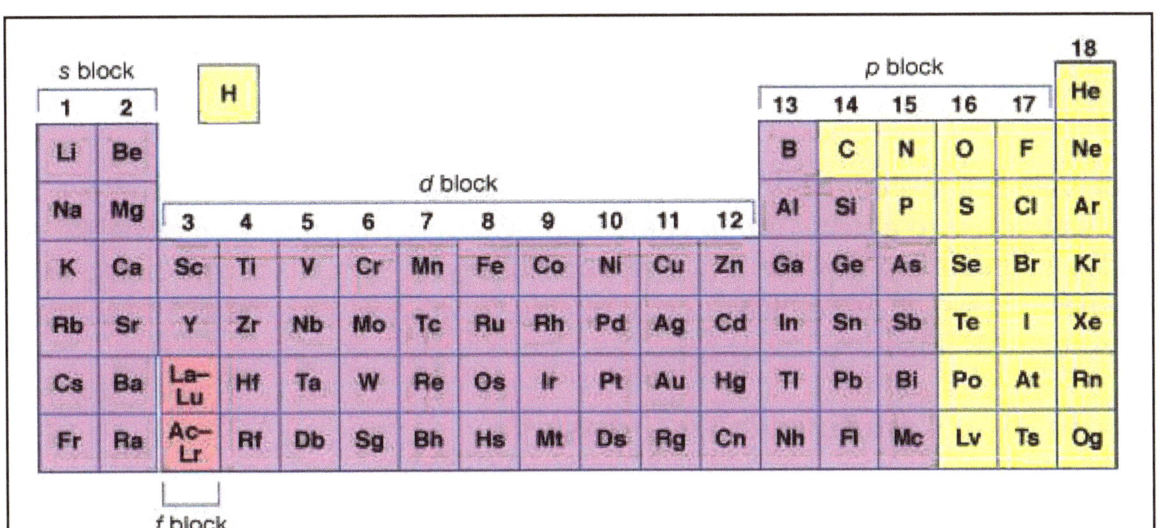

Organometallic coordination compounds, which include transition metal compounds, may be characterized by "sandwich" structures that contain two unsaturated cyclic hydrocarbons on either side of a metal atom. Organometallic compounds are found in the *p*-, *d*-, *s*-, and *f*- blocks of the periodic table (the purple-shaded blocks; the transition metals include those elements in the *d*- and *f*-blocks).

Coordination Compounds In Nature

Naturally occurring coordination compounds are vital to living organisms. Metal complexes play a variety of important roles in biological systems. Many enzymes, the naturally occurring catalysts that regulate biological processes, are metal complexes (metalloenzymes); for example, carboxypeptidase, a hydrolytic enzyme important in digestion, contains a zinc ion coordinated to several amino acid residues of the protein. Another enzyme, catalase, which is an efficient catalyst for the decomposition of hydrogen peroxide, contains iron-porphyrin complexes. In both cases, the coordinated metal ions are probably the sites of catalytic activity. Hemoglobin also contains iron-porphyrin complexes, its role as an oxygen carrier being related to the ability of the iron atoms to coordinate oxygen molecules reversibly. Other biologically important coordination compounds include chlorophyll (a magnesium-porphyrin complex) and vitamin B_{12}, a complex of cobalt with a macrocyclic ligand known as corrin.

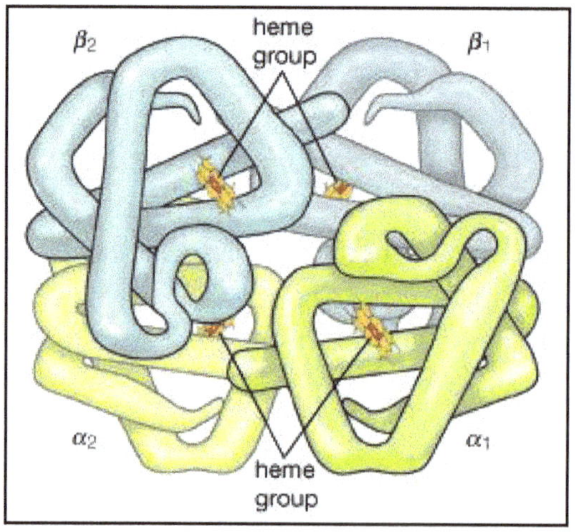

Hemoglobin is a protein made up of four polypeptide chains (α_1, α_2, β_1, and β_2). Each chain is attached to a heme group composed of porphyrin (an organic ringlike compound) attached to an iron atom. These iron-porphyrin complexes coordinate oxygen molecules reversibly, an ability directly related to the role of hemoglobin in oxygen transport in the blood.

Coordination Compounds In Industry

The applications of coordination compounds in chemistry and technology are many and varied. The brilliant and intense colours of many coordination compounds, such as Prussian blue, render them of great value as dyes and pigments. Phthalocyaninecomplexes (e.g., copper phthalocyanine), containing large-ring ligands closely related to the porphyrins, constitute an important class of dyes for fabrics.

Several important hydrometallurgical processes utilize metal complexes. Nickel, cobalt, and copper can be extracted from their ores as ammine complexes using aqueous ammonia. Differences in the stabilities and solubilities of the ammine complexes can be utilized in selective precipitation procedures that bring about separation of the metals. The purification of nickel can be effected by reaction with carbon monoxide to form the volatile tetracarbonylnickel complex, which can be distilled and thermally decomposed to deposit the pure metal. Aqueous cyanide solutions usually are employed to separate goldfrom its ores in the form of the extremely stable dicyanoaurate(−1) complex. Cyanide complexes also find application in electroplating.

There are a number of ways in which coordination compounds are used in the analysis of various substances. These include (1) the selective precipitation of metal ions as complexes—for example, nickel(2+) ion as the dimethylglyoxime complex (shown below):

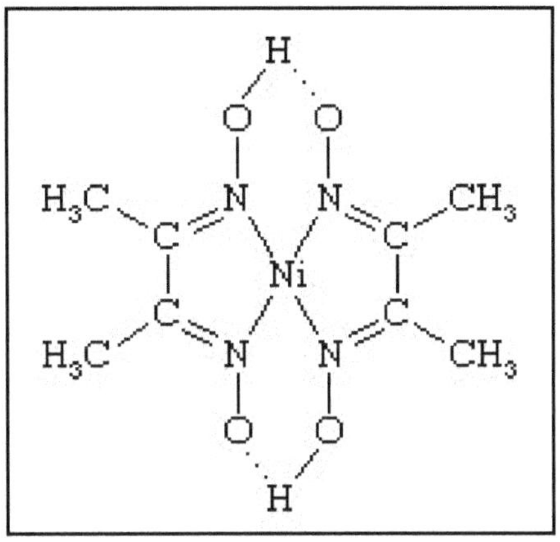

(2) the formation of coloured complexes, such as the tetrachlorocobaltate(2−) ion, which can be determined spectrophotometrically—that is, by means of their light absorption properties, and (3) the preparation of complexes, such as metal acetylacetonates, which can be separated from aqueous solution by extraction with organic solvents.

In certain circumstances, the presence of metal ions is undesirable, as, for example, in water, in which calcium (Ca^{2+}) and magnesium (Mg^{2+}) ions cause hardness. In such cases the undesirable effects of the metal ions frequently can be eliminated by "sequestering" the ions as harmless complexes through the addition of an appropriate complexing reagent. Ethylenediaminetetraacetic acid (EDTA) forms very stable complexes, and it is widely used for this purpose. Its applications include water softening (by tying up $Ca^{2+}$and Mg^{2+}) and the preservation of organic substances, such as vegetable oils and rubber, in which case it combines with traces of transition metal ions that would catalyze oxidation of the organic substances.

A technological and scientific development of major significance was the discovery in 1954 that certain complex metal catalysts—namely, a combination of titanium trichloride, or $TiCl_3$, and triethylaluminum, or $Al(C_2H_5)_3$—bring about the polymerizations of organic compounds with carbon-carbon double bonds under mild conditions to form polymersof high molecular weight and highly ordered (stereoregular) structures. Certain of these polymers are of great commercial importance because they are used to make many kinds of fibres, films, and plastics. Other technologically important processes based on metal complex catalysts include the catalysis by metal carbonyls, such as hydridotetracarbonylcobalt, of the so-called hydroformylation of olefins—i.e., of their reactions with hydrogen and carbon monoxide to form aldehydes—and the catalysis by tetrachloropalladate(2−) ions of the oxidation of ethylene in aqueous solution to acetaldehyde.

Characteristics Of Coordination Compounds

Coordination compounds have been studied extensively because of what they reveal about molecular structure and chemical bonding, as well as because of the unusual chemical nature and useful properties of certain coordination compounds. The general class of coordination compounds or complexes, as they are sometimes called—is extensive and diverse. The substances in the class may be composed of electrically neutral molecules or of positively or negatively charged species (ions).

Among the many coordination compounds having neutral molecules is uranium (+6) fluoride, or uranium hexafluoride (UF_6). The structural formula of the compound represents the actual arrangement of atoms in the molecules:

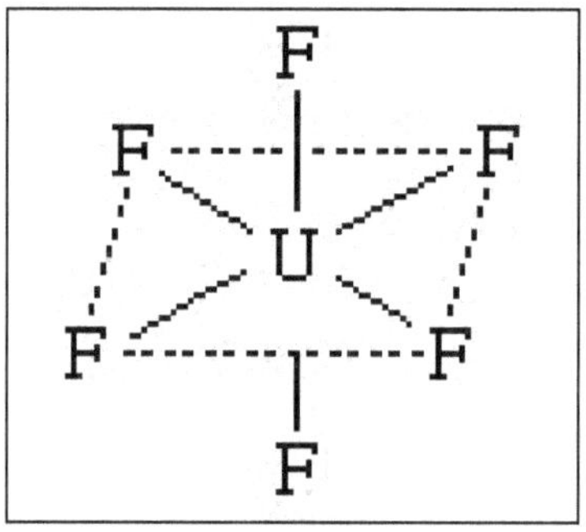

In this formula the solid lines, which represent bonds between atoms, show that four of the fluorine (F) atoms are bonded to the single atom of uranium (U) and lie in a plane with it, the plane being indicated by dotted lines (which do not represent bonds), whereas the remaining two fluorine atoms (also bonded to the uranium atom) lie above and below the plane, respectively.

An example of an ionic coordination complex is the hydrated ion of nickel, (Ni), hexaaquanickel (2+) ion, $[Ni(H_2O)_6]^{2+}$, the structure of which is shown below. In this structure, the symbols and lines are used as above, and the brackets and the "two plus" (2+) sign show that the double positive charge is assigned to the unit as a whole.

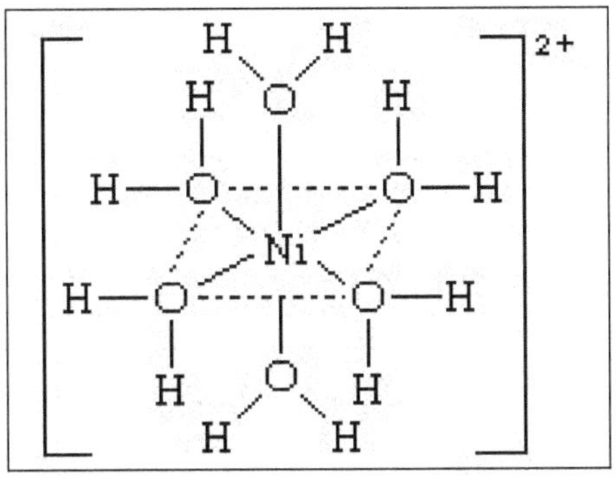

The central metal atom in a coordination compound itself may be neutral or charged (ionic). The coordinated groups—or ligands—may be neutral molecules such as water (in the above example), ammonia (NH_3), or carbon monoxide (CO); negatively charged ions (anions) such as the fluoride (in the first example above) or cyanide ion (CN^-); or, occasionally, positively charged ions (cations) such as the hydrazinium ($N_2H_5^+$) or nitrosonium (NO^+) ion.

Complex ions—that is, the ionic members of the family of coordination substances—may exist as free ions in solution, or they may be incorporated into crystalline materials (salts) with other ions of opposite charge. In such salts, the complex ion may be either the cationic (positively charged) or the anionic (negatively charged) component (or, on occasion, both). The hydrated nickel ion (above) is an example of a cationic complex. An anionic complex is the hexacyanide of the ferric iron (Fe^{+3}) ion, the hexacyanoferrate(3−) ion, $[Fe(CN)_6]^{3-}$, or

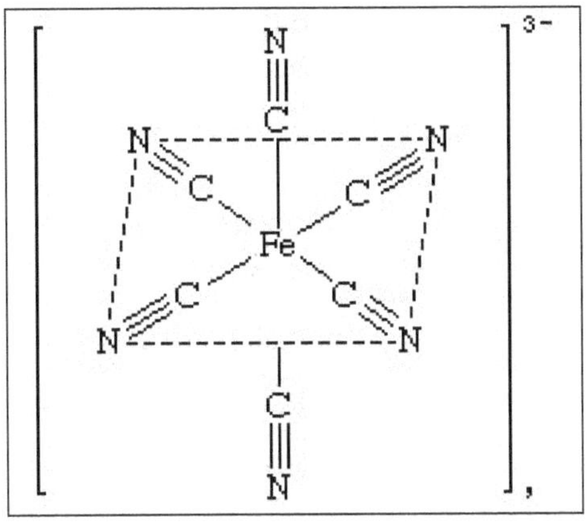

Crystalline salts containing complex ions include potassium hexacyanoferrate(3−) (potassium ferricyanide), $K_3[Fe(CN)_6]$, and the hexahydrate of nickel chloride, hexaaquanickel(2+) chloride, $[Ni(H_2O)_6]Cl_2$. In each case the charge on the complex ion is balanced by ions of opposite charge. In the case of potassium ferricyanide, three positively charged potassium ions, K^+, balance the negative charge on the complex, and in the nickel complex the positive charges are balanced by two negative chloride ions, Cl^-. The oxidation state of the central metal is determined from the charges on the ligands and the overall charge on the complex. For example, in hexaaquanickel(2+), water is electrically neutral and the charge on the complex ion is +2; thus, the oxidation state of Ni is +2. In hexacyanoferrate(3−), all six cyano ligands have a charge of −1; thus, the overall charge of −3 dictates that the oxidation state of Fe is +3.

The distinction between coordination compounds and other substances is, in fact, somewhat arbitrary. The designation coordination compound, however, is generally restricted to substances whose molecules or ions are discrete entities and in which the central atom is metal. Accordingly, molecules such as sulfur(+6) fluoride (sulfur hexafluoride; SF_6) and carbon(+4) fluoride (carbon tetrafluoride; CF_4) are not normally considered coordination compounds, because sulfur (S) and carbon (C) are nonmetallic elements. Yet there is no great difference between these compounds and, say, uranium hexafluoride. Furthermore, such simple ionic salts as sodium chloride (NaCl) or nickel(+2) fluoride (nickel difluoride; NiF_2) are not considered coordination compounds, because they consist of continuous ionic lattices rather than discrete molecules. Nevertheless, the arrangement (and bonding) of the anions surrounding the metal ions in these salts is similar to that in coordination compounds. Coordination compounds generally display a variety of distinctive physical and chemical properties, such as colour, magnetic susceptibility, solubility and volatility, an ability to undergo oxidation-reduction reactions, and catalytic activity.

A coordination compound is characterized by the nature of the central metal atom or ion, the oxidation state of the latter (that is, the gain or loss of electrons in passing from the neutral atom to the charged ion, sometimes referred to as the oxidation number), and the number, kind, and arrangement of the ligands. Because virtually all metallic elements form coordination compounds—sometimes in several oxidation states and usually with many different kinds of ligands—a large number of coordination compounds are known.

Coordination Number

Coordination number is the term proposed by Werner to denote the total number of bonds from the ligands to the metal atom. Coordination numbers generally range between 2 and 12, with 4 (tetracoordinate) and 6 (hexacoordinate) being the most common. Werner referred to the central atom and the ligands surrounding it as the coordination sphere. Coordination number should be distinguished from oxidation number (defined in the previous paragraph). The oxidation number, designated by an Arabic number with an appropriate sign (or, sometimes, by a Roman numeral in parentheses), is an index derived from a simple and formal set of rules and is not a direct indicator of electron distribution or of the charge on the central metal ion or compound as a whole. For the hexaamminecobalt(3+) ion, $[Co(NH_3)_6]^{3+}$, and the neutral moleculetriammine-trinitrocobalt(3+), $\left[Co(NO_2)_3(NH_3)_3\right]$, the coordination number of cobalt is 6 while its oxidation number is +3.

Ligands and Chelates

Each molecule or ion of a coordination compound includes a number of ligands, and, in any given substance, the ligands may be all alike, or they may be different. The term *ligand* was proposed by the German chemist Alfred Stock in 1916. Attachment of the ligands to the metal atom may be through only one atom, or it may be through several atoms. When only one atom is involved, the ligand is said to be monodentate; when two are involved, it is didentate, and so on. In general, ligands utilizing more than one bond are said to be polydentate. Because a polydentate ligand is joined to the metal atom in more than one place, the resulting complex is said to be cyclic—i.e., to contain a ring of atoms. Coordination compounds containing polydentate ligands are called chelates and their formation is termed *chelation*. Chelates are particularly stable and useful. An example of a typical chelate is bis(1,2-ethanediamine)copper(2+), the complex formed between the cupric ion (Cu^{2+}) and the organic compound ethylenediamine ($NH_2CH_2CH_2NH_2$, often abbreviated as en in formulas). The formula of the complex is,

$$\left[Cu\left(NH_2CH_2CH_2NH_2\right)_2\right]^{2+}$$

and the structural formula is

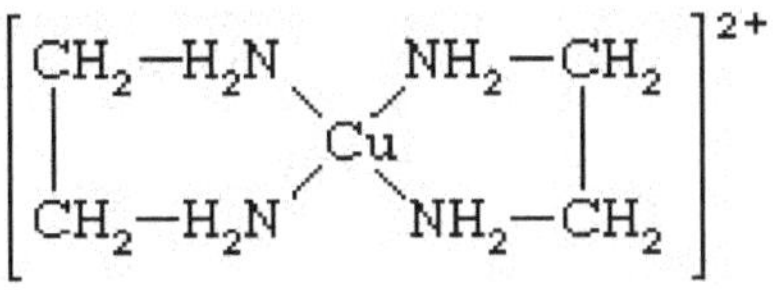

Mononuclear, Monodentate

The simplest types of coordination compounds are those containing a single metal atom or ion (mononuclear compounds) surrounded by monodentate ligands. Most of the coordination compounds already cited belong to this class. Among the ligands forming such complexes are a wide variety of neutral molecules (such as ammonia, water, carbon monoxide, and nitrogen), as well as monoatomic and polyatomic anions (such as the hydride, fluoride, chloride, oxide, hydroxide, nitrite, thiocyanate, carbonate, sulfate, and phosphate ions). Coordination of such ligands to the metal virtually always occurs through an atom possessing an unshared pair of electrons, which it donates to the metal to form a coordinate bond with the latter. Among the atoms that are known to coordinate to metals are those of virtually all the nonmetallic elements (such as hydrogen, carbon, oxygen, nitrogen, and sulfur), with the exception of the noble gases(helium [He], neon [Ne], argon [Ar], krypton [Kr], and xenon [Xe]).

Polydentate

The chelate complex of a copper ion and ethylenediamine mentioned above is an example of a compound formed between a metal ion and a didentate ligand. Two further examples of chelate complexes are shown below.

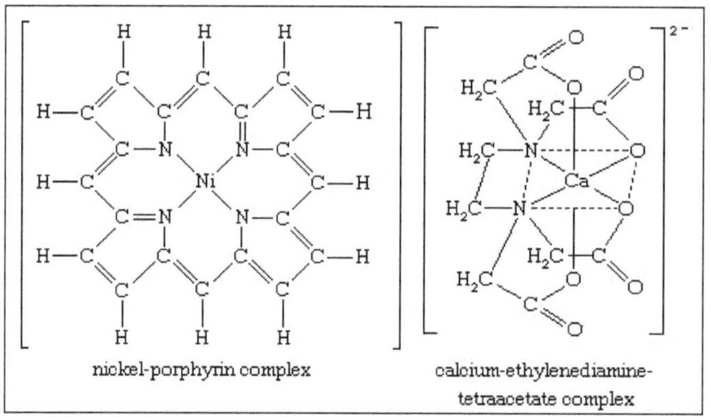

These are a nickel complex with a tetradentate large-ring ligand, known as a porphyrin, and a calcium complex with a hexadentate ligand, ethylenediaminetetraacetate (EDTA). Because metal-ligand attachment in such chelate complexes is through several bonds, such complexes tend to be very stable.

The commonest and most stable complexes of the lanthanoid metals (the series of 14 *f*-block elements following lanthanum [atomic number 57]) are those with chelating oxygen ligands, such as EDTA-type anions or hydroxo acids (e.g., tartaric or citric acids). The formation of such water-soluble complexes is employed in the separation of lanthanoids by ion-exchange chromatography. Lanthanoid β-diketonates are well known because some fluorinated β-diketonates yield volatile complexes amenable to gas-chromatographic separations. Neutral complexes can complex further to yield anionic species such as octacoordinated tetrakis(thenoyltrifluoroacetato)neodymate(1−), $[Nd(CF_3COCHCOCF_3)_4]^-$.

Certain ligands may be either monodentate or polydentate, depending on the particular compound in which they occur. The carbonate ion, $(CO_3)^{2-}$, for example, is coordinated to the cobalt (Co^{3+}) ions in two cobalt complexes, pentaamminecarbonatocobalt(+), $[Co(CO_3)(NH_3)_5]^+$, and tetraamminecarbonatocobalt(+), $[Co(CO_3)(NH_3)_4]^+$, through one and two oxygen atoms, respectively.

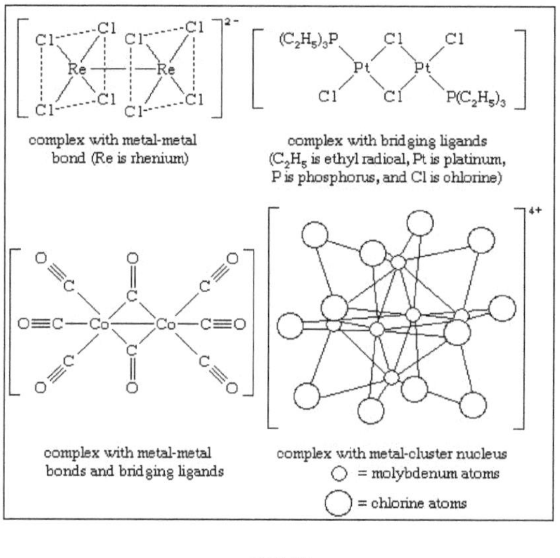

Polynuclear

Polynuclear complexes are coordination compounds containing two or more metal atoms, or ions, in a single coordination sphere. The two atoms may be held together through direct metal-metal bonds, through bridging ligands, or both. Examples of each are shown above, along with a unique metal-cluster complex having six metal atoms in its nucleus.

Nomenclature

Generally, the systematic naming of coordination compounds is carried out by rules recommended by the International Union of Pure and Applied Chemistry (IUPAC). Among the more important of these are the following:

1. Neutral and cationic complexes are named by first identifying the ligands, followed by the metal; its oxidation number may be given in Roman numerals enclosed within parentheses. Alternatively, the overall charge on the complex may be given in Arabic numbers in parentheses. This convention is generally followed here. In formulas, anionic ligands (ending in *-o*; in general, if the anion name ends in *-ide*, *-ite*, or *-ate*, the final *e* is replaced by *-o*, giving *-ido*, *-ito*, and *-ato*) are cited in alphabetical order ahead of neutral ones also in alphabetical order (multiplicative prefixes are ignored). When the complex contains more than one ligand of a given kind, the number of such ligands is designated by one of the prefixes *di-*, *tri-*, *tetra-*, *penta-*, and so on or, in the case of complex ligands, by *bis-*, *tris-*, *tetrakis-*, *pentakis-*, and so on. In names (as opposed to formulas) the ligands are given in alphabetical order without regard to charge. The oxidation number of the metal is defined in the customary way as the residual charge on the metal if all the ligands were removed together with the electron pairs involved in coordination to the metal. The following examples are illustrative (aqua is the name of the water ligand):

   $\left[Ce\left(CH_3CCHO_2CCH_3\right)_4\right]$ tetrakis(2,4-pentanedionato) cerium(IV)

   $\left[Nd\left(H_2O\right)^9\right]^{3+}$ nonaaquaneodymium (III)

2. Anionic complexes are similarly named, except that the name is terminated by the suffix *-ate*; for example:

   $\left[Er(NCS)_6\right]^{3-}$ hexakis(thiocyanato) erbate(III)

3. In the case of salts, the cation is named first and then the anion; for example:

   $(NH_4)_2\left[Ce(NO_3)_6\right]$ ammonium hexanitatocerate(IV)

4. Polynuclear complexes are named as follows, bridging ligands being identified by a prefix consisting of the Greek letter *mu* (μ-):

$[(CO)_6\,Mn - (CO)_5]$	decacarbonyldi- manganese (0) or bis (pentacarbonyl- manganese)
$[(NH_3)_5\,Cr - OH - Cr(NH_3)_5]Cl_5$	m- hydroxo - bis[penta- ammine- chormium (III)] chloride

In addition to their systematic designations, many coordination compounds are also known by names reflecting their discoverers or colours. Examples are,

$K[PtCl_3(C_2H_4)]\cdot H_2O$	Zeise's salt
$NH_4[Cr(SCN)_4(NH_3)_2]\cdot H_2O$	Reineoke's salt
$[Co(NH_3)_5(H_2O]Cl_3$	roseocobaltic chloride (red)
$KFe[Fe(CN)_6]$	Prussian blue

Structure And Bonding Of Coordination Compounds

Werner originally postulated that coordination compounds can be formed because the central atoms carry the capacity to form secondary, or coordinate, bonds, in addition to the normal, or valence, bonds. A more complete description of coordinate bonding, in terms of electron pairs, became possible in the 1920s, following the introduction of the concept that all covalent bonds consist of electron pairs shared between atoms, an idea advanced chiefly by the American physical chemist Gilbert N. Lewis. In Lewis's formulation, when both electrons are contributed by one of the atoms, as in the boron-nitrogen bond formed when the substance boron trifluoride (BF_3) combines with ammonia, the bond is called a coordinate bond:

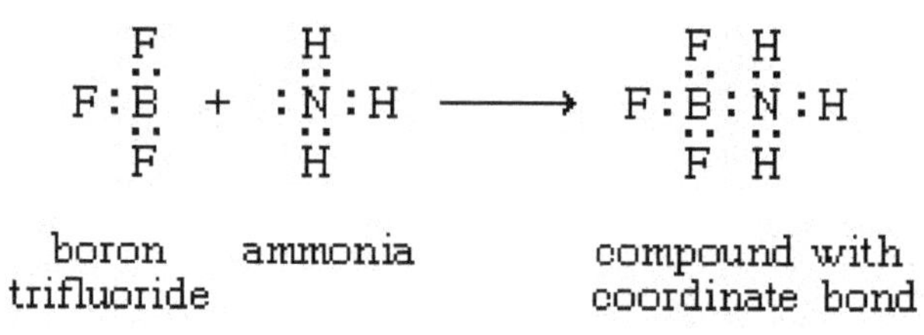

In Lewis's formulas, the valence (or bonding) electrons are indicated by dots, with each pair of dots between two atomic symbols representing a bond between the corresponding atoms.

Following Lewis's ideas, the suggestion was made that the bonds between metals and ligands were of this same type, with the ligands acting as electron donors and the metal ions as electron acceptors. This suggestion provided the first electronic interpretation of bonding in coordination compounds. The coordination reaction between silver ions and ammonia illustrates the resemblance of coordination compounds to the situation in the boron-nitrogen compound. According to this view, the metal ion can be regarded as a so-called Lewis acid and the ligands as Lewis bases:

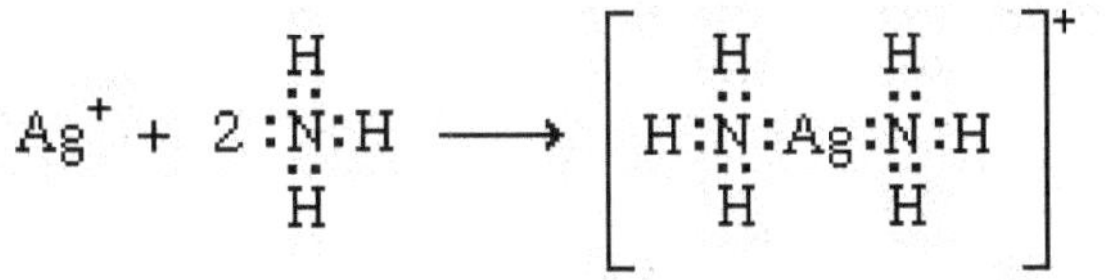

silver ion ammonia coordination complex

A coordinate bond may also be denoted by an arrow pointing from the donor to the acceptor.

Geometry

Many coordination compounds have distinct geometric structures. Two common forms are the square planar, in which four ligands are arranged at the corners of a hypothetical square around the central metal atom, and the octahedral, in which six ligands are arranged, four in a plane and one each above and below the plane. Altering the position of the ligands relative to one another can produce different compounds with the same chemical formula. Thus, a cobalt ion linked to two chloride ions and four molecules of ammonia can occur in both green and violet forms according to how the six ligands are placed. Replacing a ligand also can affect the colour. A cobalt ion linked to six ammonia molecules is yellow. Replacing one of the ammonia molecules with a water molecule turns it rose red. Replacing all six ammonia molecules with water molecules turns it purple.

Among the essential properties of coordination compounds are the number and arrangement of the ligands attached to the central metal atom or ion—that is, the coordination number and the coordination geometry, respectively. The coordination number of a particular complex is determined by the relative sizes of the metal atom and the ligands, by spatial (steric) constraints governing the shapes (conformations) of polydentate ligands, and by electronic factors, most notably the electronic configuration of the metal ion. Although coordination numbers from 1 to 16 are known, those below 3 and above 8 are rare. Possible structures and examples of species for the various coordination numbers are as follows: three, trigonal planar ($[Au\ \{P(C_6H_5)_3\}_3]^+$; four, tetrahedral ($[CoCl_4]^{2-}$) or square planar ($[PtCl_4]^{2-}$); five, trigonal bipyramid ($[CuCl_5\ |>\}]^{3-}$) or square pyramid ($VO(\text{acetylacetonate})_2$); six, octahedral ($[Co(NO_2)_6]^{3-}$) or trigonal prismatic ($[Re\ \{S_2C_2(C_6H_5)_2\}_3]$); seven, pentagonal bipyramid ($Na_5[Mo(CN)_7]10H_2O$), capped trigonal prism (cation in $[Ca(H_2O)_7]_2[Cd_6Cl_{16}(H_2O)_2]\ H_2O$), or capped octahedron (cation in $[Mo(CNC_6H_5)_7][PF_6]_2$); eight, square antiprism or dodecahedron ($[Zr(\text{acetylacetonate})_4]$; and nine, capped square antiprism ($La(NH_3)_9]^{3+}$) or tricapped trigonal prism ($[ReH_9]^{2-}$).

Coordination numbers are also affected by the 18-electron rule (sometimes called the noble gas rule), which states that coordination compounds in which the total number of valence electrons approaches but does not exceed 18 (the number of electrons in the valence shells of the noble gases) are most stable. The stabilities of 18-electron valence shells are also reflected in the coordination numbers of the stable mononuclear carbonyls of different metals that have oxidation number 0 e.g., tetracarbonylnickel, pentacarbonyliron, and hexacarbonylchromium (each of which has a valence shell of 18).

The 18-electron rule applies particularly to covalent complexes, such as the cyanides, carbonyls, and phosphines. For more ionic (also called outer-orbital) complexes, such as fluoro or aqua complexes, electronic factors are less important in determining coordination numbers, and configurations corresponding to more than 18 valence electrons are not uncommon. Several nickel (+2) complexes, for example including the hexafluoro, hexaaqua, and hexaammine complexes—each have 20 valence electrons.

Any one metal ion tends to have the same coordination number in different complexes e.g., generally six for chromium (+3) but this is not invariably so. Differences in coordination number may result from differences in the sizes of the ligands; for example, the iron (+3) ion is able to accommodate six fluoride ions in the hexafluoro complex $[FeF_6]>]^{3-}$ but only four of the larger chloride ions in the tetrachloro complex $[FeCl_4]^-$. In some cases, a metal ion and a ligand form two or more complexes with different coordination numbers e.g., tetracyanonickelate $[Ni(CN)_4]>]^{2-}$ and pentacyanonickelate $[Ni(CN)_5]>]^{3-}$, both of which contain Ni in the +2 oxidation state.

Isomerism

Coordination compounds often exist as isomers-i.e., as compounds with the same chemical composition but different structural formulas. Many different kinds of isomerism occur among coordination compounds. The following are some of the more common types.

Cis-trans Isomerism

Cis-trans (geometric) isomers of coordination compounds differ from one another only in the manner in which the ligands are distributed spatially; for example, in the isomeric pair of diamminedichloroplatinum compounds.

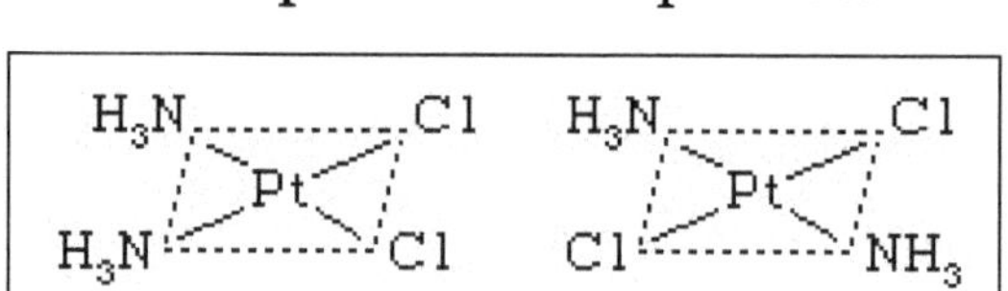

the two ammonia molecules and the two chlorine atoms are situated next to one another in one isomer, called the *cis* (Latin for "on this side") isomer, and across from one another in the other, the *trans* (Latin for "on the other side") isomer. A similar relationship exists between the *cis* and *trans* forms of the tetraamminedichlorocobalt(1+) ion:

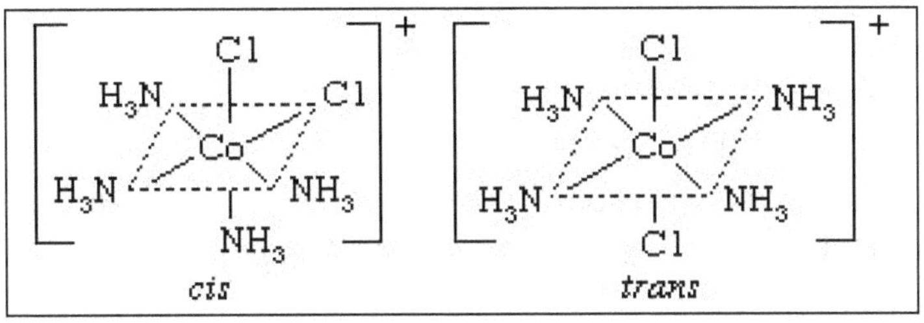

Enantiomers and Diastereomers

So-called optical isomers (or enantiomers) have the ability to rotate plane-polarized light in opposite directions. Enantiomers exist when the molecules of the substances are mirror images but are not superimposable upon one another. In coordination compounds, enantiomers can arise either from the presence of an asymmetric ligand, such as one isomer of the amino acid, alanine (aminopropionic acid),

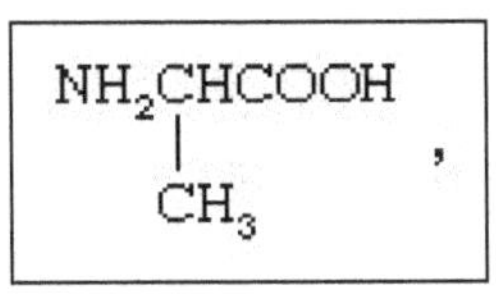

or from an asymmetric arrangement of the ligands. Familiar examples of the latter variety are octahedral complexes carrying three didentate ligands, such as ethylenediamine, $NH_2CH_2CH_2NH_2$. The two enantiomers corresponding to such a complex are depicted by the structures below.

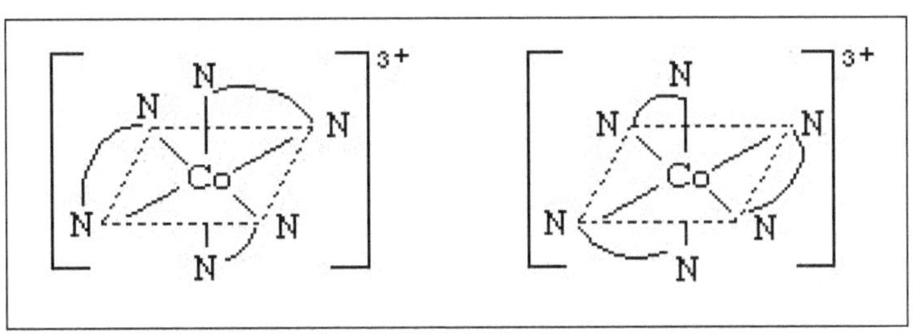

The ethylenediamine ligands above are indicated by a curved line between the symbols for the nitrogen atoms.

Diastereomers, on the other hand, are not superimposable and also are not mirror images. Using AB as an example of a chelating ligand, in which the symbol AB implies that the two ends of the chelate are different, there are six possible isomers of a complex cis-$[M(AB)_2X_2]$. For example, AB might correspond to alanine $[CH_3CH(NH_2)C(O)O]^-$, where both N and O are attached to the metal. Alternatively, AB could represent a ligand such as propylenenediamine, $[NH_2CH_2C(CH_3)HNH_2]$, where the two ends of the molecule are distinguished by the fact that one of the Hs on a C is substituted with a methyl (CH_3) group.

Ionization Isomerism

Certain isomeric pairs occur that differ only in that two ionic groups exchange positions within (and without) the primary coordination sphere. These are called ionization isomers and are exemplified by the two compounds, pentaamminebromocobalt sulfate, $[CoBr(NH_3)_5]SO_4$, and pentaamminesulfatocobalt bromide, $[Co(SO_4)(NH_3)_5]$ Br. In the former the bromide ion is coordinated to the cobalt (3+) ion, and the sulfate ion is outside the coordination sphere; in the latter the sulfate ion occurs within the coordination sphere, and the bromide ion is outside it.

Linkage Isomerism

Isomerism also results when a given ligand is joined to the central atom through different atoms of the ligand. Such isomerism is called linkage isomerism. A pair of linkage isomers are the ions $[Co(NO_2)(NH_3)_5]^{2+}$ and $[Co(ONO)(NH_3)_5]^{2+}$, in which the anionic ligand is joined to the cobalt atom through nitrogen or oxygen, as shown by designating it with the formulas NO_2^-(nitro) and ONO^-(nitrito), respectively. Another example of this variety of isomerism is given by the pair of ions $[Co(CN)_5(NCS)]^{3-}$ and $[Co(CN)_5(SCN)]^{3-}$, in which an isothiocyanate $(NCS)^-$ and a thiocyanate group $(SCN)^-$ are bonded to the cobalt(3+) ion through a nitrogen or sulfur atom, respectively.

Coordination Isomerism

Ionic coordination compounds that contain complex cations and anionscan exist as isomers if the ligands associated with the two metal atoms are exchanged, as in the pair of compounds, hexaamminecobalt(3+) hexacyanochromate(3−), $[Co(NH_3)_6][Cr(CN)_6]$, and hexaamminechromium(3+) hexacyanocobaltate(3−), $[Cr(NH_3)_6][Co(CN)_6]$. Such compounds are called coordination isomers, as are the isomeric pairs obtained by redistributing the ligands between the two metal atoms, as in the doubly coordinated pair, tetraammineplatinum(2+) hexachloroplatinate(2−), $[Pt(NH_3)_4][PtCl_6]$, and tetraamminedichloroplatinum(2+) tetrachloroplatinate(2−), $[PtCl_2(NH_3)_4][PtCl_4]$.

Ligand Isomerism

Isomeric coordination compounds are known in which the overall isomerism results from isomerism solely within the ligands groups. An example of such isomerism is shown by the ions, bis(1,3-diaminopropane)platinum(2+) and bis(1,2-diaminopropane) platinum(2+),

$$\left[Pt(NH_2CH_2CH_2CH_2NH_2)_2\right]^{2+} \text{ and } \left[Pt(NH_2\underset{\underset{CH}{|}}{C}HCH_2NH_2)_2\right]^{2+}.$$

Important Types of Reactions Of Coordination Compounds

Acid-base

Coordination to a positive metal ion usually enhances the acidity (i.e., the tendency to lose protons) of hydrogen-containing ligands, such as water and ammonia. Thus, many metal ions in aqueous solution commonly exhibit acidic behaviour. Such behaviour is exemplified by hydrolysis reactions of the type shown in the following equilibrium:

$$[M(H_2O)_x]^{n+} \rightleftharpoons [M(OH)(H_2O)_{x-1}]^{(n-1)+} + H^+,$$

in which M represents the metal ion, n its charge, and x the number of coordinated water molecules.

The acidities of such aqua ions depend on the charge, size, and electronic configuration of the metal ion. This dependence is reflected in the values of acid dissociation constants, which range from about 10^{-14}(a value only slightly larger than for pure water, for which the dissociation constant = $10^{-15.7}$) for the hydrated lithium ion, to about 10^{-2} (a value equivalent to that of a fairly strong acid) for the hydrated uranium(4+) ion. Acid-base equilibria are rapidly established in solution, generally within a fraction of a second.

In some cases, hydrolysis of a metal ion may be accompanied by polymerization to form dinuclear or polynuclear hydroxo- or oxygen-bridged complexes.

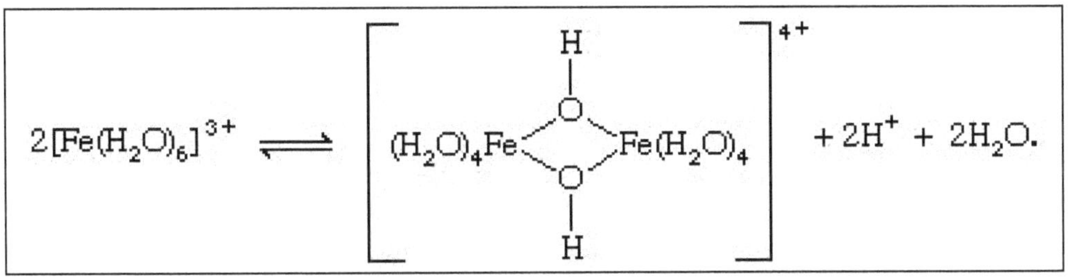

Even very weakly acidic ligands, such as ammonia, can acquire appreciable acidity through coordination to a metal ion. Thus, the hexaammineplatinum(4+) ion dissociates according to the following equilibrium:

$$[Pt(NH_3)_6]^{4+} \rightleftharpoons [Pt(NH_2)(NH_3)_5]^{3+} + H^+$$

In addition to intrinsic strength, acids and bases have other properties that determine the extent of reactions. According to the hard and soft acids and bases (HSAB) theory, the metal cation and anion are considered to be acids and bases, respectively. Hard acids and bases are small and nonpolarizable, whereas soft acids and bases are larger and more polarizable. Interactions between two hard or soft acids or bases are stronger than ones between one hard and one soft acid or base. The theory can be used to explain solubilities, formation of metallic ores, and some reactions of metal cations with ligands.

Substitution

One of the most general reactions exhibited by coordination compounds is that of substitution, or replacement, of one ligand by another. This process is depicted in a generalized manner by the equation $ML_{x-1}Y + Z \rightarrow ML_{x-1}Z + Y$ for a metal complex of coordination number *x*. The ligands L, Y, and Z may be chemically similar or different. (Charges have been omitted here for simplicity.)

A class of substitution reactions that affords the widest possible comparison of different metal ions is the replacement of water in the coordination spheres of metal-aqua complexes in aqueous solution. The substitution may be by another water molecule (which can be labeled with the isotope oxygen-18 to permit the reaction to be followed) or by a different ligand, such as the chloride ion. Reactions of both types occur as shown below (oxygen-18 is indicated by the symbol $\overset{*}{O}$).

$$[M(H_2O_x)]^{n+} + H_2\overset{*}{O} \rightarrow [M(H_2O)_{x-1}(H_2\overset{*}{O})]^{n+} + H_2O$$
$$[M(H_2O_x)]^{n+} + Cl^- \rightarrow [M(H_2O)_{x-1}Cl]^{(n-1)+} + H_2O$$

Many such reactions are extremely fast, and it has been only since 1950, following the development of appropriate experimental methods (including stopped flow, nuclear magnetic resonance, and relaxation spectrometry), that the kinetics and mechanisms of this class of reactions have been extensively investigated. Rates of substitution of metal-aqua ions have been found to span a wide range, the characteristic times required for substitution ranging from less than 10^{-9} second for monopositive ions, such as hydrated potassium ions, to several days for certain more highly charged ions, such as hexaaquachromium(3+) and hexaaquarhodium(3+). The rate of substitution parallels the ease of loss of a water molecule from the coordination sphere of the aqua complex and thus increases with increasing size and with decreasing charge of the metal ion. For transition metal ions, electronic factors also have an important influence on rates of substitution.

There are two limiting mechanisms (or pathways) through which substitution may occur—namely, dissociative and associative mechanisms. In the dissociative mechanism, a ligand is lost from the complex to give an intermediate compound of lower coordination number. This type of reaction path is typical of octahedral complexes, many aqua complexes, and metal carbonyls such as tetracarbonylnickel. An example of a dissociative reaction pathway for an octahedral complex of cobalt is as follows:

$$[Co(CN)_5(H_2O)]^{2-} \xrightarrow{-H_2O} [Co(CH)_5]^{2} \xrightarrow{+I^-} [Co(CN)_5 I]^{3-}.$$

The associative mechanism for substitution reactions, on the other hand, involves association of an extra ligand with the complex to give an intermediate of higher coordination number; one of the original ligands is then lost to restore the initial coordination number. Substitution reactions of square planar complexes, such as those of the nickel(2+), palladium(2+), and platinum(2+) ions, usually proceed through associative pathways involving intermediates with coordination number five. An example of a reaction following such a pathway is,

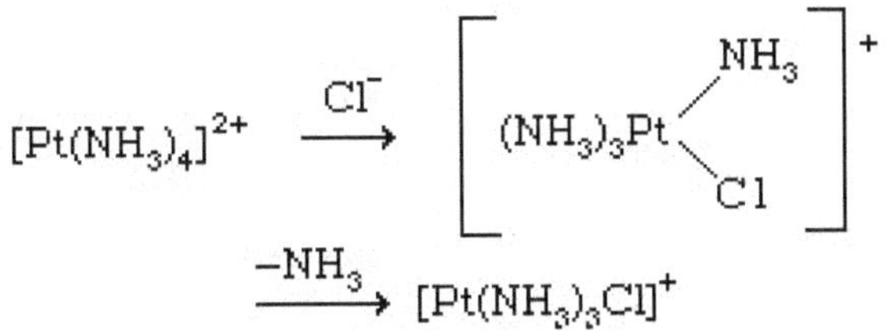

A characteristic feature of this class of reactions is the sensitivity of the rate of substitution of a given ligand to the nature of the ligand in the *trans* position. The *trans* ligand activates a ligand for replacement as follows, in decreasing order:

$$CO, CN^-, C_2H_4 > PR_3, H^- > NO_2^-, I^-, SCN^- > Br^-, Cl^- > NH_3, H_2O.$$

The *trans* effect may be used for synthetic purposes; thus, the reaction of the tetrachloroplatinate(2−) ion with ammonia yields *cis*-diamminedichloroplatinum, whereas the reaction of the tetraammineplatinum(2+) ion with the chloride ion gives the *trans* isomer, *trans*-diamminedichloroplatinum. The reactions are shown below.

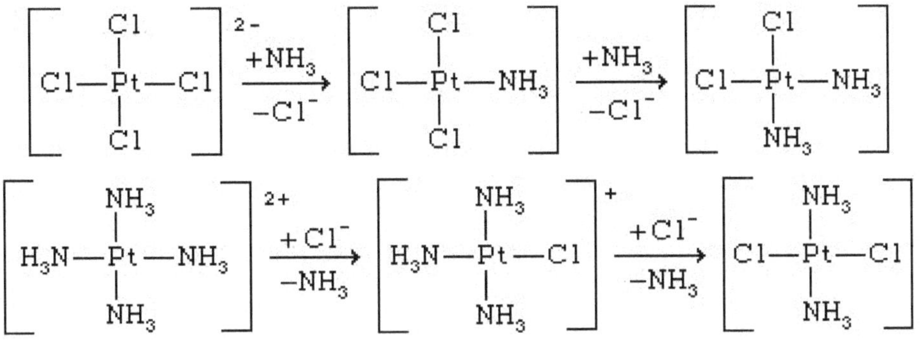

In both reactions, the *trans* effect causes the introduction of the ligand *trans* to chloride rather than *trans* to ammonia.

Lability and Inertness

In considering the mechanisms of substitution (exchange) reactions, Canadian-born American chemist Henry Taube distinguished between complexes that are labile (reacting completely in about one minute in 0.1 M solution at room temperature $[25\ °C,\ or\ 77\ °F]$) and those that are inert (under the same conditions, reacting either too slowly to measure or slowly enough to be followed by conventional techniques). These terms refer to kinetics (reaction rates) and should not be confused with the thermodynamic terms unstable and stable, which refer to equilibrium. For example, as mentioned above, most cyanide complexes are extremely stable (they possess very small dissociation constants); yet, if their rate of exchange with carbon-14-labeled cyanide, as represented in the following equation,

$$[M(CN)_x]^{y-} + x^{14}CN^- \rightleftharpoons [M(^{14}CN)_x]^{y-} + xCN^-,$$

is measured, $[Ni(CN)_4]^{2-}$ and $[Hg(CN)_4]^{2-}$ are found to be labile, whereas $[Mn(CN)_6]^{3-}$, $[Fe(CN)_6]^{4-}$, $[Fe(CN)_6]^{3-}$, and $[Cr(CN)_6]^{3-}$ are inert. On the other hand, $[Co(NH_3)_6]^{3+}$, a kinetically inert complex, is thermodynamically stable in acidic solution. Inertness may result from the lack of a suitable low-energy pathway for the reaction. In short, stable complexes possess large positive free energies of reaction (ΔG), whereas inert complexes merely possess large positive free energies of activation (ΔG^*).

While the existence of geometric or optical isomers in the solid state or in solution at nonequilibrium concentrations is evidence supporting the inertness of the complex, this does not constitute absolute proof. Conversely, the possibility of intramolecular rearrangement means that failure to isolate geometric isomers or to resolve the racemic mixture into optical isomers is not absolute proof of lability.

Taube has interpreted lability of complexes according to their electronic configuration in terms of VB theory. Labile complexes are either of the outer orbital type (outer *d* orbitals involved in bonding—e.g., sp^3d^2 as opposed to d^2sp^3 [inner orbital] for octahedral complexes) or of the inner orbital type with at least one vacant *d* orbital (available for accommodation of a seventh group during the [associative] substitution reaction).

Isomerization

Coordination compounds that exist in two or more isomeric forms may undergo reactions that convert one isomer to another. Examples are the linkage isomerization and *cis-trans* isomerization reactions depicted below.

$$[Co(ONO)(NH_3)_5]^{2+} \rightarrow [Co(NO_2)(NH_3)_5]^{2+}$$

$$\text{cis-}[Co(H_2O)_2(NH_2CH_2CH_2NH_2)_2]^{3} \rightarrow \text{trans-}[Co(H_2O)_2(NH_2CH_2CH_2NH_2)_2]^{3+}$$

The first of these has been shown to proceed intramolecularly (i.e., without dissociation of the nitrite ligand), whereas the second probably occurs through dissociation of one of the water-molecule ligands.

Oxidation-reduction

Transition metals commonly exhibit two or more stable oxidation states, and their complexes accordingly are able to undergo oxidation-reduction reactions. The simplest such reactions involve electron transfer between two complexes, with little if any accompanying rearrangement or chemical change. An example is shown below:

$$[Fe(CN)_6]^{4-} + [IrCl_6]^{2-} \rightarrow [Fe(CN)_6]^{3-} + [IrCl_6]^{3-}.$$

In other cases, oxidation-reduction is accompanied by significant chemical rearrangement. An example is,

$$[CoCl(NH_3)_5]^{2+} + [Cr(H_2O)_6]^{2+} + 5[H_3O]^{+} \rightarrow [Co(H_2O)_6]^{2+} + [CrCl(H_2O)_5]^{2+} + 5[NH_4]^{+}.$$

Two limiting mechanisms of electron transfer, commonly designated outer-sphere and inner-sphere mechanisms, have been recognized. Outer-sphere electron transfer occurs without dissociation or disruption of the coordination sphere of either complex—i.e., through both intact coordination spheres. The first reaction above is of this type. On the other hand, inner-sphere electron transfer—e.g., the second reaction above—proceeds by formation of a dinuclear complex in which the two metal ions are joined by a common bridging ligand (in this case the chloride ion) through which the electron is transferred. Such electron transfer also may occur through polyatomic bridging ligands to which the two metal ions are attached at different sites separated by several atoms, as in the reduction of pentaammine(isonicotinamide)cobalt(3+) by hexaaquachromium(2+) ion through a bridged intermediate:

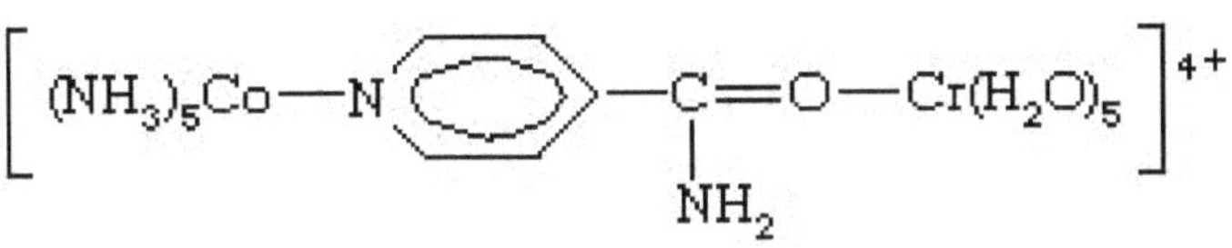

Strikingly large differences in rates of electron transfer are observed even between closely related reactions. Thus, the rate of reduction of the pentaamminebromocobalt(3+) ion by the hexaaquachromium(2+) ion is about 10^7 times higher than that of the acetatopentaamminecobalt(2+) ion by the same chromium ion.

Synthesis of Coordination Compounds

The great variety of coordination compounds is matched by the diversity of methods through which such compounds can be synthesized. Complex halides, for example, may be prepared by direct combination of two halide salts (either in the molten state or in a suitable solvent). Palladium chloride and potassium chloride, for example, react to give the complex potassium tetrachloropalladate(2−), as shown in the following equation:

$$PdCl_2 + 2KCl \rightarrow K_2[PdCl_4]$$

Another widely used route to coordination compounds is through the direct combination of a metal ion and appropriate ligands in solution. Thus, the addition of a sufficiently high concentration of ammonia to an aqueous solution of a nickel(2+) salt leads, through a series of reactions, to the formation of the hexaamminenickel(2+) ion, which can be precipitated, for example, as the sulfate salt, $[Ni(NH_3)_6]SO_4$.

Complexes of metal ions in high oxidation states are sometimes more readily formed by adding the ligands to a solution of the metal ion in a lower oxidation state in the presence of an oxidizing agent. Thus, addition of ammonia to an aqueous solution of a cobalt(2+) salt in the presence of air or oxygen leads to the formation of cobalt(3+)-ammine complexes such as hexaamminecobalt(3+), $[Co(NH_3)_6]^{3+}$, and pentaammineaquacobalt(3+), $[Co(NH_3)_5(H_2O)]^{3+}$, ions.

Complexes of metals in low oxidation states, such as the carbonyls of metals in their zero oxidation states, can sometimes be prepared by direct combination of the metal with the ligand, as, for example, in the reaction of nickel metal with carbon monoxide.

$$\mathrm{Ni(solid)} + 4\mathrm{CO} \xrightarrow[1\,\mathrm{atm}]{25^\circ} \mathrm{Ni(CO)_4}$$

More commonly, a salt of the metal is reduced in the presence of the ligand. An example of this type of synthesis is the reduction of cobalt carbonate with hydrogen in the presence of carbon monoxide to give bis(tetracarbonylcobalt).

$$2\mathrm{CoCO_3} + 2\mathrm{H_2} + 8\mathrm{CO} \xrightarrow[250-300\,\mathrm{atm}]{120-200^*} \mathrm{Co_2(CO)_8} + 2\mathrm{CO_2} + 2\mathrm{H_2O}$$

Similar procedures are applicable to the synthesis of metal sandwich compounds containing cyclopentadienyl and benzene ligands. Dibenzenechromium, for example, can be prepared from chromic chloride, benzene, and aluminum, as shown in the following equation.

$$\mathrm{CrCl_3} + 2\mathrm{C_6H_6} + \mathrm{Al} \rightarrow \mathrm{Cr(C_6H_6)_2} + \mathrm{AlCl_3}$$

Hydrido complexes of transition metals can be prepared by reactions of suitable precursors either with molecular hydrogen or with suitable reducing agents such as hydrazine or sodium borohydride; for example,

$$2[\mathrm{Co(CN)_5}]^{3-} + \mathrm{H_2} \rightarrow 2[\mathrm{CoH(CN)_5}]^{3-}.$$

Transition metal complexes containing metal-carbon bonds can be prepared by a variety of routes, some of the more important of which are illustrated by the following examples.

$$\mathrm{Mn(CO)_5\,Br} + \mathrm{LiCH_3} \rightarrow \mathrm{Mn(CO)_5(CH_3)} + \mathrm{LiBr},$$

$$[\mathrm{Mn(CO)_5}]^{-} + \mathrm{CH_3Br} \rightarrow \mathrm{Mn(CO)_5(CH_3)} + \mathrm{Br}^{-},$$

$$\mathrm{ptHCl[P(CH_3CH_2)_3]_2} + \mathrm{CH_2{=}CH_2} \rightarrow$$

$$\mathrm{pt(CH_2CH_3)Cl[p(CH_3CH_2)_3]_2}\text{, and}$$

$$2[Co(CN)_5]^{3-} + CH_3I \rightarrow [Co(CN)_5(CH_3)]^{3-} + [Co(CN)_5I]^{3-}$$

Coordination Complex

Coordination complexes are formed between a metal ion (Lewis acid) and ligands (Lewis base). The splitting of the d-orbitals (crystal field splitting) and the distribution of the d-electrons of the central transition metal ion is dependent upon the ligands that bind to the metal center. The separation of the d-orbitals is called the crystal field splitting energy and is within the visible region of the electromagnetic spectrum (400 nm – 750 nm).

Transition metal complexes have a very wide range of colors within the visible spectrum thus they are commonly found as ingredients in colored paint. A transition metal complex will absorb a photon of light which will excite an electron into a higher energy orbit. The electron then returns to the ground-state emitting a photon within the visible region of the electromagnetic spectrum. The energy of the absorbed light may be calculated as follows:

$$\Delta = E = h\nu = \frac{hc}{\lambda}$$

Spectroscopy is a method of analysis based on measuring the energy of light absorbed by a substance and relating that energy to structural characteristics. When a complex absorbs white light, the light left over is the observed color or complementary color of the coordination complex. A color wheel may be used to determine complementary colors.

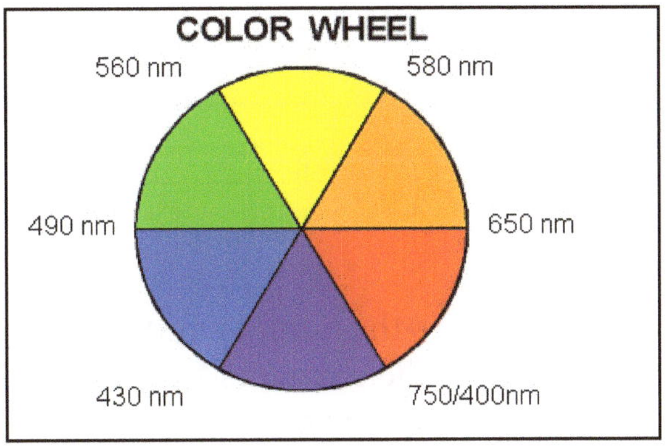

Principal Types Of Complexes

The tendency for complexes to form between a metal ion and a particular combination of ligands and the properties of the resulting complexes depend on a variety of properties of both the metal ion and the ligands. Among the pertinent properties of the metal ion are its size, charge, and electron configuration. Relevant properties of the ligand include its size and charge, the number and kinds of atoms available for coordination, the sizes of the resulting chelate rings formed (if any), and a variety of other geometric (steric) and electronic factors.

Many elements, notably certain metals, exhibit a range of oxidation states—that is, they are able to gain or lose varying numbers of electrons. The relative stabilities of these oxidation states are markedly affected by coordination of different ligands. The highest oxidation states correspond to empty or nearly empty *d* subshells (as the patterns of *d*orbitals are called). These states are generally stabilized most effectively by small negative ligands, such as fluorine and oxygen atoms, which possess unshared electron pairs. Such stabilization reflects, in part, the contribution of π bonding caused by electron donation from the ligands to empty *d* orbitals of the metal ions in the complexes. Conversely, neutral ligands, such as carbon monoxide and unsaturated hydrocarbons, which are relatively poor electron donors but which can accept π electrons from filled *d* orbitals of the metal, tend to stabilize the lowest oxidation states of metals. Intermediate oxidation states are most effectively stabilized by ligands such as water, ammonia, and cyanide ion, which are moderately good σ–electron donors but relatively poor π–electron donors or acceptors.

Chromium complexes of various oxidation states		
oxidation state	**electron configuration***	**coordination complex**
+6	d^0	$[CrF_6]$, $[CrO_4]^{2-}$
+5	d^1	$[CrO_4]^{3-}$
+4	d^2	$[CrO_4]^{4-}$, $[Cr(OR)_4]$**
+3	d^3	$[Cr(H_2O)_6]^{3+}$, $[Cr(NH_3)_6]^{3+}$
+2	d^4	$[Cr(H_2O)_6]^{2+}$
0	d^6	$[Cr(CO)_6]$, $[Cr(C_6H_6)_2]$
**R symbolizes an organic alkyl radical. *Number of d electrons indicated by superscript.		

Aqua Complexes

Few ligands equal water with respect to the number and variety of metal ions with which they form complexes. Nearly all metallic elements form aqua complexes, frequently in more than one oxidation state. Such aqua complexes include hydrated ions in aqueous solution as well as hydrated salts such as hexaaquachromium(3+) chloride, $[Cr(H_2O)_6]Cl_3$. For metal ions with partially filled *d* subshells (i.e., transition metals), the coordination numbers and geometries of the hydrated ions in solution can be inferred from their light-absorption spectra, which are generally consistent with octahedral coordination by six water molecules. Higher coordination numbers probably occur for the hydrated rare-earth ions such as lanthanum(3+).

When other ligands are added to an aqueous solution of a metal ion, replacement of water molecules in the coordination sphere may occur, with the resultant formation of other complexes. Such replacement is generally a stepwise process, as illustrated by the following series of reactions that results from the progressive addition of ammonia to an aqueous solution of a nickel(2+) salt:

$$[Ni(H_2O)_6]^{2+} + NH_3 \rightleftharpoons [Ni(NH_3)(H_2O)_5]^{2+} + H_2O$$

With increasing additions of ammonia, the equilibria are shifted toward the higher ammine complexes (those with more ammonia and less water) until ultimately the hexaamminenickel(2+) ion predominates:

$$[Ni(NH_3)_5(H_2O)]^{2+} + NH_3 \rightleftharpoons [Ni(NH_3)_6]^{2+} + H_2O$$

The tendency of metal ions in aqueous solution to form complexes with ammonia as well as with organic amines (derivatives of ammonia, with chains of carbon atoms attached to the nitrogen atom) is widespread. The stabilities of such complexes exhibit a considerable range of dependence on the nature of the metal ion as well as on that of the amine. The marked enhancement of stability that results from chelation is reflected in the equilibrium constants of the reactions values that indicate the relative proportions of the starting materials and the products at equilibrium. Complexes of hexaaquanickel(2+) ions can be formed with a series of polyamines i.e.,

$$[Ni(H_2O)_6]^{2+} + nL \rightleftharpoons [NiL_n(H_2O)_{6-n}]^{2+} + nH_2O,$$

in which L is the ligand and n the number of water molecules displaced from the complex. In this series the equilibrium constants, K_L, increase dramatically as the possibilities for chelation increase (that is, as the number of nitrogen atoms available for bonding to the metal atom increases).

It should be noted that, in the particular examples cited above, the coordination number of the metal ion is invariant throughout the substitution process, but this is not always the case. Thus, the ultimate products of the addition of the cyanide ion to an aqueous solution of hexaaquanickel(2+) ion are tetracyanonickelate(2−) and pentacyanonickelate(3−), both containing nickel in the +2 oxidation state. Similarly, addition of the chloride ion to a solution of hexaaquairon(3+) yields tetrachloroferrate(3−). Both complexes contain iron in the same oxidation state of +3.

Halo Complexes

Probably the most widespread class of complexes involving anionic ligands is that of the complexes of the halide ions—i.e., the fluoride, chloride, bromide, and iodide ions. In addition to forming simple halide salts, such as sodium chloride and nickel difluoride (in which the metal ions are surrounded by halide ions, these in a sense being regarded as coordinated to them), many metals form complex halide salts—such as potassium tetrachloroplatinate(2−), $K_2[PtCl_4]$—that contain discrete complex ions. Most metal ions also form halide complexes in aqueous solution. The stabilities of such complexes span an enormous range—from the alkali-metal ions (lithium, sodium, potassium, and so on), whose formation of halide complexes in aqueous solution can barely be detected, to extremely stable halide complexes, such as the tetraiodomercurate(2−), tetrachlorothallate(1−), and tetrachloropalladate(2−) ions, the extent of whose dissociation is extremely small.

The stabilities of halide complexes reflect a pattern by which metal ions can be divided into two general classes, designated as *A* and *B* or as hard and soft, respectively. (Generally, the electrons in the atoms of the hard elements are considered to form a compact and not easily deformable group, whereas those in the atoms of the soft elements form a looser group—that is, one more easily deformed.) For the former class which includes Be, Mg, Sc, Cr, Fe, Ni, Cu, In, and Sn the order of increasing stability of the halide complexes in aqueous solution is iodides < bromides < chlorides < fluorides. Conversely, for the class *B* (or soft) ions such as Pt, Ag, Cd, Hg, Tl, and Pb the order of increasing stability of the halide complexes is fluorides < chlorides < bromides < iodides. In contrast to class-*A* metals, those of class *B* also tend to form more stable complexes with sulfur-containing ligands than with oxygen-containing ligands and more stable complexes with phosphorus ligands than with nitrogen ligands.

Carbonyl Complexes

Following the discovery of the first metal carbonyl complex, tetracarbonylnickel, $Ni(CO)_4$, in 1890, many compounds containing carbon monoxide coordinated to transition metals have been prepared and characterized. For reasons already discussed, such compounds generally contain metal atoms or ions in low oxidation states. The following are some of the more common types of metal carbonyl compounds: (1) simple mononuclear carbonyls of metals in the zero oxidation state, such as tetracarbonylnickel, pentacarbonyliron, and hexacarbonylchromium highly toxic volatile compounds, the most stable of which have filled valence shells of 18 electrons, (2) salts of anionic and cationic carbonyls, such as tetracarbonylcobaltate(−1) and hexacarbonylmanganese(+1), (3) dinuclear and polynuclear carbonyls, such as bis(tetracarbonylcobalt), the structural formula of which was shown earlier, and (4) mixed complexes containing other ligands in addition to CO: pentacarbonylchloromanganese, tetracarbonylhydridocobalt, and tricarbonylnitrosylcobalt.

Although molecular nitrogen, N_2, is isoelectronic with carbon monoxide (that is, it has the same number and arrangement of electrons), its tendency to form complexes with metals is much smaller. The first complex containing molecular nitrogen as a ligand—i.e., pentaamminenitrogenruthenium(2+), $[Ru(NH_3)_5(N_2)]^{2+}$ was prepared in 1965, and many others have been discovered subsequently. Such complexes have attracted considerable interest because of their possible roles in the chemical and biological fixation of nitrogen.

Nitrosyl Complexes

Nitrosyl complexes can be formed by the reaction of nitric oxide (NO) with many transition metal compounds or by reactions involving species containing nitrogen and oxygen. Some of these complexes have been known for many years—e.g., pentaaquanitrosyliron(2+) ion, $[Fe(H_2O)_5NO]^{2+}$, which formed in the classical brown-ring test for the qualitative detection of nitrate ion; Roussin's red ($K_2[Fe_2S_2(NO)_4]$) and black ($K[Fe_4S_3(NO)_7]$) salts; and sodium pentacyanonitrosylferrate(3−) dihydrate (sodium nitroprusside), $Na_2[Fe(CN)_5NO]\cdot 2H_2O$. Such complexes, which can be cationic, neutral, or anionic and which are usually deeply coloured (red, brown, purple, or black), have been extensively studied because they pose unique problems of structure and bonding and because they have potential uses as homogeneous catalysts for a variety of reactions. More recently, the research field has been expanded to include organometallic species.

Because the nitrosonium ion (NO^+) is isoelectronic with carbon monoxide and because its mode of coordination to transition metals is potentially similar to that of carbon monoxide, metal nitrosyls have been recognized as similar to carbonyls and are sometimes formulated as NO^+ complexes. Carbonyl ligands can be replaced by nitric oxide in substitution reactions. Such similarities may be deceptive, however, for the additional electron in neutral nitric oxide requires a more complicated treatment of M-NO bond formation. The NO ligand exhibits several geometries of coordination—linear (e.g., $[IrH(NO)\{P(C_6H_5)_3\}_3]^+$, $[Mn(CO)_2(NO)\{P(C_6H_5)_3\}_3]$, and $Na_2[Ru(OH)(NO_2)_4(NO)]\cdot 2H_2O)$; bent (e.g., $[CoNO(NH_3)_5]^{2+}$ and $[IrCl_2(NO)\{P(C_6H_5)_3\}_2]$) ; or both (e.g., $[RuCl(NO)_2\{P(C_6H_5)_3\}_2]^+$). Like CO, NO also can act as a bridging ligand between two (e.g., $[\{Cr(\eta^5-C_5H_5)(NO)\}_2(\mu_2-NH_2)(\mu_2-NO)]$) or three (e.g., $[Mn_3(h^5-C_5H_5)_3(m_2-NO)_3(m_3-NO)]$) metal atoms. (The η^5 indicates that five carbon atoms of the $C_5H_5^-$ group are bonded to the chromium atom.)

Cyano and Isocyano Complexes

Cyano complexes, such as Prussian blue, mentioned above, are among the oldest coordination compounds. In addition to being a pseudohalide, the CN^- ion is isoelectronic with CO, RCN, RNC, N_2, and NO^+ (R is an alkyl group), and metal carbonyls and cyanide complexes are structurally similar. Also, like CO, CN^- enters into π as well as σ bonding with transition metal atoms or ions. Cyano complexes are among the most stable transition metal complexes; the extreme toxicity of CN^- (like that of CO) is due to its irreversible formation of a strong complex with hemoglobin, which prevents oxygen from binding reversibly to hemoglobin, thereby prohibiting the transport and release of oxygen in the body. Similarly, the ability of CN^- to form very stable complexes with silver ($Au(CN)_2^-$) and gold $Au(CN)_2$ is the basis for its use in the extraction and purification of these metals. As a monodentate ligand, CN^- coordinates (bonds) through carbon as the donor atom, but, as a didentate ligand, it usually coordinates at both ends (C and N) and acts as a bridging ligand (-CN-) to form infinite linear (chain) polymers as in Prussian blue, AgCN, AuCN, $Zn(CN)_2$, and $Cd(CN)_2$.

The cyanide ion forms complexes with transition metals and with zinc, cadmium, and mercury, usually by substitution in aqueous solution with no change in oxidation state. The most important complexes are anionic with the formula $[M^{n+}(CN)_x]^{(x-n)-}$, where Mn^+ represents a transition metal ion. Examples are $[Ni(CN)_4]^{2-}$, $[Pt(CN)_4]^{2-}$, $[Fe(CN)_6]^{4-\text{ or }3-}$, $[Co(CN)_6]^{3-}$, $[Pt(CN)_6]^{2-}$, and $[Mo(CN)_8]^{5-,4-,\text{or }3-}$. The free anhydrous parent acids of many of these anions for example, $H_4[Fe(CN)_6]$ and $H_3[Rh(CN)_6]$ have been isolated.

Cyanide complexes exhibit a variety of coordination numbers and configurations. Metal ions with a d^{10}structure form linear complexes of coordination number 2 as, for example, $[M(CN)_2]^-$ (where M = Hg, Ag, or Au)—while the isoelectronic complexes $[Cu(CN)_4]^{3-}$, $[Ag(CN)_4]^{3-}$, $[Zn(CN)_4]^{2-}$, $[Cd(CN)_4]^{2-}$, and $[Hg(CN)_4]^{2-}$ are tetrahedral. All the hexacoordinate complexes are octahedral, while octacoordinate complexes are cubic, dodecahedral, or square antiprismatic. (The dodecahedron and square antiprism are two structures that can be obtained by distorting the simple cube.) For d^2, d^4, d^6, d^8, and d^{10} transition metal ions, the octa-, hepta-, hexa-, penta-, and tetracoordinate complexes, respectively, are species with maximum coordination number.

Mixed complexes of type $[M(CN)_5X]n^-$ (where $X = H_2O, NH_3, CO, NO, H$, or a halogen) also exist. The cyanide ion has the ability to stabilize metal ions in low oxidation states (probably by accepting electron density into its π^*orbitals)—e.g., $[Ni(CN)_4]^{4-}$, which contains nickel in the formal oxidation state of zero. Cyanide complexes have figured prominently in numerous kinetic studies. For example, fast electron-transfer reactions between $[Fe(CN)_6]^{3-}$ and $[Fe(CN)_6]^{4-}$ and between $[Mo(CN)_8]^{3-}$ and $[Mo(CN)_8]^{4-}$ established the outer-sphere mechanism for redox reactions ; replacement of waterin $[Co(CN)_5H_2O]^{2-}$ established the dissociative mechanism for substitution at a Co(3+) ion.

Transition metals also form complexes with organic cyanides (RCN or ArCN, called nitriles) and organic isocyanides (RNC or ArNC, called isonitriles)—where R and Ar are alkyl and aryl groups, respectively—by reaction of a metal halide, carbonyl, or other complex with the nitrile or isonitrile, respectively. Nitriles and isonitriles appear to be stronger donors of σ electrons than carbon monoxide, but they are capable of extensive back acceptance of π electrons from metals in lower oxidation states—as in $Cr(CNR)_6$

or $Cr(CNAr)_6$ and $Ni(CNR)_4$ or $Ni(CNAr)_4$, which are analogous to the corresponding carbonyls $Cr(CO)_6$ and $Ni(CO)_4$, respectively. Although a bridging isonitrile group has been reported in $(p\text{-}C_5H_5)_2Fe_2(CO)_3(CNC_6H_5)$, this type of bonding is unusual.

Organometallic Complexes

Organometallic complexes are complexes formed between organic groups and metal atoms. They can be divided into two general classes: (1) complexes containing metal-carbon σ bonds and (2) π-bonded metal complexes of unsaturated hydrocarbons—that is, compounds with multiple bonds between carbon atoms.

Isopoly and Heteropoly Anions

The amphoteric metals of groups VB (vanadium, niobium, and tantalum) and VIB (chromium, molybdenum, and tungsten) in the +5 and +6 oxidation states, respectively, form weak acids that readily condense (polymerize) to form anions containing several molecules of the acid anhydride. If these condensed acids contain only one type of acid anhydride, they are called isopoly acids, and their salts are called isopoly salts. The acid anhydrides also can condense with other acids (e.g., phosphoric or silicic acids) to form heteropoly acids, which can form heteropoly salts. The condensation reactions, which occur reversibly in dilute aqueous solution, involve formation of oxo bridges by elimination of water from two molecules of the weak acid. The best-known and simplest example is the condensation of yellow chromate ion (CrO_4^{2-}) to form the orange isopoly dichromate ion ($Cr_2O_7^{2-}$), an equilibrium reaction the extent of which depends on the pH. In acidic solution the isopoly anion $Cr_2O_7^{2-}$, predominates while in basic solution the simple ion CrO_4^{2-} predominates.

$$2CrO_4^{2-} + H^+ \rightleftharpoons [O_3Cr-O-OH]^- \rightleftharpoons [O_3Cr-O-CrO_3]^{2-}\ H_2O$$

basic solution acidic solution

Heteropoly acids and their salts may be formed by coordination of the central atom with four to six oxo anions, which may be mononuclear (containing one metal ion each), as in $H_7[P(MoO_4)_6]$, or trinuclear (containing three metal ions each), as in $H_3[P(W_3O_{10})_4]$. Incomplete replacement of oxygen atoms in PO_4^{3-} ions by MoO_3 groups can result in dimers (two-molecule polymers), as, for example, $\{OP[O(MoO_3)_3]_3\}_2^{6-}$. About 70 elements can act as central (hetero) atoms in heteropoly anions. Because each element may form more than one heteropoly anion and some of these anions can contain several different heteroatoms, thousands of heteropoly acids exist. Heteroatoms may be primary (these atoms are essential to the polyanion structure and thus not susceptible to chemical exchange) or secondary (these atoms can be removed by chemical reaction from the polyanion structure without destroying it). Heteropoly anions can be regarded as coordination compounds with polyanion ligands; e.g., $[(H_3N)_5Cr(OH_2)]^{3+}$ can be considered the parent of $[(SiW_{11}O_{39})Cr(OH_2)]^{5-}$.

A variety of synthetic procedures are available for the preparation of isopoly acids and salts, which are usually less stable than heteropoly compounds. Heteropolymolybdates and heteropolytungstates are always prepared in solution, usually after acidifying and heating the theoretical amounts of reactants. In general, free heteropoly acids and salts, of which the heteropolymolybdates and heteropolytungstates are the best known, have very high molecular weights (some above 4,000) as compared with other inorganic electrolytes, are very soluble in water and organic solvents, are almost always highly hydrated with several hydrates existing, and are highly coloured. Some are strong oxidizing agents that can be reduced to stable, intensely deep blue species (heteropoly blues), which in turn can act as reducing agents, restoring the original colour on oxidation. The stoichiometry, oxidation-reduction potentials, and other characteristics of these reactions have been investigated by various methods. The free acids, which are polyprotic (contain several replaceable hydrogen ions), are fairly strong and nearly always stronger than the corresponding acids from which they are derived.

All heteropolymolybdate and heteropolytungstate anions are decomposed in strongly basic solution to form simple molybdate or tungstate ions and either an oxy anion or a hydrous metal oxide of the central metal atom, e.g.:

$$[P_2Mo_{18}O_{62}]^{6-} + 34OH^- \rightarrow 18MoO_4^{2-} + 2HPO_4^{2-} + 16H_2O\,[NiW_6O_{24}H_6]^{4-} + 8OH^- \rightarrow 6WO_4^{2-} + Ni(OH)_2 + 6H_2O$$

Throughout specific ranges of pH and other conditions, most solutions of heteropolymolybdates and heteropolytungstates appear to contain predominantly one distinct species of anion, many of which are remarkably stable and nonlabile.

The first heteropoly compound, $(NH_4)_3[PMo_{12}O_{40}]$, was obtained by the Swedish chemist Jöns Jacob Berzelius in 1826 as a yellow, crystalline precipitate, the formation of which is still used for the classical qualitative detection and quantitative estimation of phosphorus(after conversion to phosphate). By the beginning of the 20th century, hundreds of isopoly and heteropoly compounds were reported, many of which were based on incorrect analyses or failure to detect mixed crystals. Formulas were reported in terms of the old Berzelius dualistic theory as a combination of oxides, such as $3Na_2O{\cdot}Cr_2O_3{\cdot}12MoO_3{\cdot}20H_2O$ for $Na_3CrMo_6O_{24}H_6{\cdot}7H_2O$, and often merely expressed analytical results rather than structure. In addition to their use in analytical chemistry, heteropoly compounds have found use as catalysts, molecular sieves, corrosion inhibitors, photographic fixing agents, and precipitants for basic dyes.

Few structural studies of such compounds were carried out, but this lack did not prevent the elaboration of various unsuccessful theories to account for their structures. In 1907 Werner applied his coordination theory to the structure of 12-tungstosilicic acid, $H_4[SiW_{12}O_{40}]$, and its salts by assuming that the central group is an SiO_4^{4-} ion surrounded octahedrally by six $RW_2O_6^+$ groups (R = a unipositive ion), four linked by primary (ionic) and two linked by secondary (coordinate covalent) valences. Difficulties were encountered by this system as well as by the later (1910–21), more elaborate Miolati-Rosenheim theory. Modern conclusive knowledge of the structures of heteropoly compounds did not begin until 1934, with J.F. Keggin's determination of the structure of $PO_4W_{12}O_{36}]\cdot 5H_2$ O by the most direct means, X-ray diffraction.

The structures of isopoly and heteropoly compounds consist of polyhedrons sharing corners and edges with one another. In heteropolymolybdates or heteropolytungstates, each molybdenum or tungsten atom is located at the centre of an octahedron, each vertex of which is occupied by an oxygen atom. These octahedrons can share corners or edges or both with other MoO_6 or WO_6 octahedrons. In $[Mo_8O_{26}]^{4-}$ eight MoO_6octahedrons share edges. In $[PMo_{12}O_{40}]^{3-}$ the central phosphorus atom is located at the centre of a PO_4 tetrahedron, which is surrounded by 12 MoO_6 octahedrons, which share corners so that the correct number of oxygen atoms is utilized.

In coordination chemistry, the coordination number is the number of ligands attached to the central ion (more specifically, the number of donor atoms). Coordination num-bers are normally between two and nine. The number of bonds depends on the size, charge, and electron configuration of the metal ion and the ligands.

Typically the chemistry of complexes is dominated by interactions between s and p molecular orbitals of the ligands and the d orbitals of the metal ions. The s, p, and d orbitals of the metal can accommodate 18 electrons. The maximum coordination number for a certain metal is thus related to the electronic configuration of the metal ion (specifically, the number of empty orbitals) and to the ratio of the size of the ligands and the metal ion. Large metals and small ligands lead to high coordination numbers (e.g., $[Mo(CN)_8]^{4-}$). Small metals with large ligands lead to low coordination numbers (e.g., $Pt[P(CMe_3)]_2$). Due to their large size, lanthanides, actinides, and early transition metals tend to have high coordination numbers.

Coordination Number Examples

- Carbon has a coordination number of 4 in methane (CH_4) molecule since it has four hydrogen atoms bonded to it.
- In ethylene ($H_2C=CH_2$), the coordination number of each carbon is 3, where each C is bonded to 2H + 1C for a total of 3 atoms.
- The coordination number of diamond is 4, as each carbon atom rests at the center of a regular tetrahedron formed by four carbon atoms.

Calculating the Coordination Number

Here are the steps for identifying the coordination number of a coordination compound.

1. Identify the central atom in the chemical formula. Usually, this is a transition metal.
2. Locate the atom, molecule, or ion nearest the central metal atom. To do this, find the molecule or ion directly beside the metal symbol in the chemical formula of the coordination compound. If the central atom is in the middle of the formula, there will be neighboring atoms/molecules/ions on both sides.
3. Add the number of atoms of the nearest atom/molecule/ions. The central atom may only be bonded to one other element, but you still need to note the number of atoms of that element in the formula. If the central atom is in the middle of the formula, you'll need to add up the atoms in the entire molecule.
4. Find the total number of nearest atoms. If the metal has two bonded atoms, add together both numbers.

Ligands

In coordination chemistry, a ligand is an ion or molecule (functional group) that binds to a central metal atom to form a coordination complex. Virtually every molecule and every ion can serve as a ligand for (or coordinate to) metals. Denticity refers to the number of times a ligand bonds to a metal through donor atoms. Many ligands are capable of binding metal ions through multiple sites, usually because the ligands have lone pairs on more than one atom.

Monodentate ligands include virtually all anions and all simple Lewis bases. Thus, the halides and pseudohalides are important anionic ligands. Ammonia, carbon monoxide, and water are particularly common charge-neutral ligands. Simple organic species are also very common. All unsaturated molecules are also ligands, utilizing their π-electrons in forming the coordinate bond. Also, metals can bind to the σ bonds in, for example, silanes, hydrocarbons, and dihydrogen.

Ligands that bind via more than one atom are often termed polydentate or chelating. A ligand that binds through two sites is classified as bidentate, and three sites as tridentate. Chelating ligands are commonly formed by linking donor groups via organic linkers. A classic bidentate ligand is ethylenediamine, which is derived by the linking of two ammonia groups with an ethylene ($-CH_2CH_2-$) linker. A classic example of a polydentate ligand is the hexadentate chelating agent EDTA, which is able to bond through six sites, completely surrounding some metals.

There are several types of polydentate ligands which can be characterized based on how they interact with the central ion. For example, trans-spanning ligands are bidentate ligands that can span coordination positions on opposite sides of a coordination complex. Ambidentate ligands can attach to the central atom in two places but not both. A bridging ligand links two or more metal centers. Changing the size and electronic properties of ligands can be used to control catalysis of the central ion and stabilize unusual coordination sites.

Geometries

Different ligand structural arrangements result from the coordination number. Most structures follow the pattern as if the central atom were in the middle and the corners of that shape are the locations of the ligands. These shapes are defined by orbital overlap between ligand and metal orbitals and ligand-ligand repulsions, which tend to lead to certain regular geometries. However, there are many cases that deviate from regular geometry. For example, ligands of different sizes and with different electronic effects often result in irregular bond lengths.

Coordination Number 2

This arrangement is not very common for first row transition metal ion complexes and some of the best known examples are for Silver(I). In this case we have a low charge and an ion at the right hand side of the d-block indicating smaller size.

A method that was often employed for the detection of chloride ions involved the formation of the linear diamminesilver(I) complex.

The first step is:

$$Ag^{+} + Cl^{-} \rightarrow AgCl\left(\text{white ppt}\right)$$

and to ensure that the precipitate is really the chloride salt, two further tests must be done:

$$AgCl + 2\,NH_3 \rightarrow \left[Ag\left(NH_3\right)_2\right]+$$

and

$$\left[Ag\left(NH_3\right)_2\right]^{+} + HNO_3 \rightarrow AgCl\left(\text{re-ppts}\right)$$

The reaction of a bidentate ligand such as 1,2-diaminoethane with Ag(I) does not lead to chelated ring systems, but instead to linear two coordinate complexes. One reason for this is that bidentate ligands can NOT exist in *trans* arrangements, that is they are UNABLE to span 180 degrees.

Coordination Number 3

Once again, this is not very common for first row transition metal ions. Examples with three different geometries have been identified.

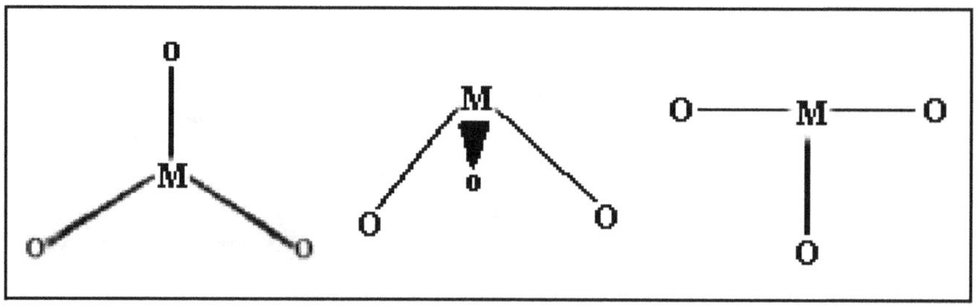

Trigonal Planar

Well known for main group species like CO_3^{2-} etc., this geometry has the four atoms in a plane with the bond angles between the ligands at 120 degrees.

Trigonal Pyramid

More common with main group ions.

T-shaped

The first example of a rare T-shaped molecule was found in 1977 however since then several further examples have been reported.

Coordination Number 4

Two different geometries are possible. The tetrahedron is the more common while the square planar is found in particular with metal ions having a d^8 electronic configuration.

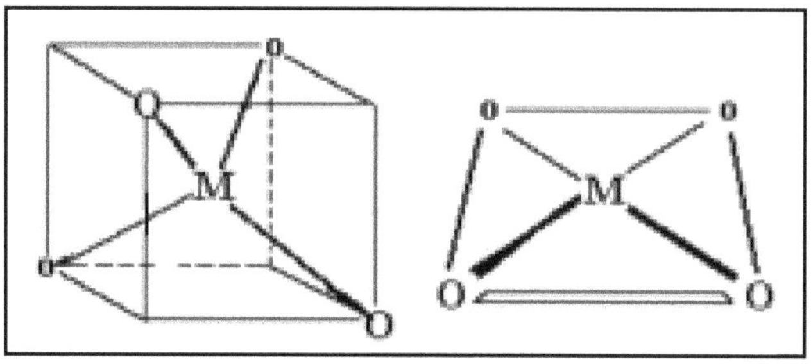

Tetrahedral, (T_d)

The chemistry of molecules centred around a tetrahedral C atom is covered in organic courses. To be politically correct, please change all occurrences of *C* to *Co*. There is a large number of tetrahedral cobalt(II) complexes known.

Square Planar, (D_{4h})

These are much less common than tetrahedral and are included here since there are some extremely important examples with this shape.

Coordination Number 5

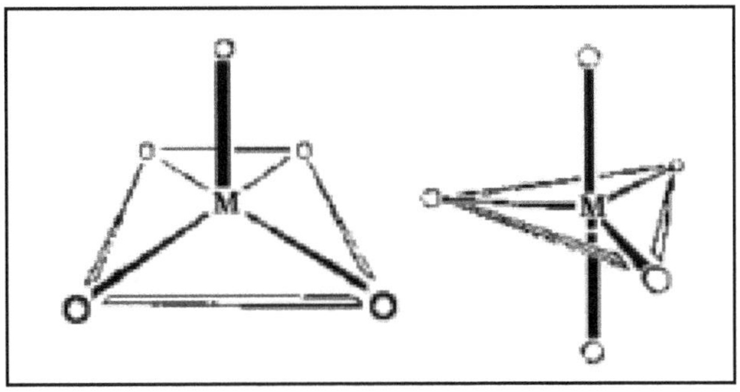

Square Pyramid, (C_{4v})

Oxovanadium salts (Vanadyl, VO^{2+}) often show square pyramidal geometry, for example, $VO(acac)_2$. Note that the Vanadium(IV) can be considered coordinatively unsaturated and addition of pyridine leads to the formation of an octahedral complex.

Trigonal Bipyramid, (D_{3h})

The structure of $[Cr(en)_3][Ni(CN)_5]_{1.5}H_2O$ was reported in 1968 to be a remarkable example of a complex exhibiting both types of geometry in the same crystal. The reaction of cyanide ion with Ni^{2+} proceeds via several steps:

$$Ni^{2+} + 2\,CN^- ® Ni(CN)_2 \text{ yellow}$$

$$Ni(CN)_2 + 2\,CN^- \rightarrow [Ni(CN)_4]^{2-} \text{ orange- red}$$

$$\log(\beta 4) = 30.1$$

$$[Ni(CN)_4]^{2-} + CN^- \rightarrow [Ni(CN)_5]^{3-} \text{ deep red}$$

Coordination Number 6

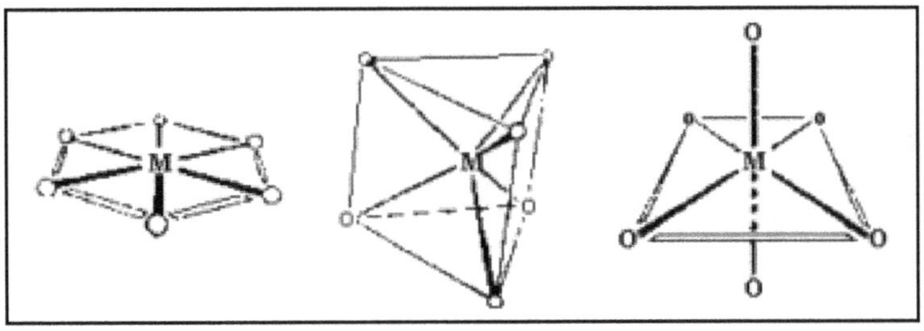

Hexagonal Planar

Unknown for first row transition metal ions, although the arrangement of six groups in a plane is found in some higher coordination number geometries.

Trigonal Prism

Most trigonal prismatic compounds have three bidentate ligands such as dithiolates or oxalates and few are known for first row transition metal ions.

Octahedral, (O_h)

Themostcommongeometryfoundforfirstrowtransitionmetalions,includingallaquaions. In some cases distortions are observed and these can sometimes be explained in terms of the Jahn-Teller Theorem.

Coordination Number 7

Three geometries are possible:

Not very common for 1st row complexes and the energy difference between the structures seems small and distortions occur so that prediction of the closest "idealised" shape is generally difficult.

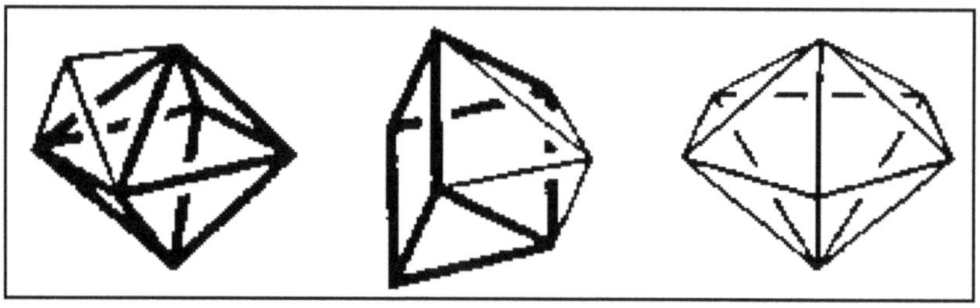

Capped octahedron, (C_{3v})

Capped trigonal prism, (C_{2v})

Pentagonal Bipyramid, (D_{5h})

Coordination Number 8

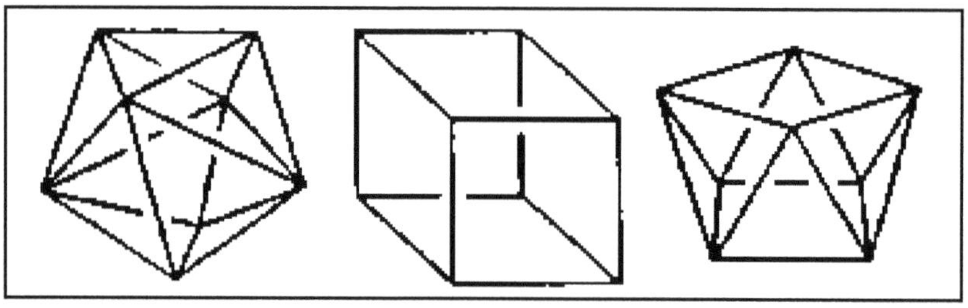

Dodecahedron, (D_{2d})

Cube, (O_h)

Square antiprism, (D_{4d})

Hexagonal bipyramid, (D_{6h})

Coordination Number 9

Three-face centred trigonal prism, (D_{3h})

Coordination Number 10

Bicapped square antiprism, (D_{4d})

Coordination Number 11

All-faced capped trigonal prism, (D_{3h})

Coordination Number 12

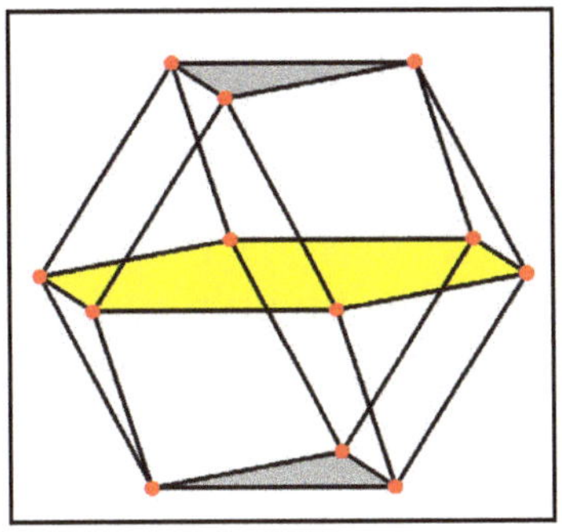

cuboctahedron, (O_h)

Coordination Sphere

In a complex compound, it usually, central metal ion and the ligands are enclosed within square bracket is called a coordination sphere. This represents a single constituent unit. The ionizable species are placed outside the square bracket.

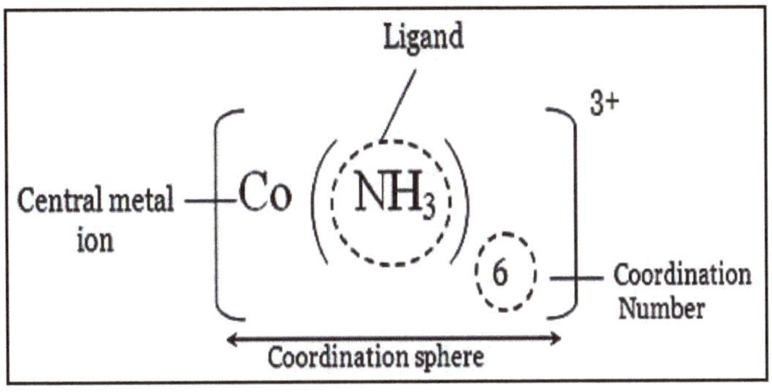

First Coordination Sphere

The first coordination sphere refers to the molecules that are attached directly to the metal. These molecules are typically solvent. The interactions between the first and second coordination spheres usually involve hydrogen-bonding. For charged complexes, ion pairing is important.

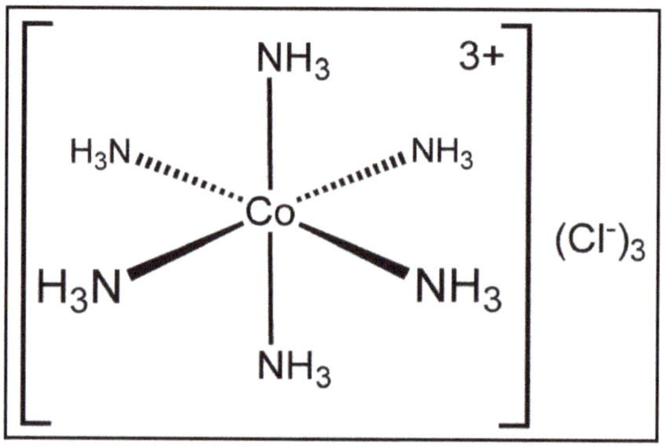

Hexamminecobalt(III) chloride is a salt of a coordination complex wherein six ammonia ("ammine") ligands occupy the first coordination sphere of the ion Co^{3+}.

In hexamminecobalt(III) chloride ($[Co(NH_3)_6]Cl_3$), the cobalt cation plus the 6 ammonia ligands comprise the first coordination sphere. The coordination sphere of this ion thus consists of a central MN_6 core "decorated" by 18 N−H bonds that radiate outwards.

Second Coordination Sphere

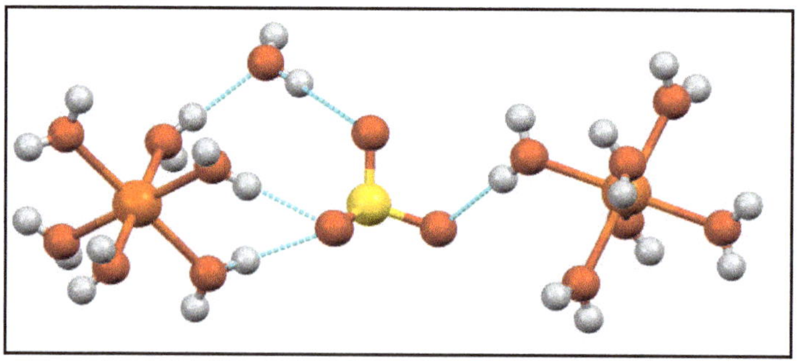

In crystalline $FeSO_4 \cdot 7H_2O$, the *first* coordination sphere of Fe^{2+} consists of six water ligands. The *second* coordination sphere consists of a water of crystallization and sulfate, which interact with the $[Fe(H_2O)_6]^{2+}$ centers.

Metal ions can be described as consisting of series of two concentric coordination spheres, the first and second. More distant from the second coordination sphere, the solvent molecules behave more like "bulk solvent." Simulation of the second coordination sphere is of interest in computational chemistry. The second coordination sphere can consist of ions (especially in charged complexes), molecules (especially those that hydrogen bond to ligands in the first coordination sphere) and portions of a ligand backbone. Compared to the first coordination sphere, the second coordination sphere has a less direct influence on the reactivity and chemical properties of the metal complex. Nonetheless the second coordination sphere is relevant to understanding reactions of the metal complex, including the mechanisms of ligand exchange and catalysis.

Role in Catalysis

Mechanisms of metalloproteins often invoke modulation of the second coordination sphere by the protein. For example, an amine cofactor in the second coordination sphere of some hydrogenase enzymes assists in the activation of dihydrogen substrate.

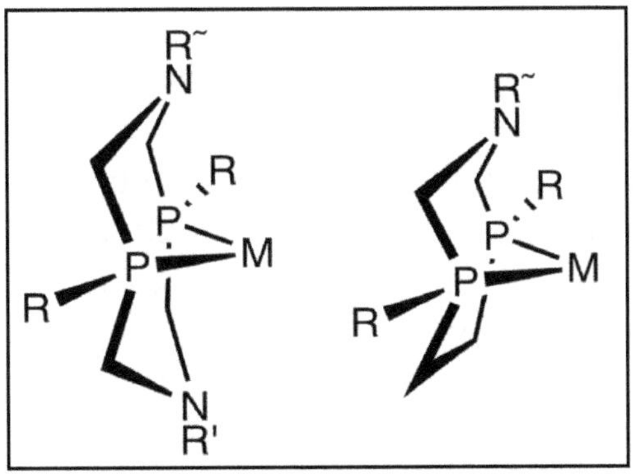

In metal complexes of 1,5-diaza-3,7-diphosphacyclooctanes and related ligands, amine groups occupy the second coordination sphere.

Role in Mechanistic Inorganic Chemistry

The rates at which ligands exchange between the first and the second coordination sphere is the first step in ligand substitution reactions. In associative ligand substitution, the entering nucleophile resides in the second coordination sphere. These effects are relevant to practical applications such as contrast agents used in MRI.

The energetics of inner sphere electron transfer reactions are discussed in terms of second coordination sphere. Some proton coupled electron transfer reactions involve atom transfer between the second coordination spheres of the reactants:

$$[Fe^*(H_2O)_6]^{2+} + [Fe(H_2O)_5(OH)]^{2+}\ [Fe(H_2O)_6]^{3+} + [Fe^*(H_2O)_5(OH)]^{2+}$$

Role in Spectroscopy

Solvent effects on colors and stability are often attributable to changes in the second coordination sphere. Such effects can be pronounced in complexes where the ligands in the first coordination sphere are strong hydrogen-bond donors and acceptors, e.g. respectively $[Co(NH_3)_6]^{3+}$ and $[Fe(CN)_6]^{3-}$. Crown-ethers bind to polyamine complexes through their second coordination sphere. Polyammonium cations bind to the nitrogen centres of cyanometallates.

Role in Supramolecular Chemistry

Macrocyclic molecules such as cyclodextrins act often as the second coordination sphere for metal complexes.

Trans and Cis Effect

The *trans* effect proper, which is often called the kinetic *trans* effect, refers to the observation that certain ligands increase the rate of ligand substitution when positioned *trans* to the departing ligand. The key word in that last sentence is "rate"—the *trans* effect proper is a kinetic effect. The *trans* influence refers to the impact of a ligand on the length of the bond *trans* to it in the ground state of a complex. The key phrase there is "ground state"—this is a thermodynamic effect, so it's sometimes called the thermodynamic *trans* effect. Adding to the insanity, *cis* effects and *cis* influences have also been observed. Evidently, ligands may also influence the kinetics or thermodynamics of their *cis* neighbors. All of these phenomena are independent of the metal center, but *do* depend profoundly on the geometry of the metal (more on that shortly).

Kinetic *trans* and *cis* effects are shown in the figure below. In both cases, we see that X^1 exhibits a stronger effect than X^2. The geometries shown are those for which each effect is most commonly observed. The metals and oxidation states shown are prototypical.

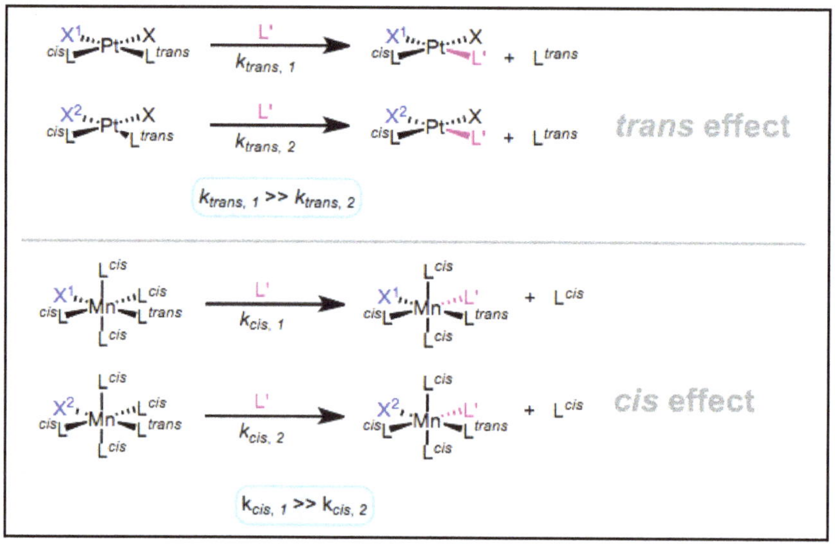

The kinetic *trans* and *cis* effects in action. X^1 is the stronger (*trans/cis*)-effect ligand in these examples.

On to the influences, which are simpler to illustrate since they're concerned with ground states, not reactions. The lengthened bonds below are exaggerated.

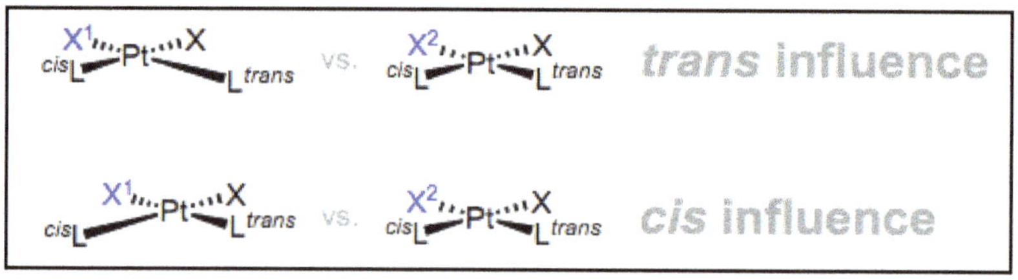

The *trans* and *cis* influences in action. Note the elongated bond lengths.

And there we have it folks, the thermodynamic and kinetic *cis/trans* effects. It's worth keeping in mind that the kinetic *trans* effect is most common for d^8 square planar complexes, and the kinetic *cis* effect is most common for d^6 octahedral complexes (particularly when the departing L is CO). But a lingering question remains: what makes for a strong *trans* effect ligand?

Origins of Effects & Influences

The *trans* effect and its cousins are all electronic, not steric effects. So, the electronic properties of the ligand dictate the strength of its *trans* effect. Let's finally dig into the *trans* effect series:

$$(weak)\ F^-,\ HO^-,\ H_2O < NH_3 < py < Cl^- < Br^- < I^-,\ SCN^-,\ NO_2^-,\ SC(NH_2)_2,$$
$$Ph^- < SO_3^{2-} < PR_3 < AsR_3,\ SR_2,\ H_3C^- < H^-,\ NO,\ CO,\ NC^-,\ C_2H_4\ (strong)$$

What's the electronic progression here? It's clear that electronegativity decreases across the series: $F^- < Cl^- < Br^- < I^- < H_3C^-$. From a bonding perspective, we can say that ligands with strong *trans* effects are strong σ-donors (or σ-bases). Yet σ-donation doesn't tell the whole story. What about ethylene and carbon monoxide, which both appear at the top of the heap? Neither of these ligands are strong σ-donors, but their π systems do interact with the metal center through backbonding. Consider the following sub-series: $S=C=N^- < PR_3 < CO$. Backbonding increases across this series, along with the strength of the *trans* effect. Strong backbonders—better known as π-acceptors or π-acids—exhibit strong *trans* effects.

$$\text{Strong trans effect} = \text{strong s-donor} + \text{strong p-acceptor}$$

Wonderful! Using these ideas we can identify ligands with strong *trans* effects. But we can dive deeper down the rabbit hole: *why* does this particular combination of electronic factors lead to a strong *trans* effect? To understand this, we need to know the mechanism of the ligand substitution reaction that's sped up by strong *trans* effect ligands. For 16-electron Pt(II) complexes, associative substitution is par for the course. The incoming ligand binds to the metal first, forming an 18-electron complex (yay!), which jettisons a ligand to yield a new 16-electron product. The mechanism in all its glory is shown in the figure below.

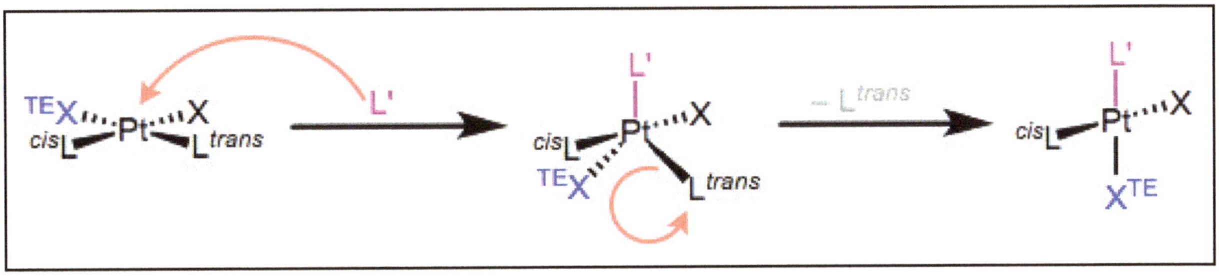

The mechanism of associative ligand substitution of Pt(II) complexes.

Some very important points about this mechanism:

- The incoming ligand always sits at an equatorial site in the trigonal bipyramidal intermediate. More on this another day, but I think of this result as governed by the principle of least motion. Consider the molecular gymnastics that would have to happen to place the incoming ligand in an axial position.
- Two ligands in the square plane are "pushed down" and become the other two equatorial ligands.
- Owing to microscopic reversibility, the leaving group must be one of the equatorial ligands.

The third point reveals that once L' has "pushed down" X^{TE} and L^{trans}, L^{trans} has no choice but to leave (assuming X^{TE} stays put). Thus, the *trans* effect has nothing to do with the second step of the mechanism, which is not rate determining anyway. The key is the first step—in particular, the "pushing down" event. Apparently, ligands with strong *trans* effects like to be pushed down. They like to occupy the equatorial plane of the TBP intermediate. Now here's the kicker: the equatorial *sites* of the TBP geometry are more *π basic* than the axial sites. The equatorial plane is just the *xy*-plane of the metal center, and the *d* orbitals in that plane (when occupied) are great electron sources for π-acidic ligands. Thus, π-acidic ligands want to occupy those equatorial sites, to receive the benefits of strong backbonding! Boom; strong π-acids encourage loss of the ligand *trans* to themselves.

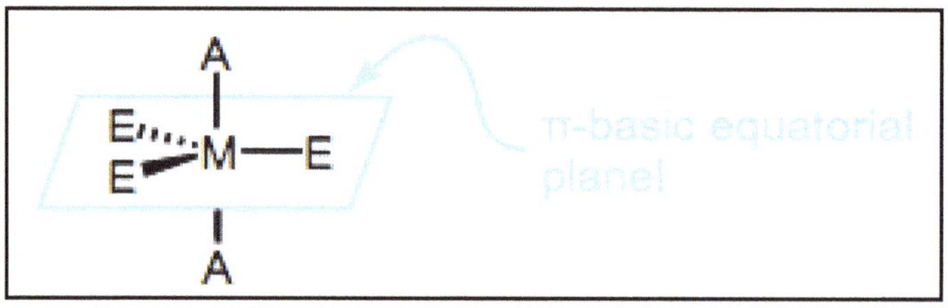

The equatorial sites of TBP metals are rich in electrons that can π bond.

What about those pesky σ donors? Well, we can imagine that in a square planar complex, a ligand and its *trans* partner are competing for donation into the same *d* orbital. Strong σ donation from a ligand should thus weaken the bond *trans* to it. Although this is the thermodynamic *trans* effect (*trans* influence) in action, the resulting destabilization of the ground state relative to the transition state is a kinetic effect. On the whole, the barrier to substitution of the *trans* ligand goes down as σ-donating strength goes up.

This idea of "competition for the metal center" is a nice heuristic to use when thinking about the *trans* and *cis*influences. The type of metallic orbital involved in M–L bonding determines the strength of L's *trans* and *cis*influences on neighboring ligands that also need that metallic orbital for bonding. For example: both influences are large if the metal's *s* orbital is a significant contributor to M–L bonding, since it's non-directional; the *trans* influence is much greater than the *cis* influence when metallic *p* orbitals are primarily involved in M–L.

Jahn–Teller Effect

The Jahn-Teller effect is a geometric distortion of a non-linear molecular system that reduces its symmetry and energy. This distortion is typically observed among octahedral complexes where the two axial bonds can be shorter or longer than those of the equatorial bonds. This effect can also be observed in tetrahedral compounds. This effect is dependent on the electronic state of the system.

In 1937, Hermann Jahn and Edward Teller postulated a theorem stating that "stability and degeneracy are not possible simultaneously unless the molecule is a linear one," in regards to its electronic state. This leads to a break in degeneracy which stabilizes the molecule and by consequence, reduces its symmetry. Since 1937, the theorem has been revised which Housecroft and Sharpe have eloquently phrased as "any non-linear molecular system in a degenerate electronic state will be unstable and will undergo distortion to form a system of lower symmetry and lower energy, thereby removing the degeneracy." This is most commonly observed with transition metal octahedral complexes, however, it can be observed in tetrahedral compounds as well.

For a given octahedral complex, the five d atomic orbitals are split into two degenerate sets when constructing a molecular orbital diagram. These are represented by the sets' symmetry labels: t_{2g} (d_{xz}, d_{yz}, d_{xy}) and e_g $(d_{z^2}$ and $d_{x^2-y^2})$. When a molecule possesses a degenerate electronic ground state, it will distort to remove the degeneracy and form a lower energy (and by consequence, lower symmetry) system. The octahedral complex will either elongate or compress the z ligand bonds as shown in Figure below:

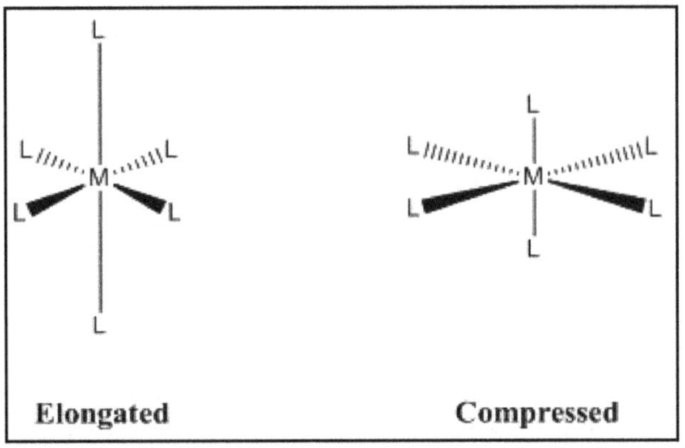

Figure: Jahn-Teller distortions for an octahedral complex.

When an octahedral complex exhibits elongation, the axial bonds are longer than the equatorial bonds. For a compression, it is the reverse; the equatorial bonds are longer than the axial bonds. Elongation and compression effects are dictated by the amount of overlap between the metal and ligand orbitals. Thus, this distortion varies greatly depending on the type of metal and ligands. In general, the stronger the metal-ligand orbital interactions are, the greater the chance for a Jahn-Teller effect to be observed.

Elongation

Elongation Jahn-Teller distortions occur when the degeneracy is broken by the stabilization (lowering in energy) of the d orbitals with a *z* component, while the orbitals without a *z* component are destabilized (higher in energy) as shown in figure below:

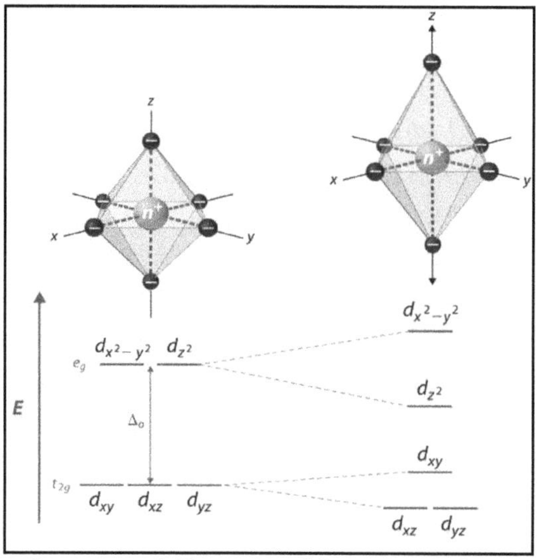

Figure: Illustration of tetragonal distortion (elongation) for an octahedral complex.

This is due to the d_{xy} and $d_{x^2-y^2}$ orbitals having greater overlap with the ligand orbitals, resulting in the orbitals being higher in energy. Since the $d_{x^2-y^2}$ orbital is antibonding, it is expected to increase in energy due to elongation. The d_{xy} orbital is still nonbonding, but is destabilized due to the interactions. Jahn-Teller elongations are well-documented for copper(II) octahedral compounds. A classic example is that of copper(II) fluoride as shown in figure below.

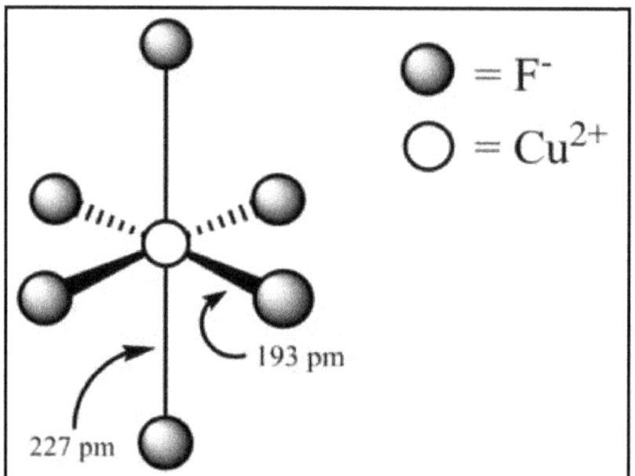

Figure: Structure of octahedral copper(II) fluoride.

Notice that the two axial bonds are both elongated and the four shorter equatorial bonds are the same length as each other. According the theorem, the orbital degeneracy is eliminated by distortion, making the molecule more stable based on the model presented in Figure.

Compression

Compression Jahn-Teller distortions occur when the degeneracy is broken by the stabilization (lowering in energy) of the d orbitals *without* a *z* component, while the orbitals with a *z* component are destabilized (higher in energy) as shown in figure below:

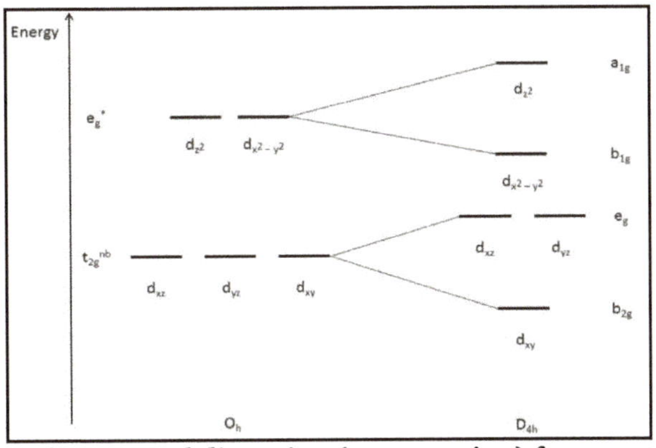

Figure: Illustration of tetragonal distortion (compression) for an octahedral complex.

This is due to the *z*-component d orbitals having greater overlap with the ligand orbitals, resulting in the orbitals being higher in energy. Since the d_{z^2} orbital is antibonding, it is expected to increase in energy due to compression. The d_{xz} and d_{yz} orbitals are still nonbonding, but are destabilized due to the interactions.

Electronic Configurations

For Jahn-Teller effects to occur in transition metals there must be degeneracy in either the t_{2g} or e_g orbitals. The electronic states of octahedral complexes are classified as either low spin or high spin. The spin of the system is dictated by the chemical environment. This includes the characteristics of the metal center and the types of ligands.

Low Spin

Figure (below) shows the various electronic configurations for low spin octahedral complexes:

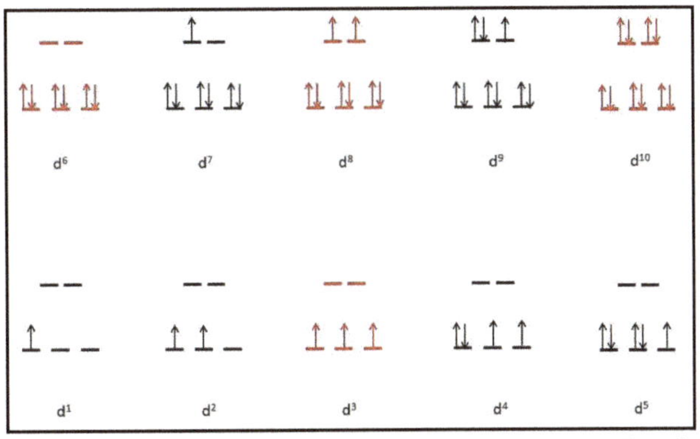

Figure: Low spin octahedral coordination diagram (red indicates no degeneracies possible, thus no Jahn-Teller effects).

The figure illustrates that low spin complexes with d^3, d^5, d^8, and d^{10} electrons that do no exhibit Jahn-Teller distortions. These electronic configurations correspond to a variety of transition metals. Many common examples include Cr^{3+}, Co^{3+}, and Ni^{2+}.

High Spin

Figure shows the various electronic configurations for high spin octahedral complexes:

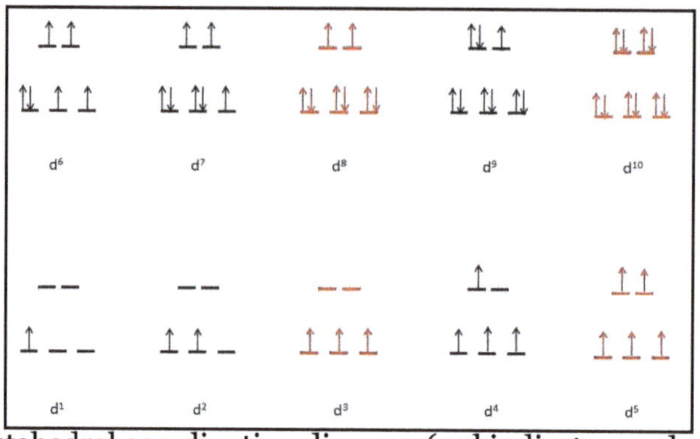

Figure: High spin octahedral coordination diagram (red indicates no degeneracies possible, thus no Jahn-Teller effects).

The figure illustrates that low spin complexes with d^3, d^5, d^8, and d^{10} electrons cannot have Jahn-Teller distortions. In general, degenerate electronic states occupying the egeg orbital set tend to show stronger Jahn-Teller effects. This is primarily caused by the occupation of these high energy orbitals. Since the system is more stable with a lower energy configuration, the degeneracy of the eg set is broken, the symmetry is reduced, and occupations at lower energy orbitals occur.

Spectroscopic Observation

Jahn-Teller distortions can be observed using a variety of spectroscopic techniques. In UV-VIS absorption spectroscopy, distortion causes splitting of bands in the spectrum due to a reduction in symmetry (O_h to D_{4h}). Consider a hypothetical molecule with octahedral symmetry showing a single absorption band. If the molecule were to undergo Jahn-Teller distortion, the number of bands would increase as shown in figure below:

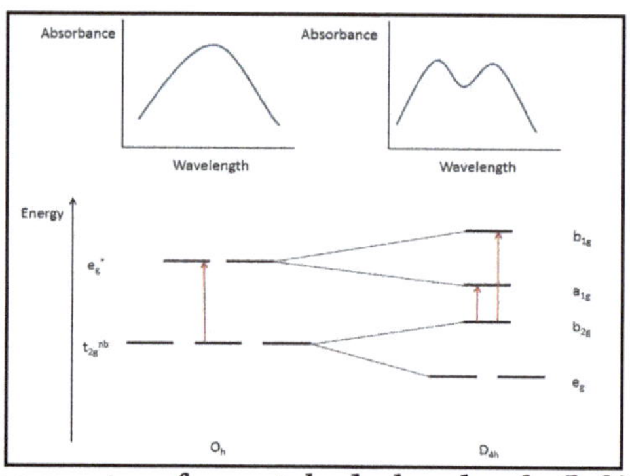

Figure: Hypothetical absorption spectra of an octahedral molecule (left) and the same molecule with Jahn-Teller elongation (right). The red arrows indicate electronic transitions.

A similar phenomenon can be seen with IR and Raman vibrational spectroscopy. The number of vibrational modes for a molecule can be calculated using the 3n - 6 rule (or 3n - 5 for linear geometry) rule. If a molecule exhibits an O_h symmetry point group, it will have fewer bands than that of a Jahn-Teller distorted molecule with D_{4h} symmetry. Thus, one could observe Jahn-Teller effects through either IR or Raman techniques. This effect can also be observed in EPR experiments as long as there is at least one unpaired electron.

Table: *Examples of Jahn-Teller distorted complexes*	
$CuBr_2$	4 Br at 240 pm 2 Br at 318 pm
$CuCl_2$	4 Cl at 230 pm 2 Cl at 295 pm
$CuCl_2.2H_2O$	2 O at 193 pm 2 Cl at 228 pm 2 Cl at 295 pm
$CsCuCl_3$	4 Cl at 230 pm 2 Cl at 265 pm
CuF_2	4 F at 193 pm 2 F at 227 pm
$CuSO_4.4NH_3.H_2O$	4 N at 205 pm 1 O at 259 pm 1 O at 337 pm
K_2CuF_4	4 F at 191 pm 2 F at 237 pm
$KCuAlF_6$	2 F at 188 pm 4 F at 220 pm
CrF_2	4 F at 200 pm 2 F at 243 pm
$KCrF_3$	4 F at 214 pm 2 F at 200 pm
MnF_3	2 F at 209 pm 2 F at 191 pm 2 F at 179 pm

The Jahn-Teller Theorem predicts that distortions should occur for *any degenerate state*, including degeneracy of the t_{2g} level, however distortions in bond lengths are much more distinctive when the degenerate electrons are in the e_g level.

References

- Definition-of-coordination-number-604956: thoughtco.com, Retrieved 29 April 2020
- Ligand-field-and-molecular-orbital-theories-277802: britannica.com, Retrieved 09 June 2020
- J. C. Fontecilla-Camps, A. Volbeda, C. Cavazza, Y. Nicolet "Structure/Function Relationships of [NiFe]- and [FeFe]-Hydrogenases" Chem. Rev. 2007, 107, 4273-4303. doi:10.1021/cr050195z
- The-transcis-effects-influences: organometallicchem.wordpress.com, Retrieved 21 March 2020
- Coordination-compound, science: britannica.com: Retrieved 26 April 2020
- What-is-coordination-sphere, chemistry: qsstudy.com, Retrieved 06 April 2020
- R. M. Supkowski, W. DeW. Horrocks Jr. "On the determination of the number of water molecules, q, coordinated to europium(III) ions in solution from luminescence decay lifetimes" Inorganic Chimica Acta 2002, Volume 340, pp. 44–48. doi:10.1016/S0020-1693(02)01022-8
- Coordination-number-ligands-and-geometries: courses.lumenlearning.com, Retrieved 14 July 2020

5

Chemical Bonding

Chemical bonding can be defined as the bond formed between molecules, atoms or ions which results in the formation of a compound. Inorganic compounds undergo chemical bonding with electrovalent or ionic bonding being the most common type of bond. This chapter covers all the topics related to chemical bonding in inorganic chemistry in a comprehensive manner.

A chemical bond is a lasting attraction between atoms that enables the formation of chemical compounds. The bond may result from the electrostatic force of attraction between atoms with opposite charges, or through the sharing of electrons as in the covalent bonds. The strength of chemical bonds varies considerably; there are “strong bonds” or “primary bond” such as metallic, covalent or ionic bonds and “weak bonds” or “secondary bond” such as Dipole-dipole interaction, the London dispersion force and hydrogen bonding.

Since opposite charges attract via a simple electromagnetic force, the negatively charged electrons that are orbiting the nucleus and the positively charged protons in the nucleus attract each other. An electron positioned between two nuclei will be attracted to both of them, and the nuclei will be attracted toward electrons in this position. This attraction constitutes the chemical bond. Due to the matter wave nature of electrons and their smaller mass, they must occupy a much larger amount of volume compared with the nuclei, and this volume occupied by the electrons keeps the atomic nuclei in a bond relatively far apart, as compared with the size of the nuclei themselves.

In general, strong chemical bonding is associated with the sharing or transfer of electrons between the participating atoms. The atoms in molecules, crystals, metals and diatomic gases—indeed most of the physical environment around us—are held together by chemical bonds, which dictate the structure and the bulk properties of matter.

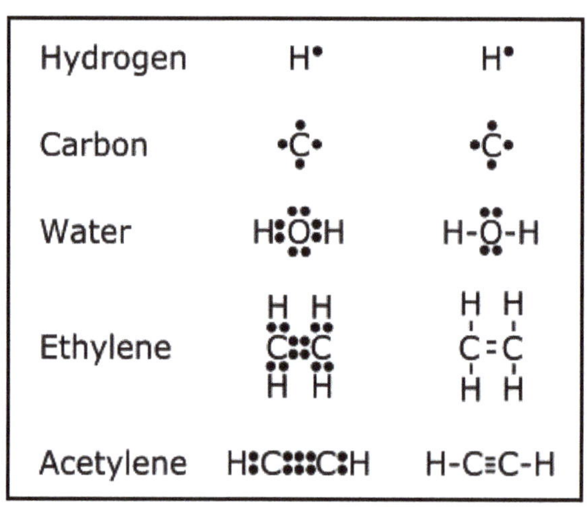

Examples of Lewis dot-style representations of chemical bonds between carbon (C), hydrogen (H), and oxygen (O). Lewis dot diagrams were an early attempt to describe chemical bonding and are still widely used today.

All bonds can be explained by quantum theory, but, in practice, simplification rules allow chemists to predict the strength, directionality, and polarity of bonds. The octet rule and VSEPR theory are two examples. More sophisticated theories are valence bond theory which includes orbital hybridization and resonance, and molecular orbital theory which includes linear combination of atomic orbitals and ligand field theory. Electrostatics are used to describe bond polarities and the effects they have on chemical substances.

Overview of Main Types Of Chemical Bonds

A chemical bond is an attraction between atoms. This attraction may be seen as the result of different behaviors of the outermost or valence electrons of atoms. These behaviors merge into each other seamlessly in various circumstances, so that there is no clear line to be drawn between them. However it remains useful and customary to differentiate between different types of bond, which result in different properties of condensed matter.

In the simplest view of a covalent bond, one or more electrons (often a pair of electrons) are drawn into the space between the two atomic nuclei. Energy is released by bond formation. This is not as a reduction in potential energy, because the attraction of the two electrons to the two protons is offset by the electron-electron and proton-proton repulsions. Instead, the release of energy (and hence stability of the bond) arises from the reduction in kinetic energy due to the electrons being in a more spatially distributed (i.e. longer de Broglie wavelength) orbital compared with each electron being confined closer to its respective nucleus. These bonds exist between two particular identifiable atoms and have a direction in space, allowing them to be shown as single connecting lines between atoms in drawings, or modeled as sticks between spheres in models.

In a polar covalent bond, one or more electrons are unequally shared between two nuclei. Covalent bonds often result in the formation of small collections of better-connected atoms called molecules, which in solids and liquids are bound to other molecules by forces that are often much weaker than the covalent bonds that hold the molecules internally together. Such weak intermolecular bonds give organic molecular substances, such as waxes and oils, their soft bulk character, and their low melting points (in liquids, molecules must cease most structured or oriented contact with each other). When covalent bonds link long chains of atoms in large molecules, however (as in polymers such as nylon), or when covalent bonds extend in networks through solids that are not composed of discrete molecules (such as diamond or quartz or the silicate minerals in many types of rock) then the structures that result may be both strong and tough, at least in the direction oriented correctly with networks of covalent bonds. Also, the melting points of such covalent polymers and networks increase greatly.

In a simplified view of an *ionic* bond, the bonding electron is not shared at all, but transferred. In this type of bond, the outer atomic orbital of one atom has a vacancy which allows the addition of one or more electrons. These newly added electrons potentially occupy a lower energy-state (effectively closer to more nuclear charge) than they experience in a different atom. Thus, one nucleus offers a more tightly bound position to an electron than does another nucleus, with the result that one atom may transfer an electron to the other. This transfer causes one atom to assume a net positive charge, and the other to assume a net negative charge. The *bond* then results from electrostatic attraction between atoms and the atoms become positive or negatively charged ions. Ionic bonds may be seen as extreme examples of polarization in covalent bonds. Often, such bonds have no particular orientation in space, since they result from equal electrostatic attraction of each ion to all ions around them. Ionic bonds are strong (and thus ionic substances require high temperatures to melt) but also brittle, since the forces between ions are short-range and do not easily bridge cracks and fractures. This type of bond gives rise to the physical characteristics of crystals of classic mineral salts, such as table salt.

A less often mentioned type of bonding is *metallic* bonding. In this type of bonding, each atom in a metal donates one or more electrons to a "sea" of electrons that reside between many metal atoms. In this sea, each electron is free (by virtue of its wave nature) to be associated with a great many atoms at once. The bond results because the metal atoms become somewhat positively charged due to loss of their electrons while the electrons remain attracted to many atoms, without being part of any given atom. Metallic bonding may be seen as an extreme example of delocalization of electrons over a large system of covalent bonds, in which every atom participates. This type of bonding is often very strong (resulting in the tensile strength of metals). However, metallic bonding is more collective in nature than other types, and so they allow metal crystals to more easily deform, because they are composed of atoms attracted to each other, but not in any particularly-oriented ways. This results in the malleability of metals. The sea of electrons in metallic bonding causes the characteristically good electrical and thermal conductivity of metals, and also their "shiny" reflection of most frequencies of white light.

History

Early speculations about the nature of the chemical bond, from as early as the 12th century, supposed that certain types of chemical species were joined by a type of chemical affinity. In 1704, Sir Isaac Newton famously outlined his atomic bonding theory, in "Query 31" of his *Opticks*, whereby atoms attach to each other by some "force". Specifically, after acknowledging the various popular theories in vogue at the time, of how atoms were reasoned to attach to each other, i.e. "hooked atoms", "glued together by rest", or "stuck together by conspiring motions", Newton states that he would rather infer from their cohesion, that "particles attract one another by some force, which in immediate contact is exceedingly strong, at small distances performs the chemical operations, and reaches not far from the particles with any sensible effect."

In 1819, on the heels of the invention of the voltaic pile, Jöns Jakob Berzelius developed a theory of chemical combination stressing the electronegative and electropositive characters of the combining atoms. By the mid 19th century, Edward Frankland, F.A. Kekulé, A.S. Couper, Alexander Butlerov, and Hermann Kolbe, building on the theory of radicals, developed the theory of valency, originally called "combining power", in which compounds were joined owing to an attraction of positive and negative poles. In 1916, chemist Gilbert N. Lewis developed the concept of the electron-pair bond, in which two atoms may share one to six electrons, thus forming the single electron bond, a single bond, a double bond, or a triple bond; in Lewis's own words, "An electron may form a part of the shell of two different atoms and cannot be said to belong to either one exclusively."

That same year, Walther Kossel put forward a theory similar to Lewis' only his model assumed complete transfers of electrons between atoms, and was thus a model of ionic bonding. Both Lewis and Kossel structured their bonding models on that of Abegg's rule (1904).

In 1927, the first mathematically complete quantum description of a simple chemical bond, i.e.

that produced by one electron in the hydrogen molecular ion, H_2^+ was derived by the Danish physicist Oyvind Burrau. This work showed that the quantum approach to chemical bonds could be fundamentally and quantitatively correct, but the mathematical methods used could not be extended to molecules containing more than one electron. A more practical, albeit less quantitative, approach was put forward in the same year by Walter Heitler and Fritz London. The Heitler-London method forms the basis of what is now called valence bond theory. In 1929, the linear combination of atomic orbitals molecular orbital method (LCAO) approximation was introduced by Sir John Lennard-Jones, who also suggested methods to derive electronic structures of molecules of F_2 (fluorine) and O_2 (oxygen) molecules, from basic quantum principles. This molecular orbital theory represented a covalent bond as an orbital formed by combining the quantum mechanical Schrödinger atomic orbitals which had been hypothesized for electrons in single atoms. The equations for bonding electrons in multi-electron atoms could not be solved to mathematical perfection (i.e., *analytically*), but approximations for them still gave many good qualitative predictions and results. Most quantitative calculations in modern quantum chemistry use either valence bond or molecular orbital theory as a starting point, although a third approach, density functional theory, has become increasingly popular in recent years.

In 1933, H. H. James and A. S. Coolidge carried out a calculation on the dihydrogen molecule that, unlike all previous calculation which used functions only of the distance of the electron from the atomic nucleus, used functions which also explicitly added the distance between the two electrons. With up to 13 adjustable parameters they obtained a result very close to the experimental result for the dissociation energy. Later extensions have used up to 54 parameters and gave excellent agreement with experiments. This calculation convinced the scientific community that quantum theory could give agreement with experiment. However this approach has none of the physical pictures of the valence bond and molecular orbital theories and is difficult to extend to larger molecules.

Bonds in Chemical Formulas

Because atoms and molecules are three-dimensional, it is difficult to use a single method to indicate orbitals and bonds. In molecular formulas the chemical bonds (binding orbitals) between atoms are indicated in different ways depending on the type of discussion. Sometimes, some details are neglected. For example, in organic chemistry one is sometimes concerned only with the functional group of the molecule. Thus, the molecular formula of ethanol may be written in conformational form, three-dimensional form, full two-dimensional form (indicating every bond with no three-dimensional directions), compressed two-dimensional form (CH_3–CH_2–OH), by separating the functional group from another part of the molecule (C_2H_5OH), or by its atomic constituents (C_2H_6O), according to what is discussed. Sometimes, even the non-bonding valence shell electrons (with the two-dimensional approximate directions) are marked, e.g. for elemental carbon ˙C˙. Some chemists may also mark the respective orbitals, e.g. the hypothetical ethene^{-4} anion (\/C=C/\ $^{-4}$) indicating the possibility of bond formation.

Strong Chemical Bonds

Strong chemical bonds are the *intramolecular* forces which hold atoms together in molecules. A strong chemical bond is formed from the transfer or sharing of electrons between atomic centers and relies on the electrostatic attraction between the protons in nuclei and the electrons in the orbitals.

The types of strong bond differ due to the difference in electronegativity of the constituent elements. A large difference in electronegativity leads to more polar (ionic) character in the bond.

Ionic Bonding

Ionic bonding is a type of electrostatic interaction between atoms which have a large electronegativity difference. There is no precise value that distinguishes ionic from covalent bonding, but a difference of electronegativity of over 1.7 is likely to be ionic, and a difference of less than 1.7 is likely to be covalent. Ionic bonding leads to separate positive and negative ions. Ionic charges are commonly between −3e to +3e. Ionic bonding commonly occurs in metal salts such as sodium chloride (table salt). A typical feature of ionic bonds is that the species form into ionic crystals, in which no ion is specifically paired with any single other ion, in a specific directional bond. Rather, each species of ion is surrounded by ions of the opposite charge, and the spacing between it and each of the oppositely charged ions near it, is the same for all surrounding atoms of the same type. It is thus no longer possible to associate an ion with any specific other single ionized atom near it. This is a situation unlike that in covalent crystals, where covalent bonds between specific atoms are still discernible from the shorter distances between them, as measured via such techniques as X-ray diffraction.

Ionic crystals may contain a mixture of covalent and ionic species, as for example salts of complex acids, such as sodium cyanide, NaCN. X-ray diffraction shows that in NaCN, for example, the bonds between sodium cations (Na^+) and the cyanide anions (CN^-) are *ionic*, with no sodium ion associated with any particular cyanide. However, the bonds between C and N atoms in cyanide are of the *covalent* type, making each of the carbon and nitrogen associated with *just one* of its opposite type, to which it is physically much closer than it is to other carbons or nitrogens in a sodium cyanide crystal.

When such crystals are melted into liquids, the ionic bonds are broken first because they are non-directional and allow the charged species to move freely. Similarly, when such salts dissolve into water, the ionic bonds are typically broken by the interaction with water, but the covalent bonds continue to hold. For example, in solution, the cyanide ions, still bound together as single CN^- ions, move independently through the solution, as do sodium ions, as Na^+. In water, charged ions move apart because each of them are more strongly attracted to a number of water molecules, than to each other. The attraction between ions and water molecules in such solutions is due to a type of weak dipole-dipole type chemical bond. In melted ionic compounds, the ions continue to be attracted to each other, but not in any ordered or crystalline way.

Covalent Bond

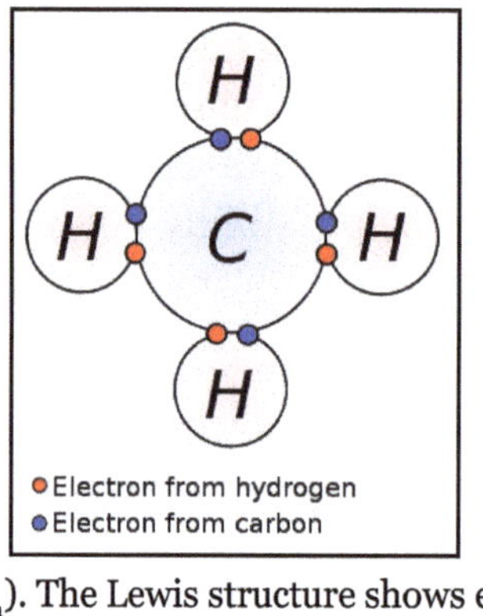

Nonpolar covalent bonds in methane (CH_4). The Lewis structure shows electrons shared between C and H atoms.

Covalent bonding is a common type of bonding, in which two atoms share two valence electrons, one from each of the atoms. In nonpolar covalent bonds, the electronegativity difference between the bonded atoms is small, typically 0 to 0.3. Bonds within most organic compounds are described as covalent. The figure shows methane (CH_4), in which each hydrogen forms a covalent bond with the carbon.

Molecules which are formed primarily from non-polar covalent bonds are often immiscible in water or other polar solvents, but much more soluble in non-polar solvents such as hexane.

A polar covalent bond is a covalent bond with a significant ionic character. This means that the two shared electrons are closer to one of the atoms than the other, creating an imbalance of charge. Such bonds occur between two atoms with moderately different electronegativities and give rise to dipole-dipole interactions. The electronegativity difference between the two atoms in these bonds is 0.3 to 1.7.

Single and Multiple Bonds

A single bond between two atoms corresponds to the sharing of one pair of electrons. The electron density of these two bonding electrons is concentrated in the region between the two atoms, which is the defining quality of a sigma bond.

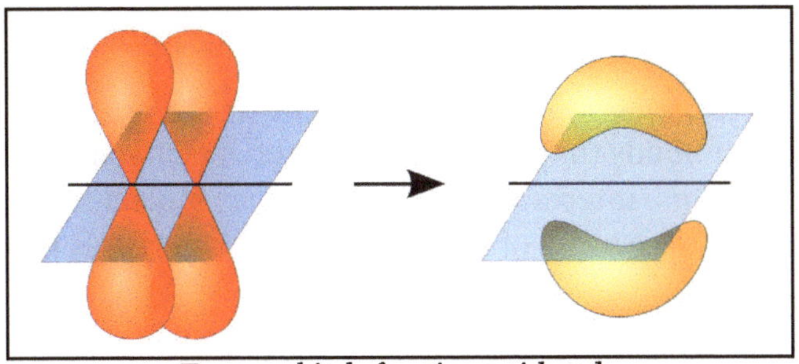

Two p-orbitals forming a pi-bond.

A double bond between two atoms is formed by the sharing of two pairs of electrons, one in a sigma bond and one in a pi bond, with electron density concentrated on two opposite sides of the internuclear axis. A triple bond consists of three shared electron pairs, forming one sigma and two pi bonds.

Quadruple and higher bonds are very rare and occur only between certain transition metal atoms.

Coordinate Covalent Bond (Dipolar Bond)

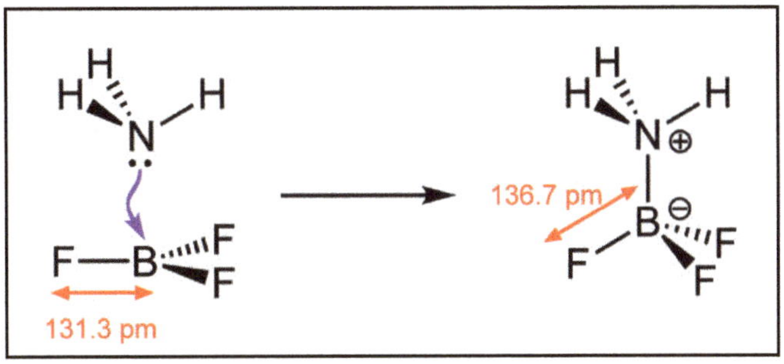

Adduct of ammonia and boron trifluoride.

A coordinate covalent bond is a covalent bond in which the two shared bonding electrons are from the same one of the atoms involved in the bond. For example, boron trifluoride (BF_3) and ammonia (NH_3) from an adduct or coordination complex $F_3B{\leftarrow}NH_3$ with a B–N bond in which a lone pair of electrons on N is shared with an empty atomic orbital on B. BF_3 with an empty orbital is described as an electron pair acceptor or Lewis acid, while NH_3 with a lone pair which can be shared is described as an electron-pair donor or Lewis base. The electrons are shared roughly equally between the atoms in contrast to ionic bonding. Such bonding is shown by an arrow pointing to the Lewis acid.

Transition metal complexes are generally bound by coordinate covalent bonds. For example, the ion Ag^+ reacts as a Lewis acid with two molecules of the Lewis base NH_3 to form the complex ion $Ag(NH_3)_2^+$, which has two Ag←N coordinate covalent bonds.

Metallic Bonding

In metallic bonding, bonding electrons are delocalized over a lattice of atoms. By contrast, in ionic compounds, the locations of the binding electrons and their charges are static. The freely-moving or delocalization of bonding electrons leads to classical metallic properties such as luster (surface light reflectivity), electrical and thermal conductivity, ductility, and high tensile strength.

Intermolecular Bonding

There are four basic types of bonds that can be formed between two or more (otherwise non-associated) molecules, ions or atoms. Intermolecular forces cause molecules to be attracted or repulsed by each other. Often, these define some of the physical characteristics (such as the melting point) of a substance.

- A large difference in electronegativity between two bonded atoms will cause a permanent charge separation, or dipole, in a molecule or ion. Two or more molecules or ions with permanent dipoles can interact within dipole-dipole interactions. The bonding electrons in a molecule or ion will, on average, be closer to the more electronegative atom more frequently than the less electronegative one, giving rise to partial charges on each atom, and causing electrostatic forces between molecules or ions.

- A hydrogen bond is effectively a strong example of an interaction between two permanent dipoles. The large difference in electronegativities between hydrogen and any of fluorine, nitrogen and oxygen, coupled with their lone pairs of electrons cause strong electrostatic forces between molecules. Hydrogen bonds are responsible for the high boiling points of water and ammonia with respect to their heavier analogues.

- The London dispersion force arises due to instantaneous dipoles in neighbouring atoms. As the negative charge of the electron is not uniform around the whole atom, there is always a charge imbalance. This small charge will induce a corresponding dipole in a nearby molecule; causing an attraction between the two. The electron then moves to another part of the electron cloud and the attraction is broken.

- A cation–pi interaction occurs between a pi bond and a cation.

Theories of Chemical Bonding

In the (unrealistic) limit of "pure" ionic bonding, electrons are perfectly localized on one of the two atoms in the bond. Such bonds can be understood by classical physics. The forces between the atoms are characterized by isotropic continuum electrostatic potentials. Their magnitude is in simple proportion to the charge difference.

Covalent bonds are better understood by valence bond theory or molecular orbital theory. The properties of the atoms involved can be understood using concepts such as oxidation number. The electron density within a bond is not assigned to individual atoms, but is instead delocalized between atoms. In valence bond theory, the two electrons on the two atoms are coupled together with the bond strength depending on the overlap between them. In molecular orbital theory, the linear combination of atomic orbitals (LCAO) helps describe the delocalized molecular orbital structures and energies based on the atomic orbitals of the atoms they came from. Unlike pure ionic bonds, covalent bonds may have directed anisotropic properties. These may have their own names, such as sigma bond and pi bond.

In the general case, atoms form bonds that are intermediate between ionic and covalent, depending on the relative electronegativity of the atoms involved. This type of bond is sometimes called polar covalent.

Bonding

The stability of the monoatomic noble gases emphasized the fact that an atomic system with eight outer most electrons (two for He) will be very stable and called as noble gas configuration. This is also known as octet rule. The above statement can be further described for the inclusion of He as, the atomic system with filled outermost shell will be stable.

Atoms are connected to another homo - or hetero - atom(s) forming molecules, while, molecules of noble gases are monoatomic. This indicates that a molecule will only be formed by the combination of either homo - or hetero - atom if it tends to a low energy, filled-configuration of the outermost shell, and more stable systems than individual atomic systems.

In a di or polyatomic molecules, atoms are held together by means of an attraction force. This force is called bond. A bond could be formed by equal sharing of electron density between atoms (covalent bonding), or uneven sharing of electron density together with coulomb interaction between the atoms (ionic bonding). Whatever, a stable electronic configuration should be achieved by atom either by releasing/accepting electron(s) or sharing electron(s).

The inter-nuclear distance between two atoms is known as bond distance. The bond dissociation energy is the enthalpy change for complete splitting the units linked by a particular chemical bond in the gaseous state. It is also defined as the amount of energy released when the bond is formed between two neutral gaseous atoms.

Types of bonds:

Elements can be subdivided mainly into two groups; electropositive and electronegative elements.

Those elements which release or accept electron(s) to achieve an inert gas configuration are called electropositive and electronegative elements, respectively. Combination of electropositive and electronegative elements forms the following compounds.

(A) Ionic compounds = Ionic bond = Electropositive elements + Electronegative elements

(B) Covalent Compounds = Covalent bond = Electronegative elements + Electronegative elements

(C) Metallic Compounds = Metallic bond = Electropositive elements + Electropositive element

Examples:

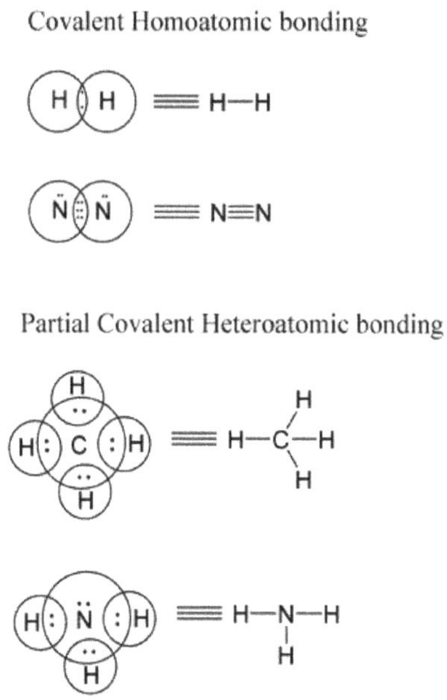

Ionic bonding

$Na(1s^2\,2s^2\,2p^6\,3s^1) \xrightarrow{-1e-} Na^+(1s^2\,2s^2\,2p^6)$

n=2 with 8 electrons
i.e. stable configuration
noble gas configuration

$Cl(1s^2\,2s^2\,2p^6\,3s^2\,3p^5) \xrightarrow{+1e-} Cl^+(1s^2\,2s^2\,2p^6\,3s^2\,3p^6)$

n=3 with 8 electrons
i.e. stable configuration,
noble gas configuration

} NaCl

Covalent, partial covalent and ionic bonding patterns maintaining octet rule.

Ionic Bonding

Ionic bonding is a type of chemical bond that involves the electrostatic attraction between oppositely charged ions, and is the primary interaction occurring in ionic compounds. The ions are atoms that have gained one or more electrons (known as anions, which are negatively charged) and atoms that have lost one or more electrons (known as cations, which are positively charged). This transfer of electrons is known as electrovalence in contrast to covalence. In the simplest case, the cation is a metal atom and the anion is a nonmetal atom, but these ions can be of a more complex nature, e.g. molecular ions like NH_4^+ or SO_4^{2-}. In simpler words, an ionic bond is the transfer of electrons from a metal to a non-metal in order to obtain a full valence shell for both atoms.

It is important to recognize that *clean* ionic bonding – in which one atom or molecule completely share an electron from another – cannot exist: all ionic compounds have some degree of covalent bonding, or electron sharing. Thus, the term "ionic bonding" is given when the ionic character is greater than the covalent character – that is, a bond in which a large electronegativity difference exists between the two atoms, causing the bonding to be more polar (ionic) than in covalent bonding where electrons are shared more equally. Bonds with partially ionic and partially covalent character are called polar covalent bonds.

Ionic compounds conduct electricity when molten or in solution, typically as a solid. Ionic compounds generally have a high melting point, depending on the charge of the ions they consist of. The higher the charges the stronger the cohesive forces and the higher the melting point. They also tend to be soluble in water. Here, the opposite trend roughly holds: the weaker the cohesive forces, the greater the solubility.

Overview

Atoms that have an almost full or almost empty valence shell tend to be very reactive. Atoms that are strongly electronegative (as is the case with halogens) often have only one or two empty orbitals in their valence shell, and frequently bond with other molecules or gain electrons to form anions. Atoms that are weakly electronegative (such as alkali metals) have relatively few valence electrons that can easily be shared with atoms that are strongly electronegative. As a result, weakly electronegative atoms tend to distort their electrons cloud and form cations.

Formation

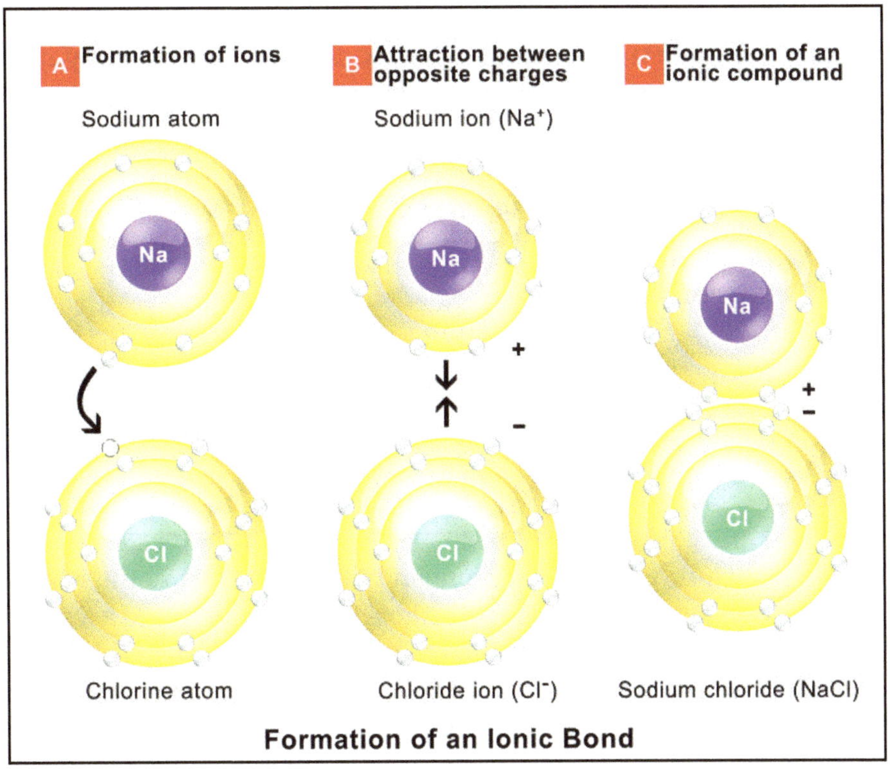

Ionic bonds in sodium chloride.

Ionic bonding can result from a redox reaction when atoms of an element (usually metal), whose ionization energy is low, give some of their electrons to achieve a stable electron configuration. In doing so, cations are formed. The atom of another element (usually nonmetal), whose electron affinity is positive, then accepts the electron(s), again to attain a stable electron configuration, and after accepting electron(s) the atom becomes an anion. Typically, the stable electron configuration is one of the noble gases for elements in the s-block and the p-block, and particular stable electron configurations for d-block and f-block elements. The electrostatic attraction between the anions and cations leads to the formation of a solid with a crystallographic lattice in which the ions are stacked in an alternating fashion. In such a lattice, it is usually not possible to distinguish discrete molecular units, so that the compounds formed are not molecular in nature. However, the ions themselves can be complex and form molecular ions like the acetate anion or the ammonium cation.

For example, common table salt is sodium chloride. When sodium (Na) and chlorine (Cl) are combined, the sodium atoms each lose an electron, forming cations (Na^+), and the chlorine atoms each gain an electron to form anions (Cl^-). These ions are then attracted to each other in a 1:1 ratio to form sodium chloride (NaCl).

$$Na + Cl \rightarrow Na^+ + Cl^- \rightarrow NaCl$$

However, to maintain charge neutrality, strict ratios between anions and cations are observed so that ionic compounds, in general, obey the rules of stoichiometry despite not being molecular compounds. For compounds that are transitional to the alloys and possess mixed ionic and metallic bonding, this may not be the case anymore. Many sulfides, e.g., do form non-stoichiometric compounds.

Many ionic compounds are referred to as salts as they can also be formed by the neutralization reaction of an Arrhenius base like NaOH with an Arrhenius acid like HCl.

$$NaOH + HCl \rightarrow NaCl + H_2O$$

The salt NaCl is then said to consist of the acid rest Cl^- and the base rest Na^+.

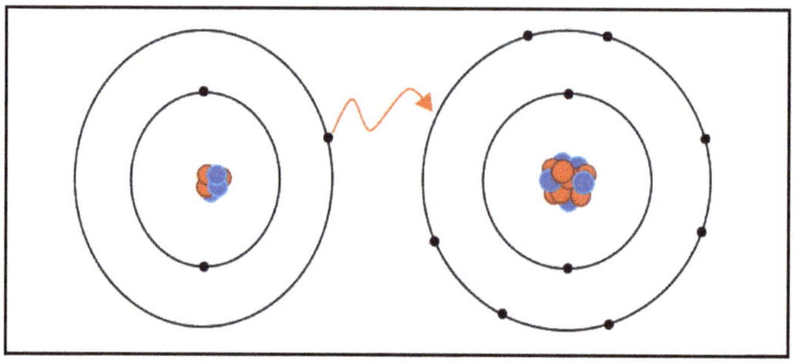

Representation of ionic bonding between lithium and fluorine to form lithium fluoride. Lithium has a low ionization energy and readily gives up its lone valence electron to a fluorine atom, which has a positive electron affinity and accepts the electron that was donated by the lithium atom. The end-result is that lithium is isoelectronic with helium and fluorine is isoelectronic with neon. Electrostatic interaction occurs between the two resulting ions, but typically aggregation is not limited to two of them. Instead, aggregation into a whole lattice held together by ionic bonding is the result.

The removal of electrons from the cation is endothermic, raising the system's overall energy. There may also be energy changes associated with breaking of existing bonds or the addition of more than one electron to form anions. However, the action of the anion's accepting the cation's valence electrons and the subsequent attraction of the ions to each other releases (lattice) energy and, thus, lowers the overall energy of the system.

Ionic bonding will occur only if the overall energy change for the reaction is favorable. In general, the reaction is exothermic, but, e.g., the formation of mercuric oxide (HgO) is endothermic. The charge of the resulting ions is a major factor in the strength of ionic bonding, e.g. a salt C^+A^- is held together by electrostatic forces roughly four times weaker than $C^{2+}A^{2-}$ according to Coulombs law, where C and A represent a generic cation and anion respectively. Of course the sizes of the ions and the particular packing of the lattice are ignored in this simple argument.

Structures

Ionic compounds in the solid state form lattice structures. The two principal factors in determining the form of the lattice are the relative charges of the ions and their relative sizes. Some structures are adopted by a number of compounds; for example, the structure of the rock salt sodium chloride is also adopted by many alkali halides, and binary oxides such as MgO. Pauling's rules provide guidelines for predicting and rationalizing the crystal structures of ionic crystals.

Bond Strength

For a solid crystalline ionic compound the enthalpy change in forming the solid from gaseous ions is termed the lattice energy. The experimental value for the lattice energy can be determined using the Born-Haber cycle. It can also be calculated (predicted) using the Born-Landé equation as the sum of the electrostatic potential energy, calculated by summing interactions between cations and anions, and a short-range repulsive potential energy term. The electrostatic potential can be expressed in terms of the inter-ionic separation and a constant (Madelung constant) that takes account of the geometry of the crystal. The further away from the nucleus the weaker the shield. The Born-Landé equation gives a reasonable fit to the lattice energy of, e.g., sodium chloride, where the calculated (predicted) value is −756 kJ/mol, which compares to −787 kJ/mol using the Born-Haber cycle.

Polarization Effects

Ions in crystal lattices of purely ionic compounds are spherical; however, if the positive ion is small and/or highly charged, it will distort the electron cloud of the negative ion, an effect summarised in Fajans' rules. This polarization of the negative ion leads to a build-up of extra charge density between the two nuclei, i.e., to partial covalency. Larger negative ions are more easily polarized, but the effect is usually important only when positive ions with charges of 3+ (e.g., Al^{3+}) are involved. However, 2+ ions (Be^{2+}) or even 1+ (Li^+) show some polarizing power because their sizes are so small (e.g., LiI is ionic but has some covalent bonding present). Note that this is not the ionic polarization effect that refers to displacement of ions in the lattice due to the application of an electric field.

Comparison with Covalent Bonding

In ionic bonding, the atoms are bound by attraction of opposite ions, whereas, in covalent bonding, atoms are bound by sharing electrons to attain stable electron configurations. In covalent bonding, the molecular geometry around each atom is determined by valence shell electron pair repulsion VSEPR rules, whereas, in ionic materials, the geometry follows maximum packing rules. One could say that covalent bonding is more *directional* in the sense that the energy penalty for not adhering to the optimum bond angles is large, whereas ionic bonding has no such penalty. There are no shared electron pairs to repel each other, the ions should simply be packed as efficiently as possible. This often leads to much higher coordination numbers. In NaCl, each ion has 6 bonds and all bond angles are 90 degrees. In CsCl the coordination number is 8. By comparison carbon typically has a maximum of four bonds.

Purely ionic bonding cannot exist, as the proximity of the entities involved in the bonding allows some degree of sharing electron density between them. Therefore, all ionic bonding has some covalent character. Thus, bonding is considered ionic where the ionic character is greater than the covalent character. The larger the difference in electronegativity between the two types of atoms involved in the bonding, the more ionic (polar) it is. Bonds with partially ionic and partially covalent character are called polar covalent bonds. For example, Na–Cl and Mg–O interactions have a few percent covalency, while Si–O bonds are usually ~50% ionic and ~50% covalent. Pauling estimated that an electronegativity difference of 1.7 (on the Pauling scale) corresponds to 50% ionic character, so that a difference greater than 50% corresponds to a bond which is predominantly ionic. Ionic character in covalent bonds can be directly measured for atoms having quadrupolar nuclei (^{2}H, ^{14}N, $^{81,79}Br$, $^{35,37}Cl$ or ^{127}I). These nuclei are generally objects of NQR nuclear quadrupole resonance and NMR nuclear magnetic resonance studies. Interactions between the nuclear quadrupole moments Q and the electric field gradients (EFG) are characterized via the nuclear quadrupole coupling constants QCC = $e^2q_{zz}Q/h$ where the eq_{zz} term corresponds to the principal component of the EFG tensor and e is the elementary charge. In turn, the electric field gradient opens the way to description of bonding modes in molecules when the QCC values are accurately determined by NMR or NQR methods.

In general, when ionic bonding occurs in the solid (or liquid) state, it is not possible to talk about a single "ionic bond" between two individual atoms, because the cohesive forces that keep the lattice together are of a more collective nature. This is quite different in the case of covalent bonding, where we can often speak of a distinct bond localized between two particular atoms. However, even if ionic bonding is combined with some covalency, the result is *not* necessarily discrete bonds of a localized character. In such cases, the resulting bonding often requires description in terms of a band structure consisting of gigantic molecular orbitals spanning the entire crystal. Thus, the bonding in the solid often retains its collective rather than localized nature. When the difference in electronegativity is decreased, the bonding may then lead to a semiconductor, a semimetal or eventually a metallic conductor with metallic bonding.

Ionic Compound

In chemistry, an ionic compound is a chemical compound composed of ions held together by electrostatic forces termed ionic bonding. The compound is neutral overall, but consists of positively charged ions called cations and negatively charged ions called anions. These can be simple ions such as the sodium (Na^+) and chloride (Cl^-) in sodium chloride, or polyatomic species such as the ammonium (NH_4^+) and carbonate (CO_3^{2-}) ions in ammonium carbonate. Individual ions within an ionic compound usually have multiple nearest neighbours, so are not considered to be part of molecules, but instead part of a continuous three-dimensional network, usually in a crystalline structure.

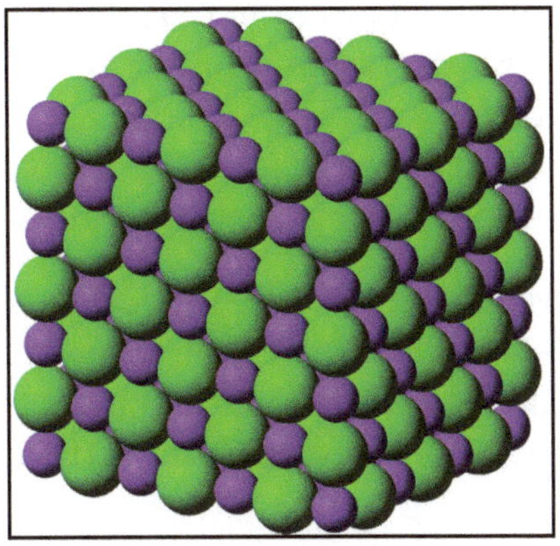

The crystal structure of sodium chloride, NaCl, a typical ionic compound. The purple spheres represent sodium cations, Na^+, and the green spheres represent chloride anions, Cl^-.

Ionic compounds containing hydrogen ions (H^+) are classified as acids, and those containing basic ions hydroxide (OH^-) or oxide (O^{2-}) are classified as bases. Ionic compounds without these ions are also known as salts and can be formed by acid–base reactions. Ionic compounds can also be produced from their constituent ions by evaporation of their solvent, precipitation, freezing, a solid-state reaction, or the electron transfer reaction of reactive metals with reactive non-metals, such as halogen gases.

Ionic compounds typically have high melting and boiling points, and are hard and brittle. As solids they are almost always electrically insulating, but when melted or dissolved they become highly conductive, because the ions are mobilized.

History of Discovery

This term was introduced by English physicist and chemist Michael Faraday in 1834 for the then-unknown species that *goes* from one electrode to the other through an aqueous medium.

X-ray spectrometer developed by Bragg.

In 1913 the crystal structure of sodium chloride was determined by William Henry Bragg and William Lawrence Bragg. This revealed that there were six equidistant nearest-neighbours for each atom, demonstrating that the constituents were not arranged in molecules or finite aggregates, but instead as a network with long-range crystalline order. Many other inorganic compounds were also found to have similar structural features. These compounds were soon described as being constituted of ions rather than neutral atoms, but proof of this hypothesis was not found until the mid-1920s, when X-ray reflection experiments (which detect the density of electrons), were performed.

Principal contributors to the development of a theoretical treatment of ionic crystal structures were Max Born, Fritz Haber, Alfred Landé, Erwin Madelung, Paul Peter Ewald, and Kazimierz Fajans. Born predicted crystal energies based on the assumption of ionic constituents, which showed good correspondence to thermochemical measurements, further supporting the assumption.

Formation

Halite, the mineral form of sodium chloride, forms when salty water evaporates leaving the ions behind.

Ionic compounds can be produced from their constituent ions by evaporation, precipitation, or freezing. Reactive metals such as the alkali metals can react directly with the highly electronegative halogen gases to form an ionic product. They can also be synthesized as the product of a high temperature reaction between solids.

If the ionic compound is soluble in a solvent, it can be obtained as a solid compound by evaporating the solvent from this electrolyte solution. As the solvent is evaporated, the ions do not go into the vapour, but stay in the remaining solution, and when they become sufficiently concentrated, nucleation occurs, and they crystallize into an ionic compound. This process occurs widely in nature, and is the means of formation of the evaporite minerals. Another method of recovering the compound from solution involves saturating a solution at high temperature and then reducing the solubility by reducing the temperature until the solution is supersaturated and the solid compound nucleates.

Insoluble ionic compounds can be precipitated by mixing two solutions, one with the cation and one with the anion in it. Because all solutions are electrically neutral, the two solutions mixed must also contain counterions of the opposite charges. To ensure that these do not contaminate the precipitated ionic compound, it is important to ensure they do not also precipitate. If the two solutions have hydrogen ions and hydroxide ions as the counterions, they will react with one another in what is called an acid–base reaction or a neutralization reaction to form water. Alternately the counterions can be chosen to ensure that even when combined into a single solution they will remain soluble as spectator ions.

If the solvent is water in either the evaporation or precipitation method of formation, in many cases the ionic crystal formed also includes water of crystallization, so the product is known as a hydrate, and can have very different chemical properties.

Molten salts will solidify on cooling to below their freezing point. This is sometimes used for the solid-state synthesis of complex ionic compounds from solid reactants, which are first melted together. In other cases, the solid reactants do not need to be melted, but instead can react through a solid-state reaction route. In this method the reactants are repeatedly finely ground into a paste, and then heated to a temperature where the ions in neighbouring reactants can diffuse together during the time the reactant mixture remains in the oven. Other synthetic routes use a solid precursor with the correct stoichiometric ratio of non-volatile ions, which is heated to drive off other species.

In some reactions between highly reactive metals (usually from Group 1 or Group 2) and highly electronegative halogen gases, or water, the atoms can be ionized by electron transfer, a process thermodynamically understood using the Born–Haber cycle.

Bonding

Ions in ionic compounds are primarily held together by the electrostatic forces between the charge distribution of these bodies, and in particular the ionic bond resulting from the long-ranged Coulomb attraction between the net negative charge of the anions and net positive charge of the cations. There is also a small additional attractive force from van der Waals interactions which contributes only around 1–2% of the cohesive energy for small ions. When a pair of ions comes close enough for their outer electron shells (most simple ions have closed shells) to overlap, a short-ranged repulsive force occurs, due to the Pauli exclusion principle. The balance between these forces leads to a potential energy well with a minimum energy when the nuclei are separated by a specific equilibrium distance.

If the electronic structure of the two interacting bodies is affected by the presence of one another, covalent interactions (non-ionic) also contribute to the overall energy of the compound formed. Ionic compounds are rarely purely ionic, i.e. held together only by electrostatic forces. The bonds between even the most electronegative/electropositive pairs such as those in caesium fluoride exhibit a small degree of covalency. Conversely, covalent bonds between unlike atoms often exhibit some charge separation and can be considered to have a partial ionic character. The circumstances under which a compound will have ionic or covalent character can typically be understood using Fajans' rules, which use only charges and the sizes of each ion. According to these rules, compounds with the most ionic character will have large positive ions with a low charge, bonded to a small negative ion with a high charge. More generally HSAB theory can be applied, whereby the compounds with the most ionic character are those consisting of hard acids and hard bases: small, highly charged ions with a high difference in electronegativities between the anion and cation. This difference in electronegativities means that the charge separation, and resulting dipole moment, is maintained even when the ions are in contact (the excess electrons on the anions are not transferred or polarized to neutralize the cations).

Structure

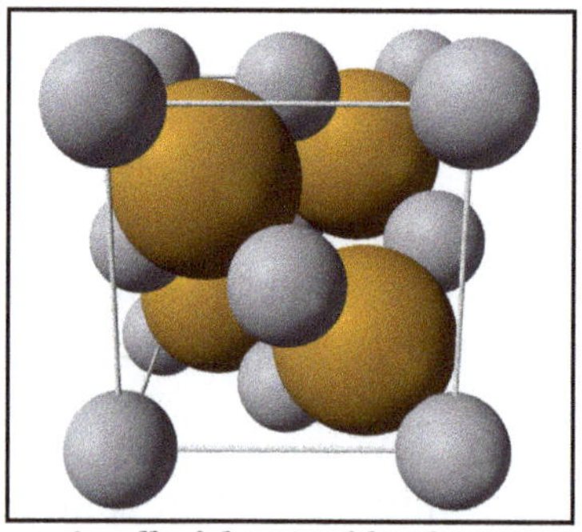

The unit cell of the zinc blende structure.

Ions typically pack into extremely regular crystalline structures, in an arrangement that minimizes the lattice energy (maximizing attractions and minimizing repulsions). The lattice energy is the summation of the interaction of all sites with all other sites. For unpolarizable spherical ions only the charges and distances are required to determine the electrostatic interaction energy. For any particular ideal crystal structure, all distances are geometrically related to the smallest internuclear distance. So for each possible crystal structure, the total electrostatic energy can be related to the electrostatic energy of unit charges at the nearest neighbour distance by a multiplicative constant called the Madelung constant that can be efficiently computed using an Ewald sum. When a reasonable form is assumed for the additional repulsive energy, the total lattice energy can be modelled using the Born–Landé equation, the Born–Mayer equation, or in the absence of structural information, the Kapustinskii equation.

Using an even simpler approximation of the ions as impenetrable hard spheres, the arrangement of anions in these systems are often related to close-packed arrangements of spheres, with the cations occupying tetrahedral or octahedral interstices. Depending on the stoichiometry of the ionic compound, and the coordination (principally determined by the radius ratio) of cations and anions, a variety of structures are commonly observed, and theoretically rationalized by Pauling's rules.

Common ionic compound structures with close-packed anions							
Stoichiometry	**Cation:anion coordination**	**Interstitial sites**		**Cubic close packing of anions**		**Hexagonal close packing of anions**	
		occupancy	**critical radius ratio**	**name**	**Madelung constant**	**name**	**Madelung constant**
MX	6:6	all octahedral	0.4142	sodium chloride	1.747565	nickeline	<1.73[a]
	4:4	alternate tetrahedral	0.2247	zinc blende	1.6381	wurtzite	1.641
MX_2	8:4	all tetrahedral	0.2247	fluorite	5.03878		
	6:3	half octahedral (alternate layers fully occupied)	0.4142	cadmium chloride	5.61	cadmium iodide	4.71

MX_3	6:2	one-third octahedral	0.4142	rhodium(III) bromide[b]	6.67[c]	bismuth iodide	8.26[d]
M_2X_3	6:4	two-thirds octahedral	0.4142			corundum	25.0312
ABO_3		two-thirds octahedral	0.4142			ilmenite	depends on charges and structure [e]
AB_2O_4		one-eighth tetrahedral and one-half octahedral	$r_A/r_O = 0.2247$, $r_B/r_O = 0.4142$[f]	spinel, inverse spinel	depends on cation site distributions	olivine	depends on cation site distributions

In some cases the anions take on a simple cubic packing, and the resulting common structures observed are:

Common ionic compound structures with simple cubic packed anions					
Stoichiometry	**Cation:anion coordination**	**Interstitial sites occupied**	**Example structure**		
			name	**critical radius ratio**	**Madelung constant**
MX	8:8	entirely filled	cesium chloride	0.7321	1.762675
MX_2	8:4	half filled	calcium fluoride		
M_2X	4:8	half filled	lithium oxide		

Some ionic liquids, particularly with mixtures of anions or cations, can be cooled rapidly enough that there is not enough time for crystal nucleation to occur, so an ionic glass is formed (with no long-range order).

Defects

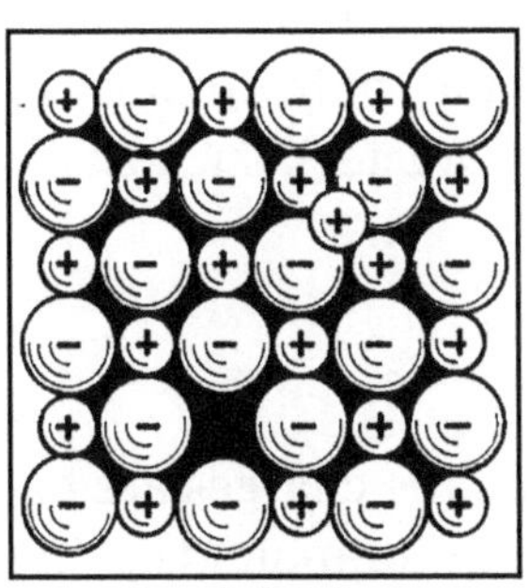

Frenkel defect.

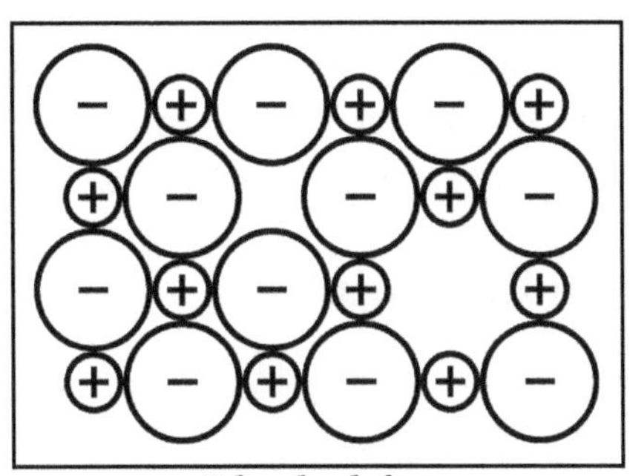

Schottky defect.

Within an ionic crystal, there will usually be some point defects, but to maintain electroneutrality, these defects come in pairs. Frenkel defects consist of a cation vacancy paired with a cation interstitial and can be generated anywhere in the bulk of the crystal, occurring most commonly in compounds with a low coordination number and cations that are much smaller than the anions. Schottky defects consist of one vacancy of each type, and are generated at the surfaces of a crystal, occurring most commonly in compounds with a high coordination number and when the anions and cations are of similar size. If the cations have multiple possible oxidation states, then it is possible for cation vacancies to compensate for electron deficiencies on cation sites with higher oxidation numbers, resulting in a non-stoichiometric compound. Another non-stoichiometric possibility is the formation of an F-center, a free electron occupying an anion vacancy. When the compound has three or more ionic components, even more defect types are possible. All of these point defects can be generated via thermal vibrations and have an equilibrium concentration. Because they are energetically costly, but entropically beneficial, they occur in greater concentration at higher temperatures. Once generated, these pairs of defects can diffuse mostly independently of one another, by hopping between lattice sites. This defect mobility is the source of most transport phenomena within an ionic crystal, including diffusion and solid state ionic conductivity. When vacancies collide with interstitials (Frenkel), they can recombine and annihilate one another. Similarly vacancies are removed when they reach the surface of the crystal (Schottky). Defects in the crystal structure generally expand the lattice parameters, reducing the overall density of the crystal. Defects also result in ions in distinctly different local environments, which causes them to experience a different crystal-field symmetry, especially in the case of different cations exchanging lattice sites. This results in a different splitting of d-electron orbitals, so that the optical absorption (and hence colour) can change with defect concentration.

Properties

Acidity/basicity

Ionic compounds containing hydrogen ions (H^+) are classified as acids, and those containing electropositive cations and basic anions ions hydroxide (OH^-) or oxide (O^{2-}) are classified as bases. Other ionic compounds are known as salts and can be formed by acid–base reactions. If the compound is the result of a reaction between a strong acid and a weak base, the result is an acidic salt. If it is the result of a reaction between a strong base and a weak acid, the result is a basic salt. If it is the result of a reaction between a strong acid and a strong base, the result is a neutral salt. Weak acids reacted with weak bases can produce ionic compounds with both the conjugate base ion and conjugate acid ion, such as ammonium acetate.

Some ions are classed as amphoteric, being able to react with either an acid or a base. This is also true of some compounds with ionic character, typically oxides or hydroxides of less-electropositive metals (so the compound also has significant covalent character), such as zinc oxide, aluminium hydroxide, aluminium oxide and lead(II) oxide.

Melting and Boiling Points

Electrostatic forces between particles are strongest when the charges are high, and the distance between the nuclei of the ions is small. In such cases, the compounds generally have very high melting and boiling points and a low vapour pressure. Trends in melting points can be even better explained when the structure and ionic size ratio is taken into account. Above their melting point ionic solids melt and become molten salts (although some ionic compounds such as aluminium chloride and iron(III) chloride show molecule-like structures in the liquid phase). Inorganic compounds with simple ions typically have small ions, and thus have high melting points, so are solids at room temperature. Some substances with larger ions, however, have a melting point below or near room temperature (often defined as up to 100 °C), and are termed ionic liquids. Ions in ionic liquids often have uneven charge distributions, or bulky substituents like hydrocarbon chains, which also play a role in determining the strength of the interactions and propensity to melt.

Even when the local structure and bonding of an ionic solid is disrupted sufficiently to melt it, there are still strong long-range electrostatic forces of attraction holding the liquid together and preventing ions boiling to form a gas phase. This means that even room temperature ionic liquids have low vapour pressures, and require substantially higher temperatures to boil. Boiling points exhibit similar trends to melting points in terms of the size of ions and strength of other interactions. When vapourized, the ions are still not freed of one another. For example, in the vapour phase sodium chloride exists as diatomic "molecules".

Brittleness

Most ionic compounds are very brittle. Once they reach the limit of their strength, they cannot deform mealleably, because the strict alignment of positive and negative ions must be maintained. Instead the material undergoes fracture via cleavage. As the temperature is elevated (usually close to the melting point) a ductile–brittle transition occurs, and plastic flow becomes possible by the motion of dislocations.

Compressibility

The compressibility of an ionic compound is strongly determined by its structure, and in particular the coordination number. For example, halides with the caesium chloride structure (coordination number 8) are less compressible than those with the sodium chloride structure (coordination number 6), and less again than those with a coordination number of 4.

Solubility

When ionic compounds dissolve, the individual ions dissociate and are solvated by the solvent and dispersed throughout the resulting solution. Because the ions are released into solution when dissolved, and can conduct charge, soluble ionic compounds are the most common class of strong electrolytes, and their solutions have a high electrical conductivity.

The solubility is highest in polar solvents (such as water) or ionic liquids, but tends to be low in nonpolar solvents (such as petrol/gasoline). This is principally because the resulting ion–dipole interactions are significantly stronger than ion-induced dipole interactions, so the heat of solution is higher. When the oppositely charged ions in the solid ionic lattice are surrounded by the opposite pole of a polar molecule, the solid ions are pulled out of the lattice and into the liquid. If the solvation energy exceeds the lattice energy, the negative net enthalpy change of solution provides a thermodynamic drive to remove ions from their positions in the crystal and dissolve in the liquid. In addition, the entropy change of solution is usually positive for most solid solutes like ionic compounds, which means that their solubility increases when the temperature increases. There are some unusual ionic compounds such as cerium(III) sulfate, where this entropy change is negative, due to extra order induced in the water upon solution, and the solubility decreases with temperature.

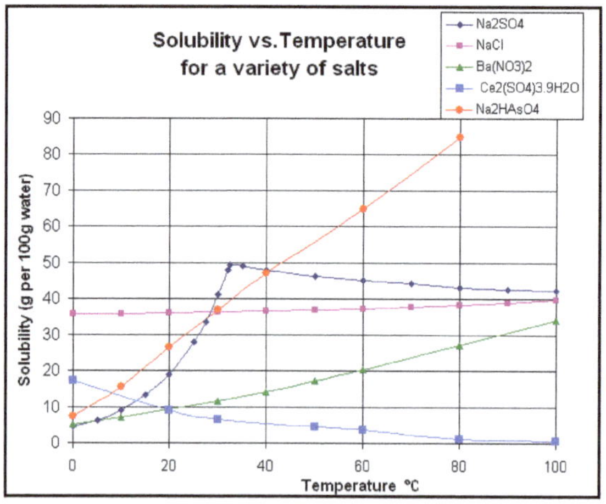

The aqueous solubility of a variety of ionic compounds as a function of temperature. Some compounds exhibiting unusual solubility behaviour have been included.

Electrical Conductivity

Although ionic compounds contain charged atoms or clusters, these materials do not typically conduct electricity to any significant extent when the substance is solid. In order to conduct, the charged particles must be mobile rather than stationary in a crystal lattice. This is achieved to some degree at high temperatures when the defect concentration increases the ionic mobility and solid state ionic conductivity is observed. When the ionic compounds are dissolved in a liquid or are melted into a liquid, they can conduct electricity because the ions become completely mobile. This conductivity gain upon dissolving or melting is sometimes used as a defining characteristic of ionic compounds.

In some unusual ionic compounds: fast ion conductors, and ionic glasses, one or more of the ionic components has a significant mobility, allowing conductivity even while the material as a whole remains solid. This is often highly temperature dependant, and may be the result of either a phase change or a high defect concentration. These materials are used in all solid-state supercapacitors, batteries, and fuel cells, and in various kinds of chemical sensors.

Colour

Anhydrous cobalt(II) chloride,
$CoCl_2$

Cobalt(II) chloride hexahydrate, $CoCl_2{\cdot}6H_2O$.

The colour of an ionic compound is often different to the colour of an aqueous solution containing the constituent ions, or the hydrated form of the same compound.

The anions in compounds with bonds with the most ionic character tend to be colourless (with an absorption band in the ultraviolet part of the spectrum). In compounds with less ionic character, their colour deepens through yellow, orange, red and black (as the absorption band shifts to longer wavelengths into the visible spectrum).

The absorption band of simple cations shift toward shorter wavelength when they are involved in more covalent interactions. This occurs during hydration of metal ions, so colourless anhydrous ionic compounds with an anion absorbing in the infrared can become colourful in solution.

Uses

Ionic compounds have long had a wide variety of uses and applications. Many minerals are ionic. Humans have processed common salt (sodium chloride) for over 8000 years, using it first as a food seasoning and preservative, and now also in manufacturing, agriculture, water conditioning, for de-icing roads, and many other uses. Many ionic compounds are so widely used in society that they go by common names unrelated to their chemical identity. Examples of this include borax, calomel, milk of magnesia, muriatic acid, oil of vitriol, saltpeter, and slaked lime.

Soluble ionic compounds like salt can easily be dissolved to provide electrolyte solutions. This is a simple way to control the concentration and ionic strength. The concentration of solutes affects many colligative properties, including increasing the osmotic pressure, and causing freezing-point depression and boiling-point elevation. Because the solutes are charged ions they also increase the electrical conductivity of the solution. The increased ionic strength reduces the thickness of the electrical double layer around colloidal particles, and therefore the stability of emulsions and suspensions.

The chemical identity of the ions added is also important in many uses. For example, fluoride containing compounds are dissolved to supply fluoride ions for water fluoridation.

Solid ionic compounds have long been used as paint pigments, and are resistant to organic solvents, but are sensitive to acidity or basicity. Since 1801 pyrotechnicians have described and widely used metal-containing ionic compounds as sources of colour in fireworks. Under intense heat, the electrons in the metal ions or small molecules can be excited. These electrons later return to lower energy states, and release light with a colour spectrum characteristic of the species present.

In chemistry, ionic compounds are often used as precursors for high-temperature solid-state synthesis.

Many metals are geologically most abundant as ionic compounds within ores. To obtain the elemental materials, these ores are processed by smelting or electrolysis, in which redox reactions occur (often with a reducing agent such as carbon) such that the metal ions gain electrons to become neutral atoms.

Nomenclature

According to the nomenclature recommended by IUPAC, ionic compounds are named according to their composition, not their structure. In the most simple case of a binary ionic compound with no possible ambiguity about the charges and thus the stoichiometry, the common name is written using two words. The name of the cation (the unmodified element name for monatomic cations) comes first, followed by the name of the anion. For example, $MgCl_2$ is named magnesium chloride, and Na_2SO_4 is named sodium sulfate (SO_4^{2-} , sulfate, is an example of a polyatomic ion). To obtain the empirical formula from these names, the stoichiometry can be deduced from the charges on the ions, and the requirement of overall charge neutrality.

If there are multiple different cations and/or anions, multiplicative prefixes (*di-*, *tri-*, *tetra-*, ...) are often required to indicate the relative compositions, and cations then anions are listed in alphabetical order. For example, $KMgCl_3$ is named magnesium potassium trichloride to distinguish it from K_2MgCl_4, magnesium dipotassium tetrachloride (note that in both the empirical formula and the written name, the cations appear in alphabetical order, but the order varies between them because the symbol for potassium is K). When one of the ions already has a multiplicative prefix within its name, the alternate multiplicative prefixes (*bis-*, *tris-*, *tetrakis-*, ...) are used. For example, $Ba(BrF_4)_2$ is named barium bis(tetrafluoridobromate).

Compounds containing one or more elements which can exist in a variety of charge/oxidation states will have a stoichiometry that depends on which oxidation states are present, to ensure overall neutrality. This can be indicated in the name by specifying either the oxidation state of the elements present, or the charge on the ions. Because of the risk of ambiguity in allocating oxidation states, IUPAC prefers direct indication of the ionic charge numbers. These are written as an arabic integer followed by the sign (... , 2−, 1−, 1+, 2+, ...) in parentheses directly after the name of the cation (without a space separating them). For example, $FeSO_4$ is named iron(2+) sulfate (with the 2+ charge on the Fe^{2+} ions balancing the 2− charge on the sulfate ion), whereas $Fe_2(SO_4)_3$ is named iron(3+) sulfate (because the two iron ions in each formula unit each have a charge of 3+, to balance the 2− on each of the three sulfate ions). Stock nomenclature, still in common use, writes the oxidation number in Roman numerals (... , −II, −I, 0, I, II, ...). So the examples given above would be named iron(II) sulfate and iron(III) sulfate respectively. For simple ions the ionic charge and the oxidation number are identical, but for polyatomic ions they often differ. For example, the uranyl(2+) ion, UO_2^{2+} , has uranium in an oxidation state of +6, so would be called a dioxouranium(VI) ion in Stock nomenclature. An even older naming system for metal cations, also still widely used, appended the suffixes *-ous* and *-ic* to the Latin root of the name, to give special names for the low and high oxidation states. For example, this scheme uses "ferrous" and "ferric", for iron(II) and iron(III) respectively, so the examples given above were classically named ferrous sulfate and ferric sulfate.

Ionic Compounds to AB Type

(A) Zinc sulfide (ZnS) structure:

In ZnS, both cation (Zn) and anion (S) are in II oxidation state. The radius ratio in ZnS is 0.40 suggests a tetrahedral arrangement where each Zn^{2+} ion is surrounded by four S^{2-} anions. The co-ordination number of both ions is 4. Therefore, the arrangement is called 4:4 arrangement.

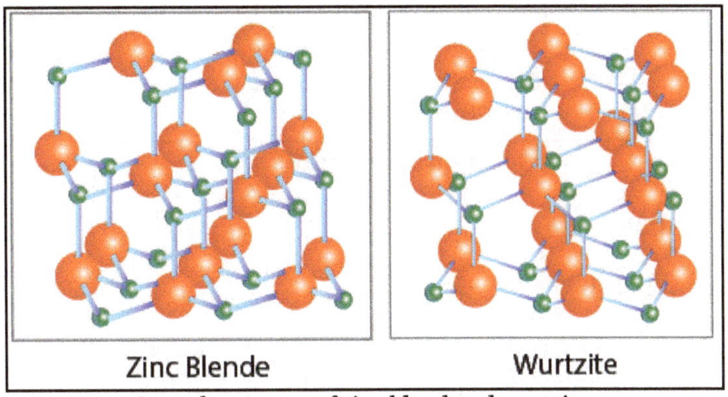

Crystal structure of zinc blend and wurtzite.

(B) Sodium chloride (NaCl) structure:

In NaCl both cation (Na) and anion (Cl) are in I state. The radius ration 0.52 suggests an octahedral arrangement around the ions. The coordination is thus 6:6.

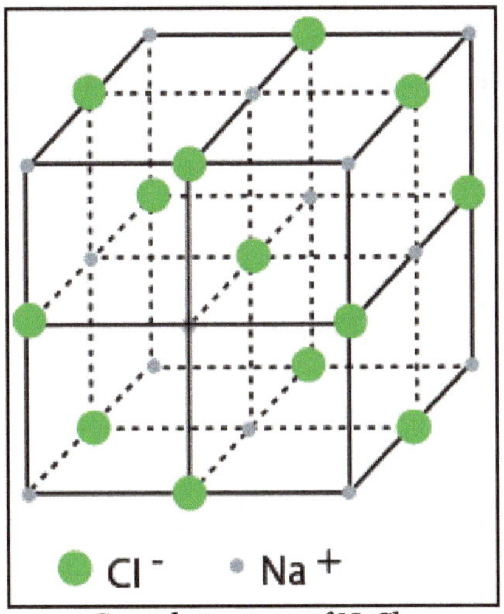

Crystal structure of NaCl.

(C) Caesium chlodire (CsCl) structure:

In CsCl both cation (Cs) and anion (Cl) are in I state. The radius ration 0.93 suggests an body centered cubic type arrangement around the ions. In this case each ion is surrounded by eight oppositely charges ions. The coordination is thus 8:8.

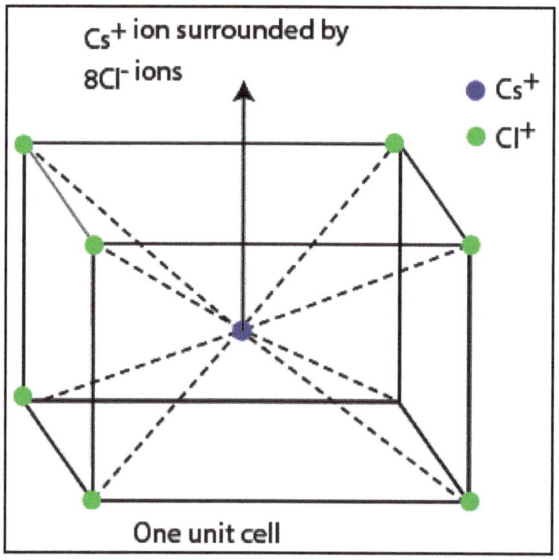

Crystal structure of CsCl.

Ionic compounds to AB_2 type:

(A) Calcium fluoride (CaF_2 , fluorite) structure:

In CaF_2 the radius ratio is 0.73 or above. It has a body centered cubic arrangement of fluoride ions around calcium ion. The ratio between Ca^{2+} and F^- is 2:1. Since, there is twice as many fluoride ions as calcium ion, the coordination number of the both ions will be different. A 8:4 F^-:Ca^{2+} arrangement has been found for CaF_2 where each Ca^{2+} ions are tetrahedrally arranged around each F^- ion.

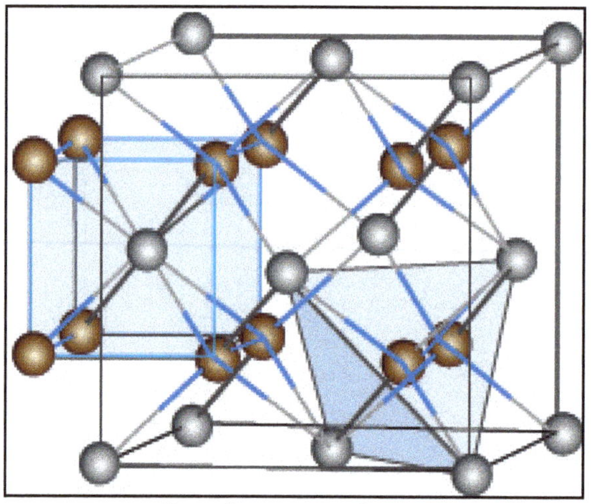
Crystal structure of CaF_2 .

(B) TiO_2 (rutile) structure:

TiO_2 exists in three different forms; (i) anatase, (ii) brookite, and (iii) rutile. In many crystals rutile structure is found and the radius ration is between 0.41 and 0.73. This indicates a coordination number is 6. Therefore the other will be 3 (as it is AB_2 kind structure). Rutile form has a 3:6 structure where each Ti^{4+} is octahedrally surrounded by six O^{2-} ions and each O^{2-} ion has three Ti^{4+} ions around in a plane triangular fashion.

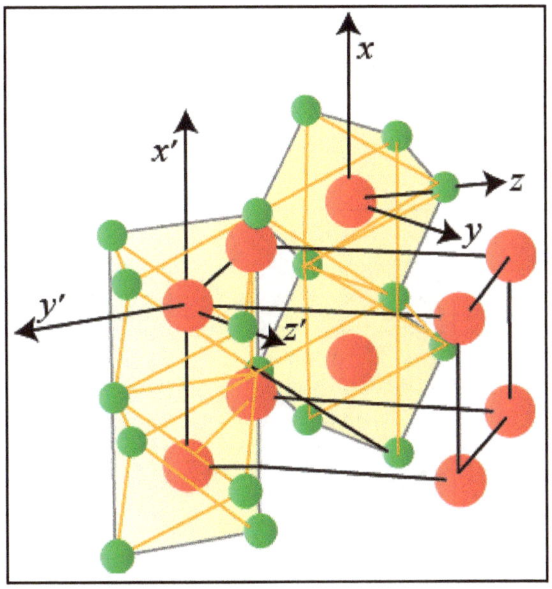

Crystal structure of TiO_2 (rutile).

The rutile structure is not close packed. The unit cell is not a cube as one of the three axes is about 30% shorter. It is convenient to describe as distorted cube.

Limitations of the radius ratio rule:

The structure adapted by the alkali metal halides may be considered to demonstrate the failure of structural prediction from the radius ratio rule.

LiCl, LiBr, and LiI, the radius ratio lies around 0.3 suggesting fourfold coordination. In case of KF, KCl, RbF, RbCl, RbBr, and CsF, the radius ratio is 0.73, hence coordination number shoud be minimum 8. In fact, all the alkali metal halides adapt 6:6 (NaCl type) coordination environment except CsCl, CsBr, and CsI.

It is said in the radius ratio rule that with increase of the ratio value the coordination number will increase. Let consider a AB type ionic compound, as we know that with increase of radius ration value the number of counter ions around a particular ion will increase. As the system is AB type, therefore, each cation will be surrounded by one anion. So, the gain in electrostatic attraction force will be cancelled by large repulsion force by the same charge ions as the coordination number is high. Therefore, the stability of ionic solid cannot be answered.

This rule provides a rough guide to the structure of ionic solids. Ultimately, the reason why any particular crystal structure is formed is that it gives the most favorable lattice energy.

Lattice Energy

The lattice energy of a crystalline solid is usually defined as the energy of formation of the crystals from infinitely-separated ions and as such is invariably positive. The precise value of the lattice energy may not be determined experimentally, because of the impossibility of preparing an adequate amount of gaseous ions or atoms and measuring the energy released during their condensation to form the solid. However, the value of the lattice energy may either be derived theoretically from electrostatics or from a thermodynamic cycling reaction, the Born–Haber cycle.

Historical Development

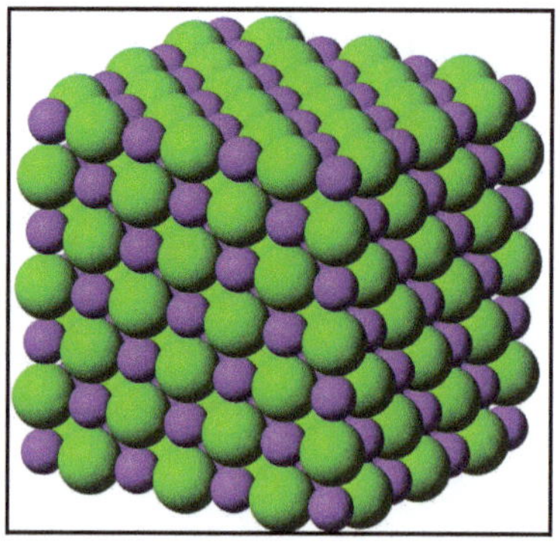

Sodium chloride crystal lattice.

The concept of lattice energy was originally developed for rocksalt-structured and sphalerite-structured compounds like NaCl and ZnS, where the ions occupy high-symmetry crystal lattice sites. In the case of NaCl,lattice energy is the energy released by the reaction:

$$Na^{+}(g) + Cl^{-}(g) \rightarrow NaCl(s)$$

which would amount to -786 kJ/mol.

Some older textbooks define lattice energy with the opposite sign, i.e. the energy required to convert the crystal into infinitely separated gaseous ions in vacuum, an endothermic process. Following this convention, the lattice energy of NaCl would be +786 kJ/mol. The lattice energy for ionic crystals such as sodium chloride, metals such as iron, or covalently linked materials such as diamond is considerably greater in magnitude than for solids such as sugar or iodine, whose neutral molecules interact only by weaker dipole-dipole or van der Waals forces.

Theoretical Treatments

The relationship between the molar lattice energy and the molar lattice enthalpy is given by the following equation:

$$\Delta_G U = \Delta_G H - p\Delta V_m$$

where $\Delta_G U$ is the molar lattice energy, $\Delta_G H$ the molar lattice enthalpy and ΔV_m the change of the volume per mol. Therefore, the lattice enthalpy further takes into account that work has to be performed against an outer pressure p. Lattice Energy of an ionic compound depends upon charge of the ion and size of the ions. Moreover, factors such as packing of ions doesn't matter efficiently.

Born–Landé Equation

In 1918 Born and Landé proposed that the lattice energy could be derived from the electric potential of the ionic lattice and a repulsive potential energy term.

$$E = -\frac{N_A M z^+ z^- q_e^2}{4\pi\varepsilon_0 r_0}\left(1-\frac{1}{n}\right),$$

where,

N_A is the Avogadro constant;

M is the Madelung constant, relating to the geometry of the crystal;

z^+ is the charge number of cation;

z^- is the charge number of anion;

q_e is the elementary charge, equal to 1.6022×10^{-19} C;

ε_0 is the permittivity of free space, equal to 8.854×10^{-12} C^2 J^{-1} m^{-1};

r_0 is the distance to closest ion; and

n is the Born exponent, a number between 5 and 12, determined experimentally by measuring the compressibility of the solid, or derived theoretically.

The Born–Landé equation gives a reasonable fit to the lattice energy.

Compound	Calculated Lattice Energy	Experimental Lattice Energy
NaCl	−756 kJ/mol	−787 kJ/mol
LiF	−1007 kJ/mol	−1046 kJ/mol
$CaCl_2$	−2170 kJ/mol	−2255 kJ/mol

From the Born–Landé equation it can be seen that the lattice energy of a compound is dependent on a number of factors:

- As the charges on the ions increase the lattice energy increases (becomes more negative),
- When ions are closer together the lattice energy increases (becomes more negative).

Barium oxide (BaO), for instance, which has the NaCl structure and therefore the same Madelung constant, has a bond radius of 275 picometers and a lattice energy of -3054 kJ/mol, while sodium chloride (NaCl) has a bond radius of 283 picometers and a lattice energy of -786 kJ/mol.

Kapustinskii Equation

The Kapustinskii equation can be used as a simpler way of deriving lattice energies where high precision is not required.

Effect of Polarisation

For ionic compounds with ions occupying lattice sites with crystallographic point groups C_1, C_{1h}, C_n or C_{nv} (n = 2, 3, 4 or 6) the concept of the lattice energy and the Born–Haber cycle has to be extended. In these cases the polarization energy E_{pol} associated with ions on polar lattice sites has to be included in the Born–Haber cycle and the solid formation reaction has to start from the already polarized species. As an example, one may consider the case of iron-pyrite FeS_2, where sulfur ions occupy lattice site of point symmetry group C_3. The lattice energy defining reaction then reads,

$$Fe^{2+}\,(g) + 2\text{ pol }S^-\,(g) \rightarrow FeS_2\,(s)$$

where pol S^- stands for the polarized, gaseous sulfur ion. It has been shown that the neglection of the effect led to 15% difference between theoretical and experimental thermodynamic cycle energy of FeS_2 that reduced to only 2%, when the sulfur polarization effects were included.

Lattice Defects

Why does not $NaCl_2$ exist?

The composition of $NaCl_2$ will be Na^{2+} and $2Cl^-$.

Process		Related energy terms
Na(s)	Na(g)	+S (enthalpy of atomization)
Na(g)	Na^{2+}(g)	$+I = +(I_1+I_2)$ (I_1 and I_2 are respective first and second ionization energy)
Cl_2(g)	2Cl(g)	D (enthalpy of bond dissociation)
2Cl(g)	$2Cl^-$(g)	−2E.A (electron affinity)
$Na^{2+}(g) + 2Cl^-(g)$	$NaCl_2$(s)	$-U_0$ (lattice energy)
$Na(s) + Cl_2(g)$	$NaCl_2$(s)	$+S + I + D - 2E.A - U_0 = -H_f$ (heat of formation of NaCl) H_f is negative as heat release on the formation of NaCl.

Therefore, we can write;

$$+S+(I_1+I_2)+D-2E.A-U_0=H_f$$

$NaCl_2$ would crystallize in fluorite (CaF_2) structure with A = 2.52. Theoretically, the U_0 for $NaCl_2$ obtained is 2180 kJmol $^{-1}$.

Hence,

$$\begin{aligned} H_f &= +S+(I_1+I_2)+D-2E.A-U_0 \\ &= +108+(496+5462)+242-2\times348-2180 \\ &= 2530\,\text{kJmol}^{-1} \end{aligned}$$

As the H_f is positive, hence $NaCl_2$ cannot be stabilized. Here, the very large I_2 (5462 kJmol $^{-1}$) of Na cannot be compensate by the stabilizing lattice energy.

Why does not CaCl exist?

Process		Related energy terms
Ca(s)	Ca(g)	+S (enthalpy of atomization)
Ca(g)	$Ca^{+}(g)$	+I (ionization energy)
½ F(g)	F(g)	½ D (enthalpy of bond dissociation)
F(g)	$F^{-}(g)$	E.A (electron affinity)
$Ca^{2+}(g) + Cl^{-}(g)$	CaF(s)	$-U_0$ (lattice energy)
Na(s) + ½ F_2(g)	NaF(s)	$+S + I + ½ D - E.A - U_0 = -H_f$(heat of formation of NaCl) H_f is negative as heat release on the formation of NaCl.

Assume that CaF will be crystallized in the same geometry as KF. Therefore, the calculated lattice energy will be $U_0 = -795$ kJmol^{-1}.

$$\text{Hence, } H_f = +S+I+1/2\,D\text{-}E.A\text{-}U_0$$
$$= +178 + 590 + 79 - 328 - 795$$
$$= -276\,\text{kJmol}^{-1}$$

As the H_f value is not large (276 kJmol^{-1}), the CaF is unstable and disproportionate to CaF_2 and Ca.

Lattice defects:

In reality, lattices do not acquire a perfectly ordered arrangement of the constituting ions. A defective lattice might not conform to the stoichiometric cation and anion ratio as represented by its chemical formula. Hence, lattices can be divided in (i) stoichiometric defect and (ii) non- stoichiometric defect.

Stoichiometric defect:

Frenkel defect and Schottky defect belong to this type of lattice defect.

Frenkel defect:

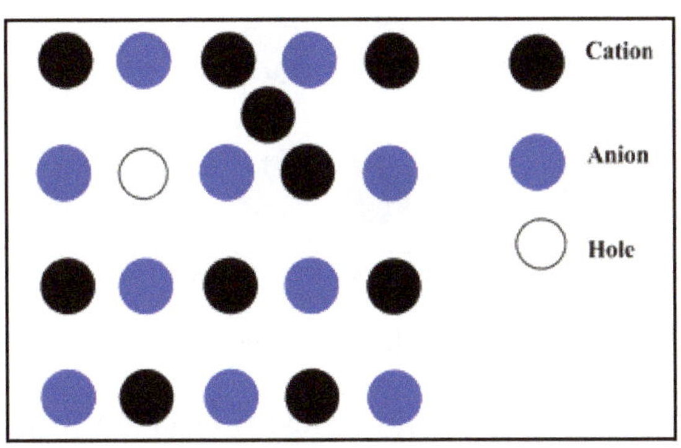

In Frenkel defect, an ion is missing from its lattice site. Therefore, a hole is generated in the lattice. The missing ion occupies some interstitial position. Crystals with small positive ions and large negative ion, like AgCl, AgBr, AgI, ZnS, etc shows this kind of defect.

Schottky defect:

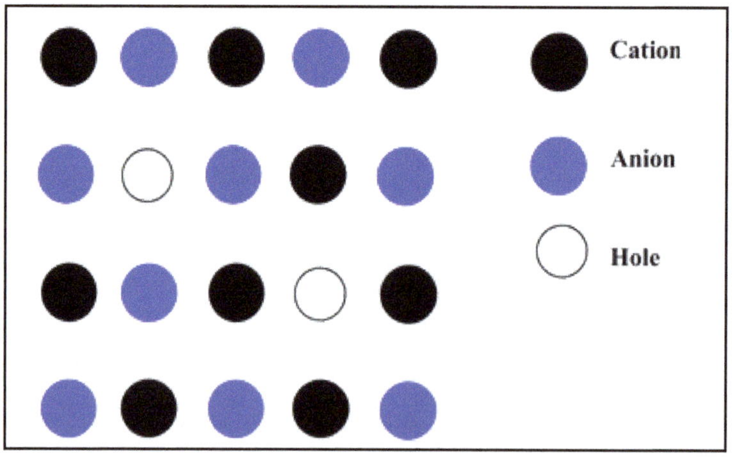

In Schottky defect, both a positive ion and a negative ion are missing from its lattice site. Therefore, two holes are generated in the lattice. Some examples are NaCl, KCl, KBr etc.

Crystalline solid with these kinds of defect show some sort of electric conductivity via ion migration to the hole(s).

Non-stoichiometric Defects:

In the non-stoichiometric defect the ratio of cation and anion differ from that of the indicated chemical formula. The charge is balanced is either by extra electron or by cation according to necessary. In this type of compounds, either the matal or the non-metal atoms may be in excess.

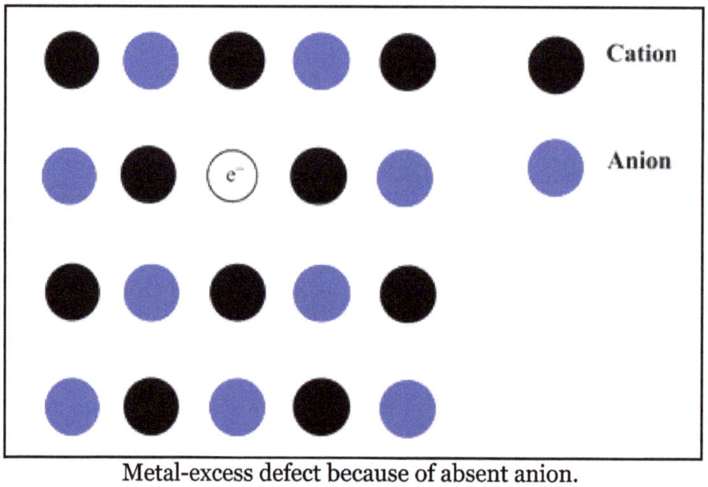

Metal-excess defect because of absent anion.

The excess of metal can be obtained by the following ways;

(A) Missing of anion from the lattice site and an extra electron in present there to balance the charge.

Example: NaCl treated with Na-vapor forms such a yellow non-stoichiometric lattice. KCl with K-vapour becomes bule.

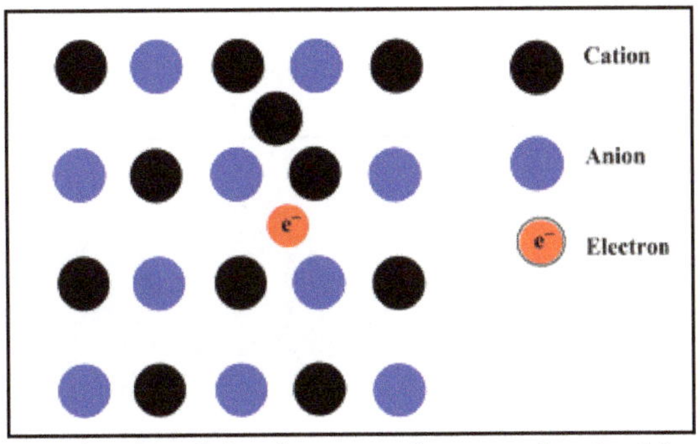

Metal excess defect because of excess of cation(s).

(B) Presence of an extra metal ion in interstitial position and extra electron in some other interstitial position to balance the charge of the lattice.

Example: ZnO, CdO, Fe_2O_3, etc shows such type of lattice defect. ZnO under normal room temperature is white in color but when it is heated it becomes yellow. This is because it the presence of extra electrons, at the interstitial position of ZnO lattice, undergo to higher energy excited states by absorbing heat.

The metal deficiency can be obtained by the following ways;

(C) Positive ion may absent in a lattice and a doubly charged cation maintains the charge balance.

Example: FeO, CuO, NiO, Fes, etc.

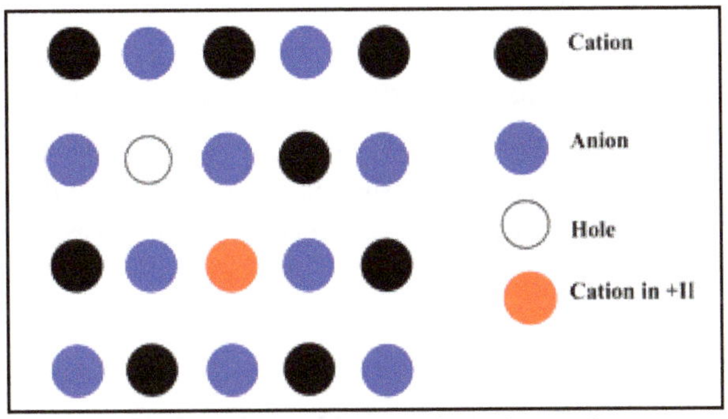

Metal deficiency because of mission cation.

(D) A non-stoichiometric defect might be possible as shown in case (C) with a additional anion in the interstitial position. No example of crystals containing this kind of anion in the interstitial position of a lattice is known.

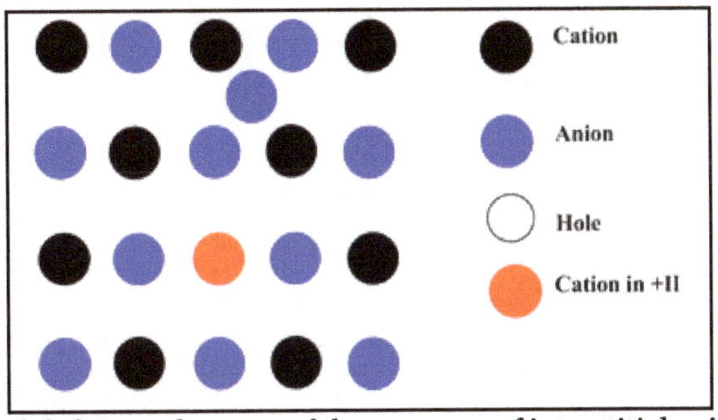

Metal deficiency because of the presence of interstitial anion.

Covalent Bond

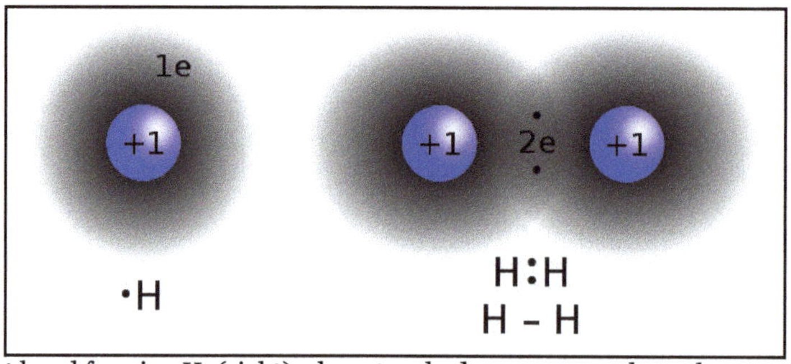

A covalent bond forming H_2 (right) where two hydrogen atoms share the two electrons.

A covalent bond, also called a molecular bond, is a chemical bond that involves the sharing of electron pairs between atoms. These electron pairs are known as shared pairs or bonding pairs, and the stable balance of attractive and repulsive forces between atoms, when they share electrons, is known as covalent bonding. For many molecules, the sharing of electrons allows each atom to attain the equivalent of a full outer shell, corresponding to a stable electronic configuration.

Covalent bonding includes many kinds of interactions, including σ-bonding, π-bonding, metal-to-metal bonding, agostic interactions, bent bonds, and three-center two-electron bonds. The term *covalent bond* dates from 1939. The prefix *co-* means *jointly, associated in action, partnered to a lesser degree,* etc.; thus a "co-valent bond", in essence, means that the atoms share "valence", such as is discussed in valence bond theory.

In the molecule H_2, the hydrogen atoms share the two electrons via covalent bonding. Covalency is greatest between atoms of similar electronegativities. Thus, covalent bonding does not necessarily require that the two atoms be of the same elements, only that they be of comparable electronegativity. Covalent bonding that entails sharing of electrons over more than two atoms is said to be delocalized.

History

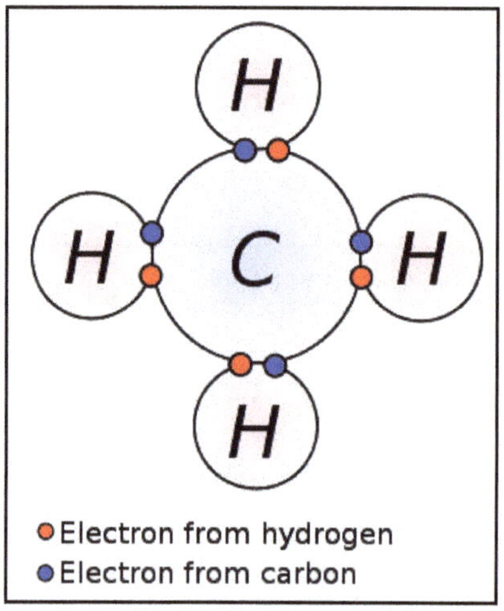

Early concepts in covalent bonding arose from this kind of image of the molecule of methane. Covalent bonding is implied in the Lewis structure by indicating electrons shared between atoms.

The term *covalence* in regard to bonding was first used in 1919 by Irving Langmuir in a *Journal of the American Chemical Society* article entitled "The Arrangement of Electrons in Atoms and Molecules". Langmuir wrote that "we shall denote by the term *covalence* the number of pairs of electrons that a given atom shares with its neighbors."

The idea of covalent bonding can be traced several years before 1919 to Gilbert N. Lewis, who in 1916 described the sharing of electron pairs between atoms. He introduced the *Lewis notation* or *electron dot notation* or *Lewis dot structure*, in which valence electrons (those in the outer shell) are represented as dots around the atomic symbols. Pairs of electrons located between atoms represent covalent bonds. Multiple pairs represent multiple bonds, such as double bonds and triple bonds. An alternative form of representation, not shown here, has bond-forming electron pairs represented as solid lines.

Lewis proposed that an atom forms enough covalent bonds to form a full (or closed) outer electron shell. In the diagram of methane shown here, the carbon atom has a valence of four and is, therefore, surrounded by eight electrons (the octet rule), four from the carbon itself and four from the hydrogens bonded to it. Each hydrogen has a valence of one and is surrounded by two electrons (a duet rule) – its own one electron plus one from the carbon. The numbers of electrons correspond to full shells in the quantum theory of the atom; the outer shell of a carbon atom is the $n = 2$ shell, which can hold eight electrons, whereas the outer (and only) shell of a hydrogen atom is the $n = 1$ shell, which can hold only two.

While the idea of shared electron pairs provides an effective qualitative picture of covalent bonding, quantum mechanics is needed to understand the nature of these bonds and predict the structures and properties of simple molecules. Walter Heitler and Fritz London are credited with the first successful quantum mechanical explanation of a chemical bond (molecular hydrogen) in 1927. Their work was based on the valence bond model, which assumes that a chemical bond is formed when there is good overlap between the atomic orbitals of participating atoms.

Types of Covalent Bonds

Atomic orbitals (except for s orbitals) have specific directional properties leading to different types of covalent bonds. Sigma (σ) bonds are the strongest covalent bonds and are due to head-on overlapping of orbitals on two different atoms. A single bond is usually a σ bond. Pi (π) bonds are weaker and are due to lateral overlap between p (or d) orbitals. A double bond between two given atoms consists of one σ and one π bond, and a triple bond is one σ and two π bonds.

Covalent bonds are also affected by the electronegativity of the connected atoms which determines the chemical polarity of the bond. Two atoms with equal electronegativity will make nonpolar covalent bonds such as H–H. An unequal relationship creates a polar covalent bond such as with H–Cl. However polarity also requires geometric asymmetry, or else dipoles may cancel out resulting in a non-polar molecule.

Covalent Structures

There are several types of structures for covalent substances, including individual molecules, molecular structures, macromolecular structures and giant covalent structures. Individual molecules have strong bonds that hold the atoms together, but there are negligible forces of attraction between molecules. Such covalent substances are usually gases, for example, HCl, SO_2, CO_2, and CH_4. In molecular structures, there are weak forces of attraction. Such covalent substances are low-boiling-temperature liquids (such as ethanol), and low-melting-temperature solids (such as iodine and solid CO_2). Macromolecular structures have large numbers of atoms linked by covalent bonds in chains, including synthetic polymers such as polyethylene and nylon, and biopolymers such as proteins and starch. Network covalent structures (or giant covalent structures) contain large numbers of atoms linked in sheets (such as graphite), or 3-dimensional structures (such as diamond and quartz). These substances have high melting and boiling points, are frequently brittle, and tend to have high electrical resistivity. Elements that have high electronegativity, and the ability to form three or four electron pair bonds, often form such large macromolecular structures.

One- and Three-electron Bonds

Bonds with one or three electrons can be found in radical species, which have an odd number of electrons. The simplest example of a 1-electron bond is found in the dihydrogen cation, H_2^+. One-electron bonds often have about half the bond energy of a 2-electron bond, and are therefore called "half bonds". However, there are exceptions: in the case of dilithium, the bond is actually stronger for the 1-electron Li_2^+ than for the 2-electron Li_2. This exception can be explained in terms of hybridization and inner-shell effects.

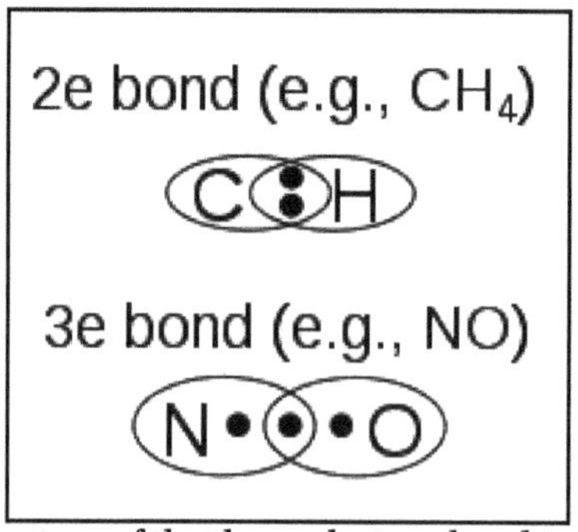

Comparison of the electronic structure of the three-electron bond to the conventional covalent bond.

The simplest example of three-electron bonding can be found in the helium dimer cation, He_2^+. It is considered a "half bond" because it consists of only one shared electron (rather than two); in molecular orbital terms, the third electron is in an anti-bonding orbital which cancels out half of the bond formed by the other two electrons. Another example of a molecule containing a 3-electron bond, in addition to two 2-electron bonds, is nitric oxide, NO. The oxygen molecule, O_2 can also be regarded as having two 3-electron bonds and one 2-electron bond, which accounts for its paramagnetism and its formal bond order of 2. Chlorine dioxide and its heavier analogues bromine dioxide and iodine dioxide also contain three-electron bonds.

Molecules with odd-electron bonds are usually highly reactive. These types of bond are only stable between atoms with similar electronegativities.

Resonance

There are situations whereby a single Lewis structure is insufficient to explain the electron configuration in a molecule, hence a superposition of structures are needed. The same two atoms in such molecules can be bonded differently in different structures (a single bond in one, a double bond in another, or even none at all), resulting in a non-integer bond order. The nitrate ion is one such example with three equivalent structures. The bond between the nitrogen and each oxygen is a double bond in one structure and a single bond in the other two, so that the average bond order for each N–O interaction is $\frac{2+1+1}{3}=\frac{4}{3}$.

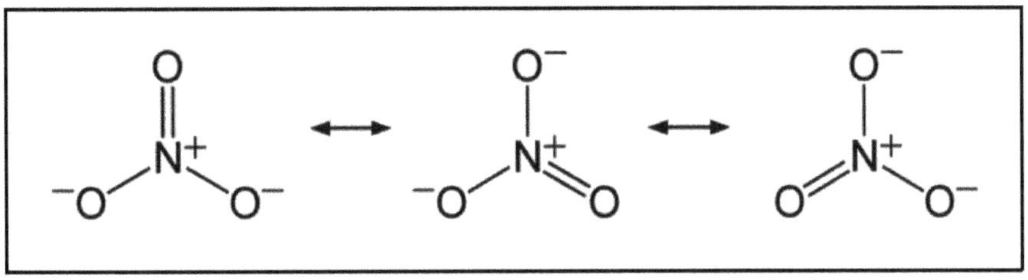

Aromaticity

In organic chemistry, when a molecule with a planar ring obeys Hückel's rule, where the number of π electrons fit the formula $4n + 2$ (where n is an integer), it attains extra stability and symmetry. In benzene, the prototypical aromatic compound, there are 6 π bonding electrons ($n = 1$, $4n + 2 = 6$). These occupy three delocalized π molecular orbitals (molecular orbital theory) or form conjugate π bonds in two resonance structures that linearly combine (valence bond theory), creating a regular hexagon exhibiting a greater stabilization than the hypothetical 1,3,5-cyclohexatriene.

In the case of heterocyclic aromatics and substituted benzenes, the electronegativity differences between different parts of the ring may dominate the chemical behaviour of aromatic ring bonds, which otherwise are equivalent.

Hypervalence

Certain molecules such as xenon difluoride and sulfur hexafluoride have higher co-ordination numbers than would be possible due to strictly covalent bonding according to the octet rule. This is explained by the three-center four-electron bond ("3c–4e") model which interprets the molecular wavefunction in terms of non-bonding highest occupied molecular orbitals in molecular orbital theory and ionic-covalent resonance in valence bond theory.

Electron-deficiency

In three-center two-electron bonds ("3c–2e") three atoms share two electrons in bonding. This type of bonding occurs in electron deficient compounds like diborane. Each such bond (2 per molecule in diborane) contains a pair of electrons which connect the boron atoms to each other in a banana shape, with a proton (nucleus of a hydrogen atom) in the middle of the bond, sharing electrons with both boron atoms. In certain cluster compounds, so-called four-center two-electron bonds also have been postulated.

Quantum Mechanical Description

After the development of quantum mechanics, two basic theories were proposed to provide a quantum description of chemical bonding: valence bond (VB) theory and molecular orbital (MO) theory. A more recent quantum description is given in terms of atomic contributions to the electronic density of states.

Covalency from Atomic Contribution to the Electronic Density of States

In COOP, COHP and BCOOP, evaluation of bond covalency is dependent on the basis set. To overcome this issue, an alternative formulation of the bond covalency can be provided in this way.

The center mass $cm(n,l,m_l,m_s)$ of an atomic orbital $|n,l,m_l,m_s\rangle$, with quantum numbers n, l, m_l, m_s, for atom A is defined as:

$$cm^{\mathrm{A}}(n,l,m_l,m_s) = \frac{\int_{E_0}^{E_1} E g^{\mathrm{A}}_{|n,l,m_l,m_s\rangle}(E)\,dE}{\int_{E_0}^{E_1} g^{\mathrm{A}}_{|n,l,m_l,m_s\rangle}(E)\,dE}$$

where $g^{\mathrm{A}}_{|n,l,m_l,m_s\rangle}(E)$ is the contribution of the atomic orbital $|n,l,m_l,m_s\rangle$ of the atom A to the total electronic density of states $g(E)$ of the solid:

$$g(E) = \sum_{\mathrm{A}} \sum_{n,l} \sum_{m_l,m_s} g^{\mathrm{A}}_{|n,l,m_l,m_s\rangle}(E)$$

where the outer sum runs over all atoms A of the unit cell. The energy window $[E_0,E_1]$ is chosen in such a way that it encompasses all relevant bands participating in the bond. If the range to select is unclear, it can be identified in practice by examining the molecular orbitals that describe the electron density along the considered bond.

The relative position $C_{n_{\mathrm{A}}l_{\mathrm{A}},n_{\mathrm{B}}l_{\mathrm{B}}}$ of the center mass of $|n_{\mathrm{A}},l_{\mathrm{A}}\rangle$ levels of atom A with respect to the center mass of $|n_{\mathrm{B}},l_{\mathrm{B}}\rangle$ levels of atom B is given as:

$$C_{n_{\mathrm{A}}l_{\mathrm{A}},n_{\mathrm{B}}l_{\mathrm{B}}} = -\left|cm^{\mathrm{A}}(n_{\mathrm{A}},l_{\mathrm{A}}) - cm^{\mathrm{B}}(n_{\mathrm{B}},l_{\mathrm{B}})\right|$$

where the contributions of the magnetic and spin quantum numbers are summed. According to this definition, the relative position of the A levels with respect to the B levels is:

$$C_{\mathrm{A,B}} = -\left|cm^{\mathrm{A}} - cm^{\mathrm{B}}\right|$$

where, for simplicity, we may omit the dependence from the principal quantum number n in the notation referring to $C_{n_{\mathrm{A}}l_{\mathrm{A}},n_{\mathrm{B}}l_{\mathrm{B}}}$.

In this formalism, the greater the value of $C_{\mathrm{A,B}}$, the higher the overlap of the selected atomic bands, and thus the electron density described by those orbitals gives a more covalent A–B bond. The quantity $C_{\mathrm{A,B}}$ is denoted as the *covalency* of the A–B bond, which is specified in the same units of the energy E.

A covalent bond is a chemical bonding formed by sharing of pairs of electrons between two atoms.

Lewis Theory (Octet rule):

Rule: A stable arrangement is attended when the atom is surrounded by eight electrons. This octet can be made up by own electrons and some electrons which are shared. Thus, an atom continues to form bonds until an octet of electrons is made. This is known as octet rule by Lewis.

Example:

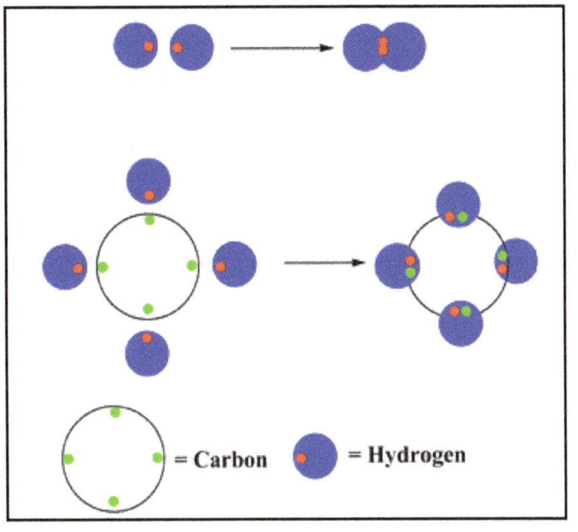

(i) Normally two electrons pairs up and forms a bond. Ex: H_2

(ii) For most atoms there will be a maximum of eight electrons in the valence shell (octet structure). Ex: CH_4

Exception of the octet rule:

(A) PF 3 maintains octet rule, while PF 5 does not. PF 5 has ten outer electrons.

(B) Molecules comprised of odd number of electrons do not follow octet rule. Ex: NO, ClO_2 , etc. This rule even does not explain the origin of paramagnetic character in O_2 molecule.

(C) For atoms like Be and B which have less than four outer electrons. Even if all the outer electrons used to form bonds an octet will not be resulted for Be and B.

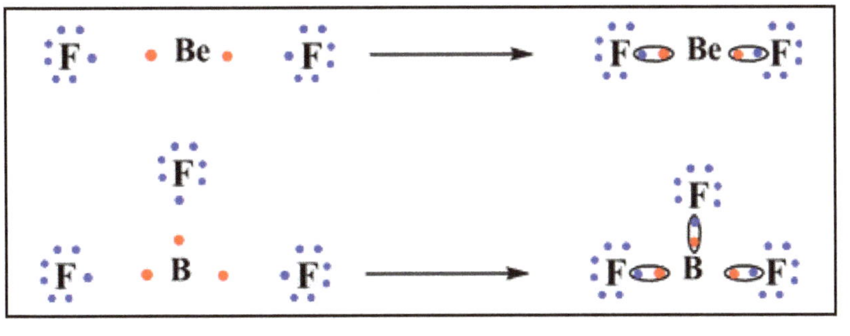

The octet rule has been further extended to include the exceptional part. Hence, the following rules are added to the octet rules.

(iii) For elements with available d orbitals, the valence shell can be expanded beyond an octet.

This rule explained the formation of PF 5. Here P has low-energy d orbitals.

(iv) The molecule will seek the lowest overall energy. This means the maximum number of bonds will form, that the strongest possible bonds will form, and that the arrangement of the atoms in molecule will be in such as to minimize adverse repulsion energy.

Valence Bond Theory:

Atoms containing unpaired electron(s) tend to combine with other atom(s) which also possess(es) unpaired electron(s). In this process unpaired electron(s) paired up and attained a stable noble gas configuration. Two electrons shared by two atoms constitute a bond. The number of bonds formed by an atom is same as the number of unpaired electrons in the ground state (lowest energy state). However, in some cases the atom may form more bonds than that of its acquired unpaired electron. This takes place by the excitation of an atom that promotes paired-electrons in the ground state to the excited state in a suitable empty orbital as unpaired electron.

The shape of a molecule is preliminary determined by the direction in which the participating orbitals point. Electrons in the valence shell of the original orbital are known as lone pairs.

A covalent bond is resulted by pairing of electrons from each atom. In this case the spin of the electrons must be opposite to each other.

Limitations:

(i) Relative stability of different molecules cannot be explained.

(ii) Paramagnetic nature of complexes or molecules cannot be explained.

(iii) Different shape of molecules cannot be explained.

Hybridization:

Hybridization is the mixing of atomic orbitals prior to overlap that change the partial distribution of orbitals.

Due to this hybridization mainly two different kind of bonds forms, (i) sigma (σ) and (ii) pi (π) bond.

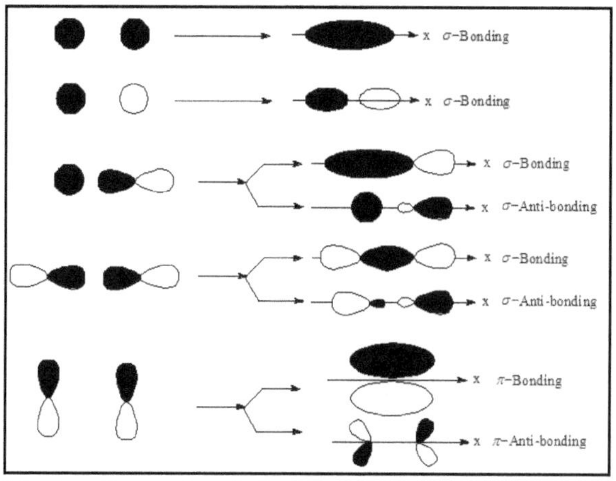

Sigma (σ) and pi (π) bonding and anti - bonding by the combination of s and p orbitals.

Sigma (σ) bond: A covalent bond established between two atoms having the maximum density of the electron cloud along the axis connecting the centers of the two participating atoms is called sigma (σ) bond.

Pi (π) bond: A bond is formed by the lateral overlap between two atomic orbitals possessing maximum electron density on both sides of the overlapping axis is known as pi (π) bond.

Therefore, from the definition it is clear that sigma (σ) bonds are more strong than pi (π) bonds.

Sigma (σ) bonds may arises from the overlap between s, p, and d orbitals, like (i) s - s orbitals, (ii) s - p prbitals, (iii) p - p orbitals, (iv) s - d orbitals, so on.

s orbitals are spherical, while p , d orbitals are dumbbell shaped. Hence, s - s over lap is weaker compared to s - p , p - p overlaps.

**Herein, it is important to note that due to hybridization of two atomic orbitals two molecular orbitals formed. One is called bonding orbital and the other is called anti - bonding orbital. Hence, n number of atomic orbitals forms n /2 bonding and n /2 anti - bonding orbitals.

Examples:

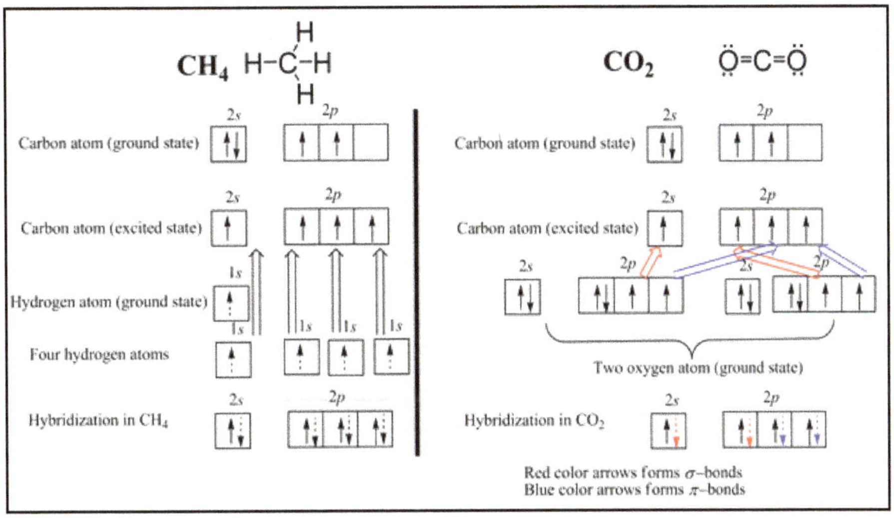

In the formation of CH_4 one s orbital and three p orbital of the central carbon atom participate.

Similarly, in the CO_2 molecule one s orbital and three p orbital of the central carbon atom participate but here two p orbitals of carbon atom forms two π - bonds. Therefore, CH_4 is sp^3 hybridized and CO_2 is sp^2 hybridized. NOTE: Orbitals participate in π - bond formation is not counted to the hybridization.

VSEPR Theory

Valence shell electron pair repulsion (VSEPR) theory is a model used in chemistry to predict the geometry of individual molecules from the number of electron pairs surrounding their central atoms. It is also named the Gillespie-Nyholm theory after its two main developers. The acronym "VSEPR" is pronounced either "*ves*-pur" or "vuh-*seh*-per."

Example of bent electron arrangement. Shows location of unpaired electrons, bonded atoms, and bond angles. (Water molecule) The bond angle for water is 104.5°.

The premise of VSEPR is that the valence electron pairs surrounding an atom tend to repel each other and will, therefore, adopt an arrangement that minimizes this repulsion, thus determining the molecule's geometry. Gillespie has emphasized that the electron-electron repulsion due to the Pauli exclusion principle is more important in determining molecular geometry than the electrostatic repulsion.

VSEPR theory is based on observable electron density rather than mathematical wave functions and hence unrelated to orbital hybridisation, although both address molecular shape. While it is mainly qualitative, VSEPR has a quantitative basis in quantum chemical topology (QCT) methods such as the electron localization function and the quantum theory of atoms in molecules (QTAIM).

History

The idea of a correlation between molecular geometry and number of valence electrons (both shared and unshared) was originally proposed in 1939 by Ryutaro Tsuchida in Japan, and was independently presented in a Bakerian Lecture in 1940 by Nevil Sidgwick and Herbert Powell of the University of Oxford. In 1957, Ronald Gillespie and Ronald Sydney Nyholm of University College London refined this concept into a more detailed theory, capable of choosing between various alternative geometries.

Overview

VSEPR theory is used to predict the arrangement of electron pairs around non-hydrogen atoms in molecules, especially simple and symmetric molecules, where these key, central atoms participate in bonding to two or more other atoms; the geometry of these key atoms and their non-bonding electron pairs in turn determine the geometry of the larger whole.

The number of electron pairs in the valence shell of a central atom is determined after drawing the Lewis structure of the molecule, and expanding it to show all bonding groups and lone pairs of electrons. In VSEPR theory, a double bond or triple bond are treated as a single bonding group. The sum of the number of atoms bonded to a central atom and the number of lone pairs formed by its nonbonding valence electrons is known as the central atom's steric number.

The electron pairs (or groups if multiple bonds are present) are assumed to lie on the surface of a sphere centered on the central atom and tend to occupy positions that minimize their mutual repulsions by maximizing the distance between them. The number of electron pairs (or groups), therefore, determines the overall geometry that they will adopt. For example, when there are two electron pairs surrounding the central atom, their mutual repulsion is minimal when they lie at opposite poles of the sphere. Therefore, the central atom is predicted to adopt a *linear* geometry. If there are 3 electron pairs surrounding the central atom, their repulsion is minimized by placing them at the vertices of an equilateral triangle centered on the atom. Therefore, the predicted geometry is *trigonal*. Likewise, for 4 electron pairs, the optimal arrangement is *tetrahedral*.

Degree of Repulsion

The overall geometry is further refined by distinguishing between *bonding* and *nonbonding* electron pairs. The bonding electron pair shared in a sigma bond with an adjacent atom lies further from the central atom than a nonbonding (lone) pair of that atom, which is held close to its positively charged nucleus. VSEPR theory therefore views repulsion by the lone pair to be greater than the repulsion by a bonding pair. As such, when a molecule has 2 interactions with different degrees of repulsion, VSEPR theory predicts the structure where lone pairs occupy positions that allow them to experience less repulsion. Lone pair–lone pair (lp–lp) repulsions are considered stronger than lone pair–bonding pair (lp–bp) repulsions, which in turn are considered stronger than bonding pair–bonding pair (bp–bp) repulsions, distinctions that then guide decisions about overall geometry when 2 or more non-equivalent positions are possible. For instance, when 5 valence electron pairs surround a central atom, they adopt a trigonal bipyramidal molecular geometry with two collinear *axial* positions and three *equatorial* positions. An electron pair in an axial position has three close equatorial neighbors only 90° away and a fourth much farther at 180°, while an equatorial electron pair has only two adjacent pairs at 90° and two at 120°. The repulsion from the close neighbors at 90° is more important, so that the axial positions experience more repulsion than the equatorial positions; hence, when there are lone pairs, they tend to occupy equatorial positions.

The difference between lone pairs and bonding pairs may also be used to rationalize deviations from idealized geometries. For example, the H_2O molecule has four electron pairs in its valence shell: two lone pairs and two bond pairs. The four electron pairs are spread so as to point roughly towards the apices of a tetrahedron. However, the bond angle between the two O–H bonds is only 104.5°, rather than the 109.5° of a regular tetrahedron, because the two lone pairs (whose density or probability envelopes lie closer to the oxygen nucleus) exert a greater mutual repulsion than the two bond pairs.

An advanced-level explanation replaces the above distinction with two rules:

- Bent's rule: An electron pair of a more electropositive ligand constitutes greater repulsion. This explains why the Cl in $PClF_4$ prefers the equatorial position and why the bond angle in oxygen difluoride (103.8°) is smaller than that of water (104.5°). Lone pairs are then considered to be a special case of this rule, held by a "ghost ligand" in the limit of electropositivity.

- A higher bond order constitutes greater repulsion. This explains why in phosgene, the oxygen–chlorine bond angle (124.1°) is larger than the chlorine–chlorine bond angle (111.8°)

even though chlorine is more electropositive than oxygen. In the carbonate ion, all three bond angles are equivalent due to resonance.

AXE Method

The "AXE method" of electron counting is commonly used when applying the VSEPR theory. The *A* represents the central atom and always has an implied subscript one. The *X* represents each of ligands (atoms bonded to A). The *E* represents the number of lone electron *pairs* surrounding the central atom. The sum of *X* and *E* is known as the steric number.

Based on the steric number and distribution of *X*s and *E*s, VSEPR theory makes the predictions in the following tables. Note that the geometries are named according to the atomic positions only and not the electron arrangement. For example, the description of AX_2E_1 as a bent molecule means that the three atoms AX_2 are not in one straight line, although the lone pair helps to determine the geometry.

Steric number	Molecular geometry 0 lone pairs	Molecular geometry 1 lone pair	Molecular geometry 2 lone pairs	Molecular geometry 3 lone pairs
2	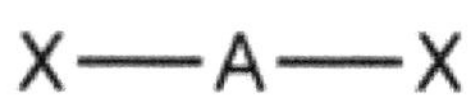 Linear (CO_2)			
3	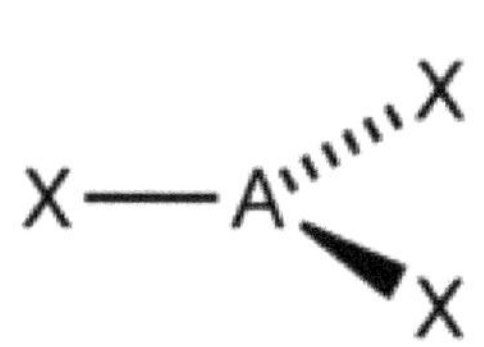 Trigonal planar (BCl_3)	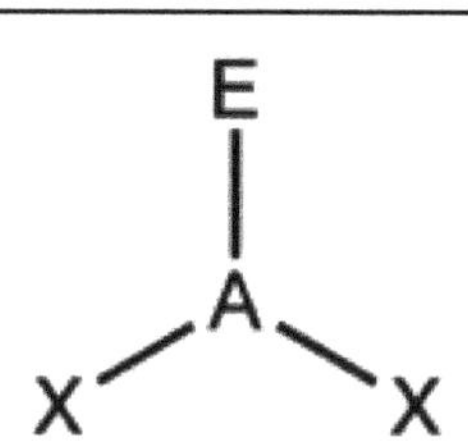 V-shape (SO_2)		
4	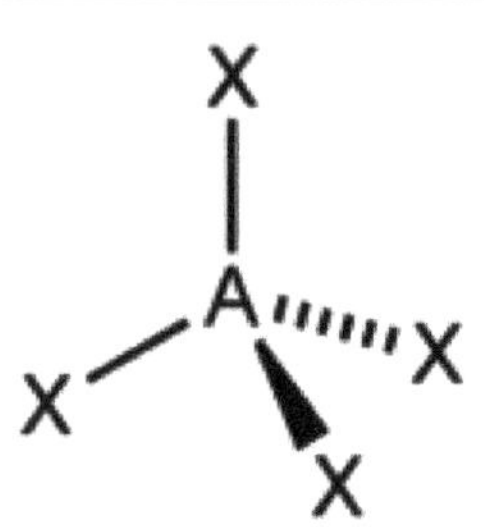 Tetrahedral (CH_4)	 Trigonal pyramidal (NH_3)	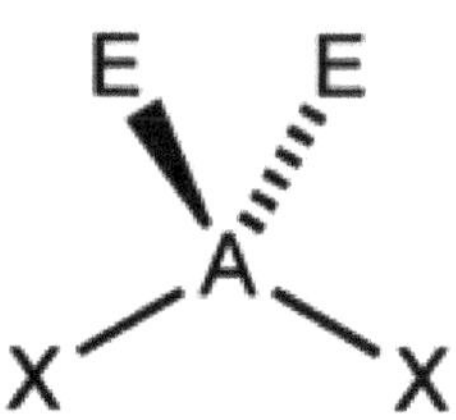 Bent (H_2O)	
5	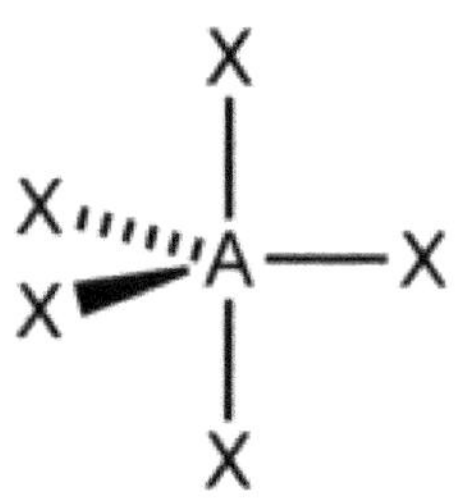 Trigonal bipyramidal (PCl_5)	 Seesaw (SF_4)	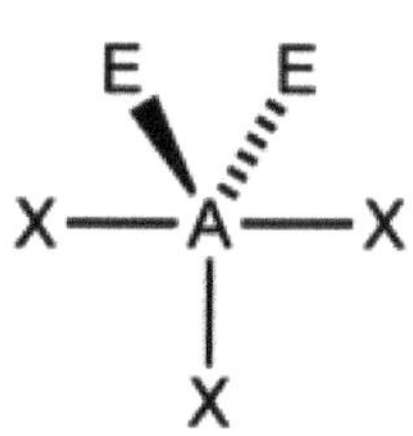 T-shaped (ClF_3)	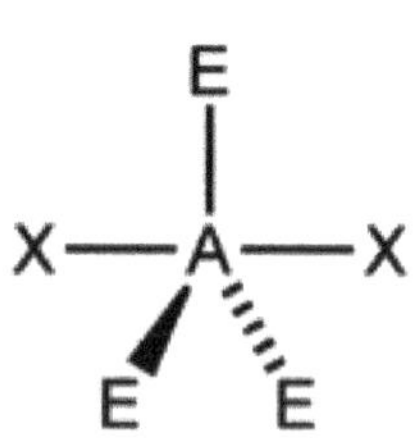 Linear (I– 3)

6	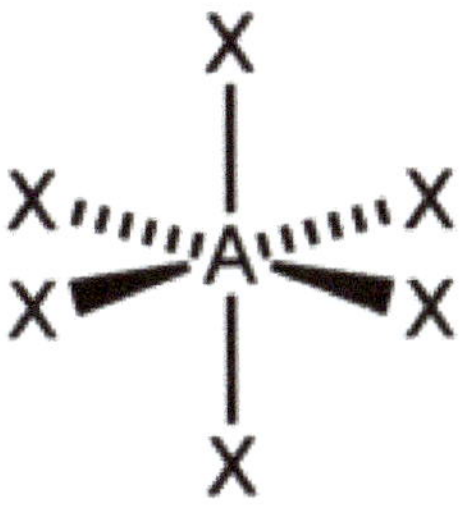 Octahedral (SF_6)	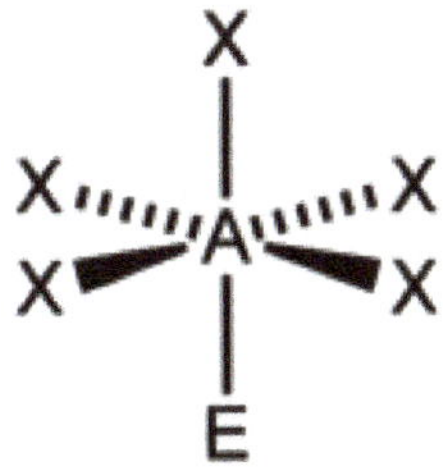 Square pyramidal (BrF_5)	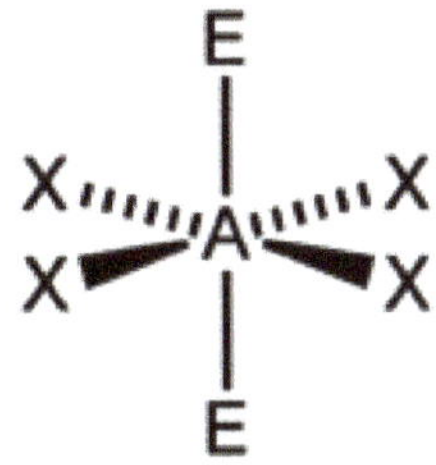 Square planar (XeF_4)	
7	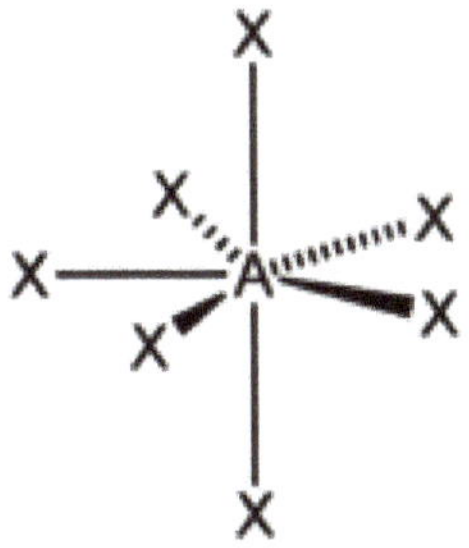 Pentagonal bipyramidal (IF_7)	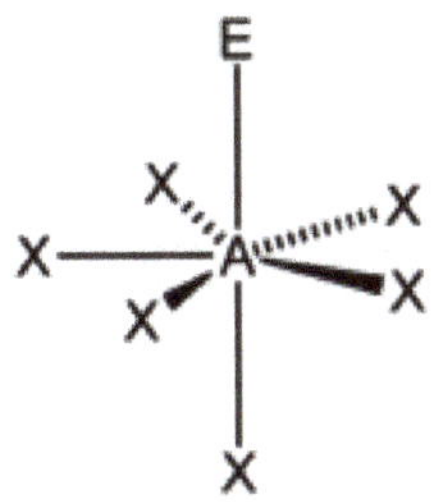 Pentagonal pyramidal ($XeOF_5^-$)	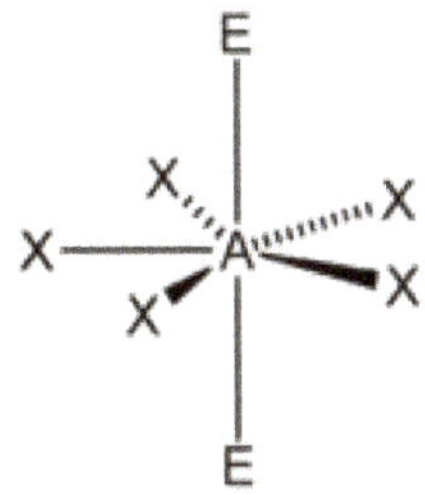 Pentagonal planar (XeF_5^-)	

Molecule type	Shape	Electron arrangement including lone pairs, shown in pale yellow	Geometry excluding lone pairs	Examples
AX_2E_0	Linear			$BeCl_2$, $HgCl_2$, CO_2
AX_2E_1	Bent		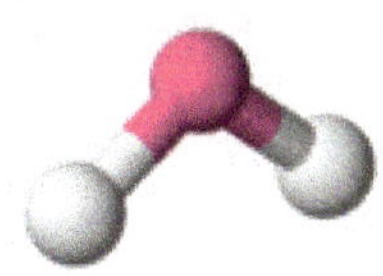	NO_2^-, SO_2, O_3, CCl_2
AX_2E_2	Bent	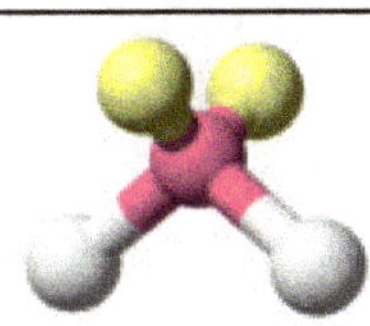	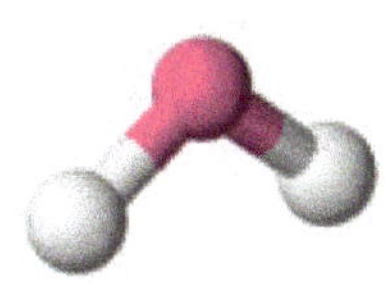	H_2O, OF_2
AX_2E_3	Linear	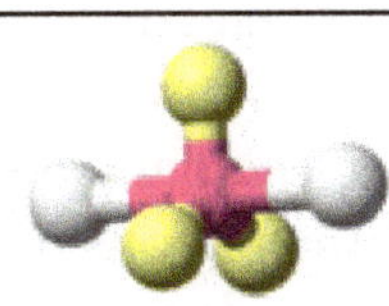	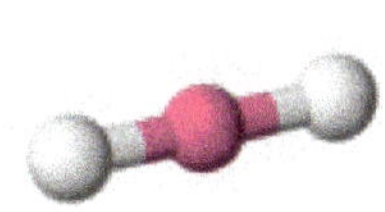	XeF_2, I_3^-, $XeCl_2$
AX_3E_0	Trigonal planar	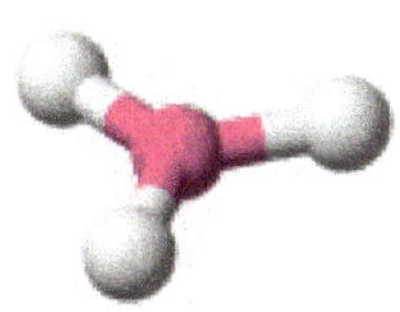	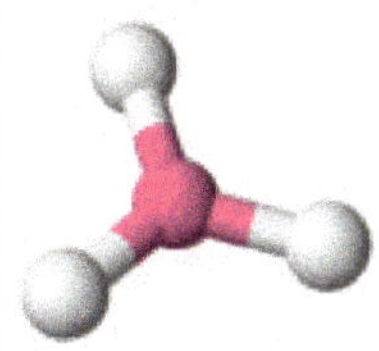	BF_3, CO_3^{2-}, NO_3^-, SO_3

AX_6E_0	Octahedral	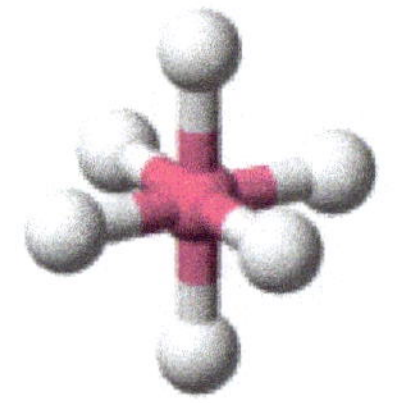	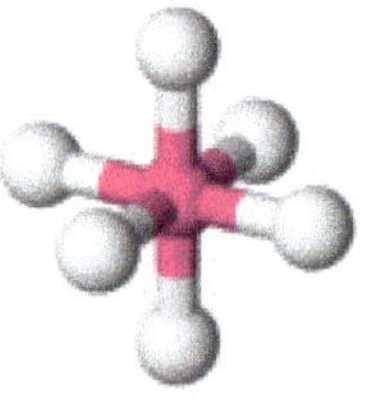	SF_6, WCl_6
AX_6E_1	Pentagonal pyramidal		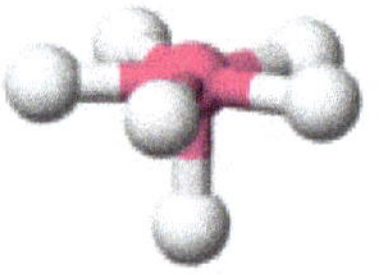	$XeOF_5^-$, IOF_5^{2-}

When the substituent (X) atoms are not all the same, the geometry is still approximately valid, but the bond angles may be slightly different from the ones where all the outside atoms are the same. For example, the double-bond carbons in alkenes like C_2H_4 are AX_3E_0, but the bond angles are not all exactly 120°. Likewise, $SOCl_2$ is AX_3E_1, but because the X substituents are not identical, the X–A–X angles are not all equal.

As a tool in predicting the geometry adopted with a given number of electron pairs, an often used physical demonstration of the principle of minimal electron pair repulsion utilizes inflated balloons. Through handling, balloons acquire a slight surface electrostatic charge that results in the adoption of roughly the same geometries when they are tied together at their stems as the corresponding number of electron pairs. For example, five balloons tied together adopt the trigonal bipyramidal geometry, just as do the five bonding pairs of a PCl_5 molecule (AX_5) or the two bonding and three non-bonding pairs of a XeF_2 molecule (AX_2E_3). The molecular geometry of the former is also trigonal bipyramidal, whereas that of the latter is linear.

Examples

The methane molecule (CH_4) is tetrahedral because there are four pairs of electrons. The four hydrogen atoms are positioned at the vertices of a tetrahedron, and the bond angle is $\cos^{-1}(-\frac{1}{3}) \approx 109° 28'$. This is referred to as an AX_4 type of molecule. As mentioned above, A represents the central atom and X represents an outer atom.

The ammonia molecule (NH_3) has three pairs of electrons involved in bonding, but there is a lone pair of electrons on the nitrogen atom. It is not bonded with another atom; however, it influences the overall shape through repulsions. As in methane above, there are four regions of electron density. Therefore, the overall orientation of the regions of electron density is tetrahedral. On the other hand, there are only three outer atoms. This is referred to as an AX_3E type molecule because the lone pair is represented by an E. By definition, the molecular shape or geometry describes the geometric arrangement of the atomic nuclei only, which is trigonal-pyramidal for NH_3.

Steric numbers of 7 or greater are possible, but are less common. The steric number of 7 occurs in iodine heptafluoride (IF_7); the base geometry for a steric number of 7 is pentagonal bipyramidal. The most common geometry for a steric number of 8 is a square antiprismatic geometry. Examples of this include the octacyanomolybdate ($Mo(CN)_8^{4-}$) and octafluorozirconate (ZrF_8^{4-}) anions.

The nonahydridorhenate ion (ReH_9^{2-}) in potassium nonahydridorhenate is a rare example of a compound with a steric number of 9, which has a tricapped trigonal prismatic geometry. Another example is the octafluoroxenate ion (XeF_8^{2-}) in nitrosonium octafluoroxenate(VI), although in this case one of the electron pairs is a lone pair, and therefore the molecule actually has a distorted square antiprismatic geometry.

Possible geometries for steric numbers of 10, 11, 12, or 14 are bicapped square antiprismatic (or bicapped dodecadeltahedral), octadecahedral, icosahedral, and bicapped hexagonal antiprismatic, respectively. No compounds with steric numbers this high involving monodentate ligands exist, and those involving multidentate ligands can often be analysed more simply as complexes with lower steric numbers when some multidentate ligands are treated as a unit.

Exceptions

There are groups of compounds where VSEPR fails to predict the correct geometry.

Some AX_2E_0 Molecules

The gas phase structures of the triatomic halides of the heavier members of group 2, (i.e., calcium, strontium and barium halides, MX_2), are not linear as predicted but are bent, (approximate X–M–X angles: CaF_2, 145°; SrF_2, 120°; BaF_2, 108°; $SrCl_2$, 130°; $BaCl_2$, 115°; $BaBr_2$, 115°; BaI_2, 105°). It has been proposed by Gillespie that this is caused by interaction of the ligands with the electron core of the metal atom, polarising it so that the inner shell is not spherically symmetric, thus influencing the molecular geometry. Ab initio calculations have been cited to propose that contributions from d orbitals in the shell below the valence shell are responsible. Disilynes are also bent, despite having no lone pairs.

Some AX_2E_2 Molecules

One example of the AX_2E_2 geometry is molecular lithium oxide, Li_2O, a linear rather than bent structure, which is ascribed to its bonds being essentially ionic and the strong lithium-lithium repulsion that results. Another example is $O(SiH_3)_2$ with an Si–O–Si angle of 144.1°, which compares to the angles in Cl_2O (110.9°), $(CH_3)_2O$ (111.7°), and $N(CH_3)_3$ (110.9°). Gillespie and Robinson rationalize the Si–O–Si bond angle based on the observed ability of a ligand's lone pair to most greatly repel other electron pairs when the ligand electronegativity is greater than or equal to that of the central atom. In $O(SiH_3)_2$, the central atom is more electronegative, and the lone pairs are less localized and more weakly repulsive. The larger Si–O–Si bond angle results from this and strong ligand-ligand repulsion by the relatively large $-SiH_3$ ligand. Burford et al showed through X-ray diffraction studies that $Cl_3Al–O–PCl_3$ has a linear Al–O–P bond angle and is therefore a non-VSEPR molecule.

Some AX_6E_1 and AX_8E_1 Molecules

Some AX_6E_1 molecules, e.g. xenon hexafluoride (XeF_6) and the Te(IV) and Bi(III) anions, $TeCl_6^{2-}$, $TeBr_6^{2-}$, $BiCl_6^{3-}$, $BiBr_6^{3-}$ and BiI_6^{3-}, are octahedra, rather than pentagonal pyramids, and the lone pair does not affect the geometry to the degree predicted by VSEPR. One rationalization is that steric crowding of the ligands allows little or no room for the non-bonding lone pair; another rationalization is the inert pair effect.

Xenon hexafluoride, which has a distorted octahedral geometry.

Transition Metal Molecules

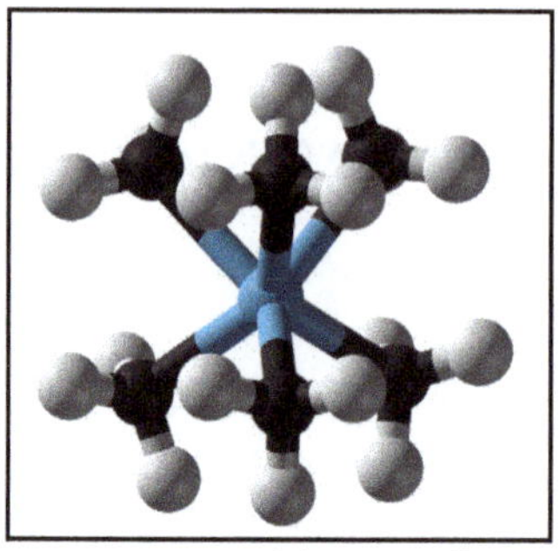

Hexamethyltungsten, a transition metal compound whose geometry is different from main group coordination.

Many transition metal compounds have unusual geometries, which can be ascribed to ligand bonding interaction with the d subshell and to absence of valence shell lone pairs. Gillespie suggested that this interaction can be weak or strong. Weak interaction is dealt with by the Kepert model, while strong interaction produces bonding pairs that also occupy the respective antipodal points of the sphere. This is similar to predictions based on sd hybrid orbitals using the VALBOND theory. The repulsion of these bidirectional bonding pairs leads to a different prediction of shapes.

Molecule type	Shape	Geometry	Examples
AX_2	Bent	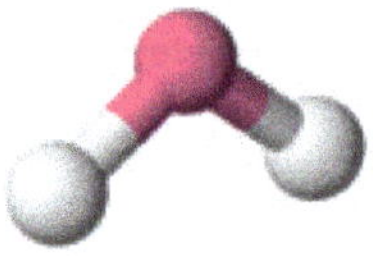	VO_2^+
AX_3	Trigonal pyramidal	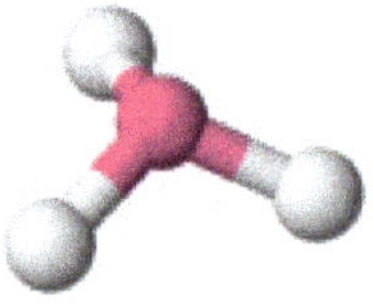	CrO_3
AX_4	Tetrahedral	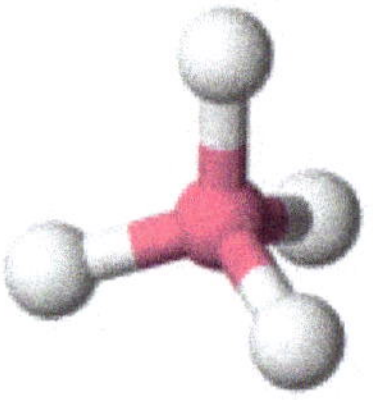	$TiCl_4$

AX_5	Square pyramidal	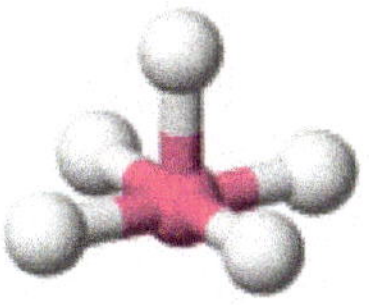	$Ta(CH_3)_5$
AX_6	Trigonal prismatic	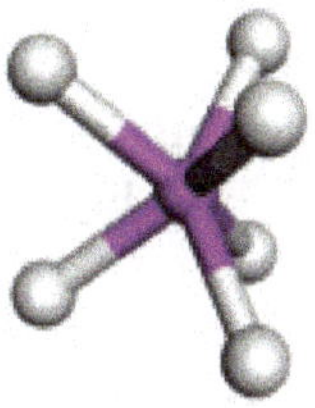	$W(CH_3)_6$

The square planar shape associated with a d^8 electronic configuration is an exception to the Kepert model. This can be rationalized by considering the increased crystal field stabilization energy as compared to a tetrahedral geometry.

Odd-electron Molecules

The VSEPR theory can be extended to molecules with an odd number of electrons by treating the unpaired electron as a "half electron pair" — for example, Gillespie and Nyholm suggested that the decrease in the bond angle in the series NO_2^+ (180°), NO_2 (134°), NO_2^- (115°) indicates that a given set of bonding electron pairs exert a weaker repulsion on a single non-bonding electron than on a pair of non-bonding electrons. In effect, they considered nitrogen dioxide as an $AX_2E_{0.5}$ molecule, with a geometry intermediate between NO_2^+ and NO_2^-. Similarly, chlorine dioxide (ClO_2) is an $AX_2E_{1.5}$ molecule, with a geometry intermediate between ClO_2^+ and ClO_2^-.

Finally, the methyl radical (CH_3) is predicted to be trigonal pyramidal like the methyl anion (CH_3^-), but with a larger bond angle (as in the trigonal planar methyl cation (CH_3^+)). However, in this case, the VSEPR prediction is not quite true, as CH_3 is actually planar, although its distortion to a pyramidal geometry requires very little energy.

Valence Shell Electron Pair Repulsion (VSEPR) Theory:

(i) Lone pairs of electrons (lp) repel each other more strongly than that of bond pair (bp) of electrons. The decreasing order of repulsion is lp - lp > lp - bp > bp - bp.

Examples:

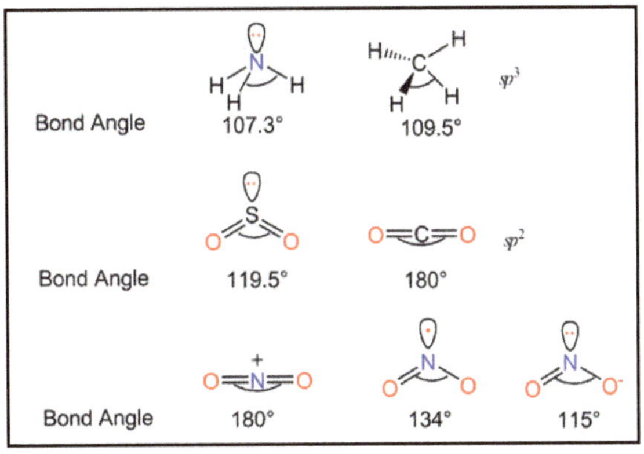

- In CH_4 only bp-bp repulsion exist while, in NH_3 two types of repulsive force present, lp - bp repulsion and bp - bp repulsion. lp - bp repulsion dominates over bp - bp repulsion. Because of this the bonds in NH_3 is more compressed than that in CH_4 which provides lower band angle in NH_3.
- In CO_2 the central C atom does not occupy any lone pair of electron, hence in CO_2 only bp - bp repulsion exists. The central S atom in SO_2 contains one lone pair of electron. Therefore, in SO_2 lp - bp repulsion and bp - bp repulsion present. As we know that lp - bp repulsion dominates over bp - bp repulsion, the O-S-O bonds is compressed by dominating lp - bp.
- NO_2^+ does not contain any paired or unpaired electron on the central N atom. Here, only bp - bp repulsion exists. In NO_2 the central N atom contains one unpaired electron, while in NO_2^- the N atom contains one lone pair of electrons. As in NO_2 only one unpaired electron present, the lp - bp repulsion is lower compared to lp - bp repulsion in NO_2^-. Hence, the dominating lp - bp is more pronounced in NO_2^- compared to NO_2.

(ii) Repulsion between bond pairs decreases as the electronegativity of the atom bound to the central atom increases.

A bond is formed by the sharing of electrons. When the difference in electronegativity between the two connecting atoms increases the electron pair is shifted towards the more electronegative atom. With increases of electronegativity this shift is more to the electronegative atom. This decreases the repulsion effect between two bond pairs.

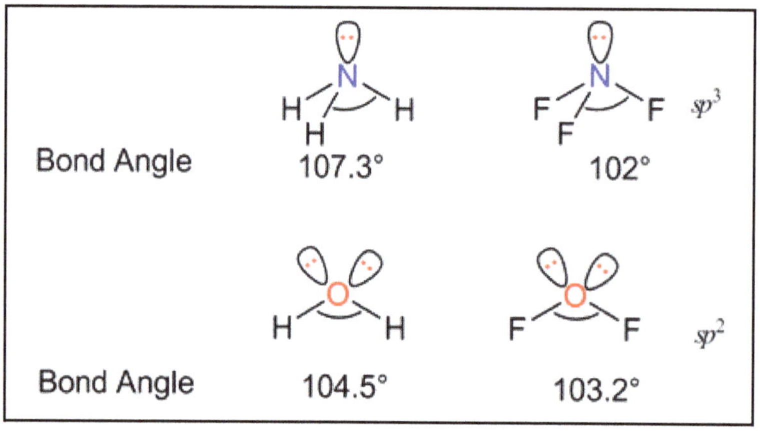

- F is more electronegative that that of H. Therefore, compared to NH_3, in NF_3 the bond forming lone pair of electron is more shifted to the F atom. This decreases bp - bp repulsion. Therefore, in NF_3 the lp - bp repulsion is more predominating that that of NH_3 and makes F-N-F bond angle 102° compared to H-N-H bond angle 107.3°.
- Same (above) argument is valid for having higher H-O-H bond angle in H_2O compared to F-O-F bond angle in OF_2.

(iii) Electronpairs in filled shell repel stronger that electronpairs in incomplete shell.

When the electron pair containing atom contains low-energy lying suitable vacant orbitals the electron pair of electrons can undergoes diffusion to the vacant orbitals. Hence, the lp - bp repulsion diminishes dramatically and bp - bp repulsion dominates. In this case, the bond angles around the central metal increase.

Example:

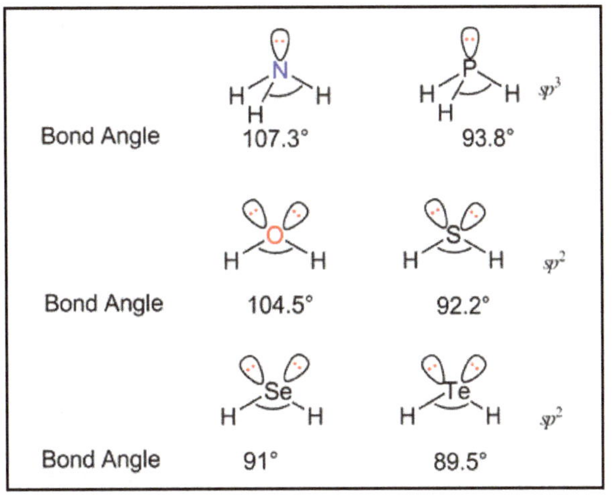

- On going from O to Te the size of the central atom increases. Therefore, diffusion is more predominating which further decreases lp - bp repulsion.

(IV) Lone pair of electron can be transferred from a filled shell to a energetically suitable empty shell of the other bonded atom.

Due to transfer of lone pair of electron the bond accumulate some sort of double bond character. This enhances the bp - bp repulsion and consequently, causing a bigger bond angle.

Example:

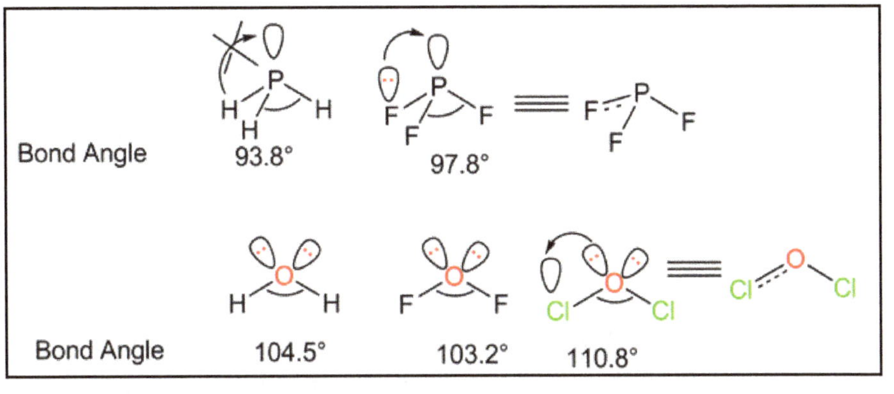

- P and Cl contain low-lying vacant d orbitals. Lone pair of electron from the filled orbitals of F or O can be transferred to the vacant orbital. This causes a partial - multiple bond that enhances the bp - bp repulsion. Hence, the bond angle increases.

(V) Multiple bond orbitals repel each other more strongly than single bond orbitals.

Examples:

O = PF_3 (F-P-F bond angle = 103°), while in O = PBr_3 (Br-P-Br bond angle = 108°).

Molecular Orbital Theory

In chemistry, molecular orbital (MO) theory is a method for determining molecular structure in which electrons are not assigned to individual bonds between atoms, but are treated as moving under the influence of the nuclei in the whole molecule. The spatial and energetic properties of electrons within atoms are fixed by quantum mechanics to form orbitals that contain these electrons. While atomic orbitals contain electrons ascribed to a single atom, molecular orbitals, which surround a number of atoms in a molecule, contain valence electrons between atoms. Molecular orbital theory, which was proposed in the early twentieth century, revolutionized the study of bonding by approximating the positions of bonded electrons—the molecular orbitals—as linear combinations of atomic orbitals (LCAO). These approximations are now made by applying the density functional theory (DFT) or Hartree–Fock (HF) models to the Schrödinger equation.

Quantitative Applications

In this theory, each molecule has a set of molecular orbitals, in which it is assumed that the molecular orbital wave function ψ_j can be written as a simple weighted sum of the n constituent atomic orbitals χ_i, according to the following equation:

$$\psi_j = \sum_{i=1}^{n} c_{ij}\chi_i$$

One may determine c_{ij} coefficients numerically by substituting this equation into the Schrödinger's equation and applying the variational principle. The variational principle is a mathematical technique used in quantum mechanics to build up the coefficients of each atomic orbital basis. A larger coefficient means that the orbital basis is composed more of that particular contributing atomic orbital—hence, the molecular orbital is best characterized by that type. This method of quantifying orbital contribution as Linear Combinations of Atomic Orbitals is used in computational chemistry. An additional unitary transformation can be applied on the system to accelerate the convergence in some computational schemes. Molecular orbital theory was seen as a competitor to valence bond theory in the 1930s, before it was realized that the two methods are closely related and that when extended they become equivalent.

History

Molecular orbital theory was developed, in the years after valence bond theory had been established (1927), primarily through the efforts of Friedrich Hund, Robert Mulliken, John C. Slater, and John Lennard-Jones. MO theory was originally called the Hund-Mulliken theory. According to German physicist and physical chemist Erich Hückel, the first quantitative use of molecular orbital theory was the 1929 paper of Lennard-Jones. This paper notably predicted a triplet ground state for the dioxygen molecule which explained its paramagnetism before valence bond theory, which came up with its own explanation in 1931. The word *orbital* was introduced by Mulliken in 1932. By 1933, the molecular orbital theory had been accepted as a valid and useful theory.

Erich Hückel applied molecular orbital theory to unsaturated hydrocarbon molecules starting in 1931 with his Hückel molecular orbital (HMO) method for the determination of MO energies for pi electrons, which he applied to conjugated and aromatic hydrocarbons. This method provided an explanation of the stability of molecules with six pi-electrons such as benzene.

The first accurate calculation of a molecular orbital wavefunction was that made by Charles Coulson in 1938 on the hydrogen molecule. By 1950, molecular orbitals were completely defined as eigenfunctions (wave functions) of the self-consistent field Hamiltonian and it was at this point that molecular orbital theory became fully rigorous and consistent. This rigorous approach is known as the Hartree–Fock method for molecules although it had its origins in calculations on atoms. In calculations on molecules, the molecular orbitals are expanded in terms of an atomic orbital basis set, leading to the Roothaan equations. This led to the development of many ab initio quantum chemistry methods. In parallel, molecular orbital theory was applied in a more approximate manner using some empirically derived parameters in methods now known as semi-empirical quantum chemistry methods.

The success of Molecular Orbital Theory also spawned ligand field theory, which was developed during the 1930s and 1940s as an alternative to crystal field theory.

Types of Orbitals

Molecular orbital (MO) theory uses a linear combination of atomic orbitals (LCAO) to represent molecular orbitals resulting from bonds between atoms. These are often divided into *bonding* orbitals, anti-bonding orbitals, and non-bonding orbitals. A bonding orbital concentrates electron density in the region *between* a given pair of atoms, so that its electron density will tend to attract each of the two nuclei toward the other and hold the two atoms together. An anti-bonding orbital concentrates electron density "behind" each nucleus (i.e. on the side of each atom which is farthest from the other atom), and so tends to pull each of the two nuclei away from the other and actually weaken the bond between the two nuclei. Electrons in non-bonding orbitals tend to be associated with atomic orbitals that do not interact positively or negatively with one another, and electrons in these orbitals neither contribute to nor detract from bond strength.

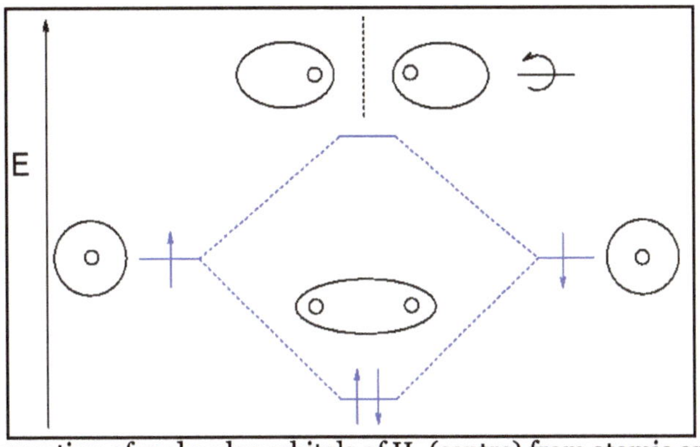

MO diagram showing the formation of molecular orbitals of H_2 (centre) from atomic orbitals of two H atoms. The lower-energy MO is bonding with electron density concentrated between the two H nuclei. The higher-energy MO is anti-bonding with electron density concentrated behind each H nucleus.

Molecular orbitals are further divided according to the types of atomic orbitals they are formed from. Chemical substances will form bonding interactions if their orbitals become lower in energy when they interact with each other. Different bonding orbitals are distinguished that differ by electron configuration (electron cloud shape) and by energy levels.

The molecular orbitals of a molecule can be illustrated in molecular orbital diagrams.

Overview

MO theory provides a global, delocalized perspective on chemical bonding. In MO theory, *any* electron in a molecule may be found *anywhere* in the molecule, since quantum conditions allow electrons to travel under the influence of an arbitrarily large number of nuclei, as long as they are in eigenstates permitted by certain quantum rules. Thus, when excited with the requisite amount of energy through high-frequency light or other means, electrons can transition to higher-energy molecular orbitals. For instance, in the simple case of a hydrogen diatomic molecule, promotion of a single electron from a bonding orbital to an antibonding orbital can occur under UV radiation. This promotion weakens the bond between the two hydrogen atoms and can lead to photodissociation—the breaking of a chemical bond due to the absorption of light.

Although in MO theory *some* molecular orbitals may hold electrons that are more localized between specific pairs of molecular atoms, *other* orbitals may hold electrons that are spread more uniformly over the molecule. Thus, overall, bonding is far more delocalized in MO theory, which makes it more applicable to resonant molecules that have equivalent non-integer bond orders than valence bond (VB) theory. This makes MO theory more useful for the description of extended systems.

An example is the MO description of benzene, C_6H_6, which is an aromatic hexagonal ring of six carbon atoms and three double bonds. In this molecule, 24 of the 30 total valence bonding electrons—24 coming from carbon atoms and 6 coming from hydrogen atoms—are located in 12 σ (sigma) bonding orbitals, which are located mostly between pairs of atoms (C-C or C-H), similarly to the electrons in the valence bond description. However, in benzene the remaining six bonding electrons are located in three π (pi) molecular bonding orbitals that are delocalized around the ring. Two of these electrons are in an MO that has equal orbital contributions from all six atoms. The other four electrons are in orbitals with vertical nodes at right angles to each other. As in the VB theory, all of these six delocalized π electrons reside in a larger space that exists above and below the ring plane. All carbon-carbon bonds in benzene are chemically equivalent. In MO theory this is a direct consequence of the fact that the three molecular π orbitals combine and evenly spread the extra six electrons over six carbon atoms.

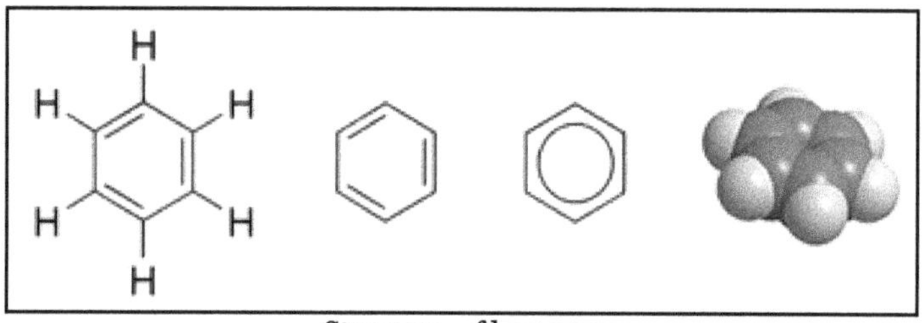

Structure of benzene.

In molecules such as methane, CH_4, the eight valence electrons are found in four MOs that are spread out over all five atoms. However, it is possible to transform the MOs into four localized

sp^3 orbitals. Linus Pauling, in 1931, hybridized the carbon 2s and 2p orbitals so that they pointed directly at the hydrogen 1s basis functions and featured maximal overlap. However, the delocalized MO description is more appropriate for predicting ionization energies and the positions of spectral absorption bands. When methane is ionized, a single electron is taken from the valence MOs, which can come from the s bonding or the triply degenerate p bonding levels, yielding two ionization energies. In comparison, the explanation in VB theory is more complicated. When one electron is removed from an sp^3 orbital, resonance is invoked between four valence bond structures, each of which has a single one-electron bond and three two-electron bonds. Triply degenerate T_2 and A_1 ionized states (CH_4^+) are produced from different linear combinations of these four structures. The difference in energy between the ionized and ground state gives the two ionization energies.

As in benzene, in substances such as beta carotene, chlorophyll, or heme, some electrons in the π orbitals are spread out in molecular orbitals over long distances in a molecule, resulting in light absorption in lower energies (the visible spectrum), which accounts for the characteristic colours of these substances. This and other spectroscopic data for molecules are well explained in MO theory, with an emphasis on electronic states associated with multicenter orbitals, including mixing of orbitals premised on principles of orbital symmetry matching. The same MO principles also naturally explain some electrical phenomena, such as high electrical conductivity in the planar direction of the hexagonal atomic sheets that exist in graphite. This results from continuous band overlap of half-filled p orbitals and explains electrical conduction. MO theory recognizes that some electrons in the graphite atomic sheets are completely delocalized over arbitrary distances, and reside in very large molecular orbitals that cover an entire graphite sheet, and some electrons are thus as free to move and therefore conduct electricity in the sheet plane, as if they resided in a metal.

In the Valence bond theory, the individuality of atomic properties of the constituting atoms forming a molecule is retained too some extent. Whereas, in the molecular orbital theory it is considered that molecular orbitals are formed by the linear combination of the two constituting atomic orbitals´ wave functions.

Let us consider, ψ(A) and ψ(B) are two atomic orbitals which are combining to form molecular orbitals for A - B bond. Molecular orbitals will also be wave functions anf they can be represented as;

$\psi(MO_b) = \psi(A) + \psi(B)$; $y\ (MO_b)$ = bonding MO wave function

and

$\psi(MO_{ab}) = \psi(A) - \psi(B)$; $\psi(MO_{ab})$ = anti-bonding MO wave function

Conditions for the favorable combination of atomic orbitals:

(A) Atomic orbitals must have comparable energies.

(B) They must have comparable energies.

(C) They must have same symmetry with respect to the bonding molecular axis, i.e. if the atomic orbitals are perpendicular to each other there will be no overlap or linear combination between them.

Bond order:

Bond order (BO) of a bond is defined as;

BO = [No of electron in bonding MO - no of electron in anti - bonding MO]/2

Starting from hydrogen to fluorine atom the energy difference between the 2 s and 2 p orbitals increases due to increase in electronegativity. It is important to note that up to nitrogen, the energy difference between the 2s and 2p orbital is sufficient for sp type hybridization. Therefore, up to nitrogen nitrogen molecule the energy level follow the order 1σ , 1σ *, 2π , 2σ , 2σ *, 2π * (i.e. energy of 2σ *> 2π*) and for oxygen and nitrogen molecules the order is 1σ , 1σ *, 2σ , 2π , 2π *, 2σ *are (i.e. energy of 2σ *<2π *).

MOs of H_2 Molecule:

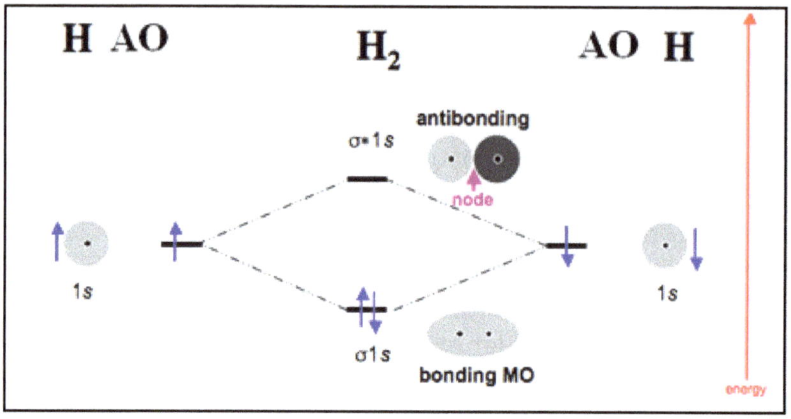

Bond Order = [2-0]/2 = 2/2 = 1

Here * represents anti-bonding orbital.

MOs of He_2 Molecule:

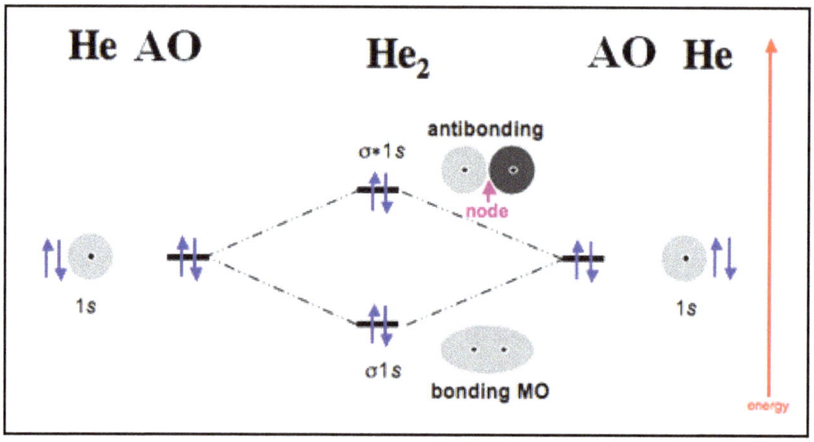

Bond Order = [2-2]/2 = 0/2 = 0.

As the BO is zero, there is no bond between He_2 molecule. This means He is monoatomic molecule.

MOs of Li_2 Molecule:

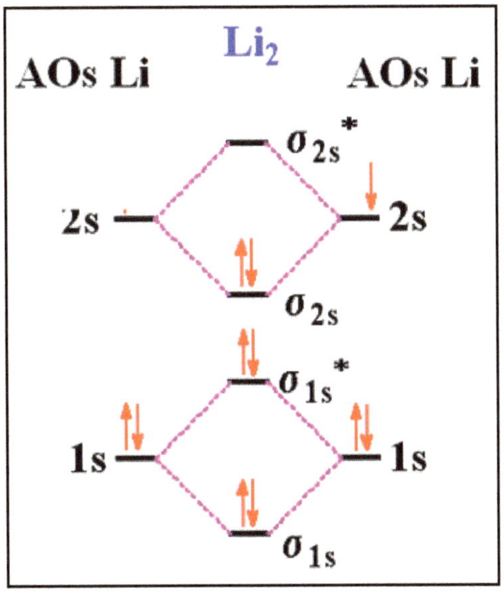

Bond Order = [4-2]/2 = 2/2 = 1

MOs of Be_2 Molecule:

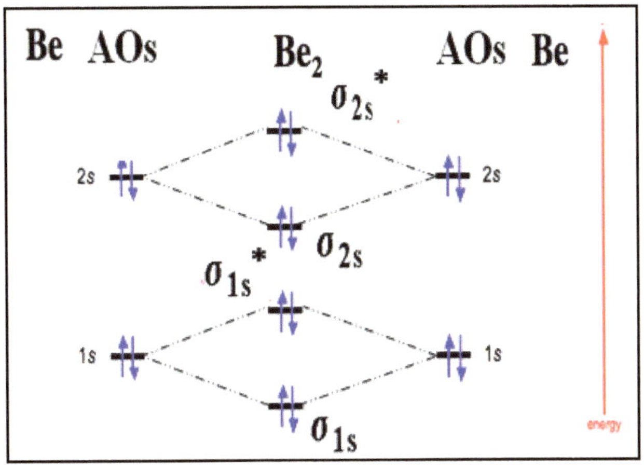

Bond Order = [4-4]/2 = 0/2 = 0

That is no bond between Be atom. Be molecule is monoatomic.

MOs of B_2 Molecule:

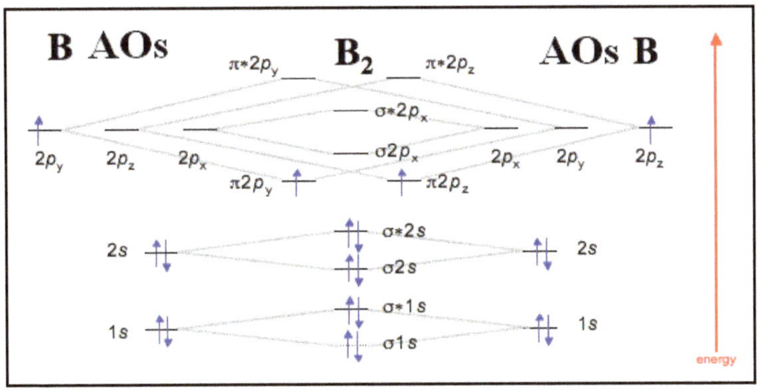

Bond Order = [6-4]/2 = 2/2 = 1

Carbon and nitrogen have same MO energy distribution as in B_2 .

In Carbon and nitrogen the electron distributions in MO are;

	Carbon	**Nitrogen**
σ1S	2	2
σ*1S	2	2
σ2S	2	2
σ*2S	2	2
$\pi 2p_y$	2	2
π2pz	2	2
$\sigma 2P_x$	0	2
Bond order	[8-4]/2 = 4/2 = 2	[10-4]/2 = 6/2 = 3

MOs of B_2 Molecule:

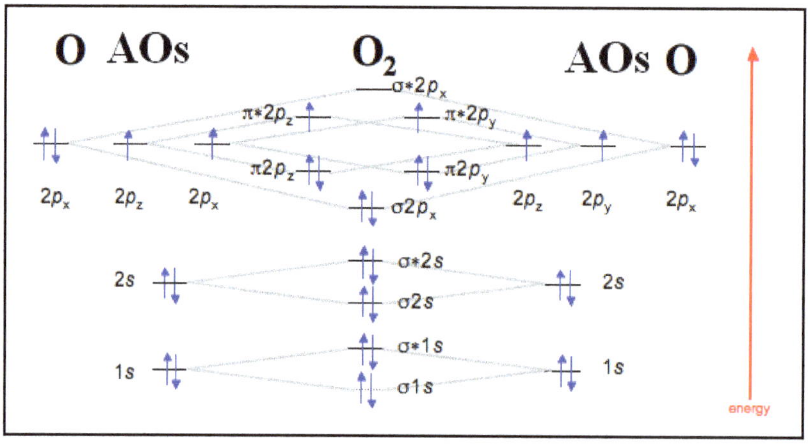

References

- Gillespie, R. J. (2008). "Fifty years of the VSEPR model". Coord. Chem. Rev. 252: 1315–1327. doi:10.1016/j.ccr.2007.07.007
- Hall, G.G. (7 August 1950). "The Molecular Orbital Theory of Chemical Valency. VI. Properties of Equivalent Orbitals" (pdf). Proc. Roy. Soc. A. 202 (1070): 336–344. Bibcode:1950RSPSA.202..336H. doi:10.1098/rspa.1950.0104
- Campbell, Neil A.; Williamson, Brad; Heyden, Robin J. (2006). Biology: Exploring Life. Boston, MA: Pearson Prentice Hall. ISBN 0-13-250882-6. Retrieved 01, June 2020
- Hanson, Robert M. (1995). Molecular origami: precision scale models from paper. University Science Books. ISBN 0-935702-30-X
- Langmuir, Irving (1919-06-01). "The Arrangement of Electrons in Atoms and Molecules". Journal of the American Chemical Society. 41 (6): 868–934. doi:10.1021/ja02227a002

- Anslyn, E. V.; Dougherty, D. A. (2006). Modern Physical Organic Chemistry. University Science Books. p. 57. ISBN 978-1891389313
- Lewis, Gilbert N. (1916-04-01). "The atom and the molecule". Journal of the American Chemical Society. 38 (4): 762–785. doi:10.1021/ja02261a002

6

Applied Inorganic Chemistry

Inorganic chemistry finds its use in a wide variety of applications. Due to the high melting points and specific high or low electrical conductivity of inorganic compounds, they are used as catalysts and pigments in many industrial processes. Materials science is another field where inorganic chemistry is applied. This chapter deals with the various applications of inorganic chemistry to provide an extensive understanding of the subjects.

Catalysis

Catalysis is the increase in the rate of a chemical reaction due to the participation of an additional substance called a catalyst. With a catalyst, reactions occur faster because they require less activation energy. Furthermore since they are not consumed in the catalyzed reaction, catalysts can continue to act repeatedly. Often only tiny amounts are required in principle.

An air filter that utilizes low-temperature oxidation catalyst used to convert carbon monoxide to less toxic carbon dioxide at room temperature. It can also remove formaldehyde from the air.

Technical Perspective

In the presence of a catalyst, less free energy is required to reach the transition state, but the total free energy from reactants to products does not change. A catalyst may participate in multiple chemical transformations. The effect of a catalyst may vary due to the presence of other substances known as inhibitors or poisons (which reduce the catalytic activity) or promoters (which increase the activity). The opposite of a catalyst, a substance that reduces the rate of a reaction, is an inhibitor.

Catalyzed reactions have a lower activation energy (rate-limiting free energy of activation) than the corresponding uncatalyzed reaction, resulting in a higher reaction rate at the same temperature and for the same reactant concentrations. However, the detailed mechanics of catalysis is complex. Catalysts may affect the reaction environment favorably, or bind to the reagents to polarize bonds, e.g. acid catalysts for reactions of carbonyl compounds, or form specific intermediates that are not produced naturally, such as osmate esters in osmium tetroxide-catalyzed dihydroxylation of alkenes, or cause dissociation of reagents to reactive forms, such as chemisorbed hydrogen in catalytic hydrogenation.

Kinetically, catalytic reactions are typical chemical reactions; i.e. the reaction rate depends on the frequency of contact of the reactants in the rate-determining step. Usually, the catalyst participates in this slowest step, and rates are limited by amount of catalyst and its "activity". In heterogeneous catalysis, the diffusion of reagents to the surface and diffusion of products from the surface can be rate determining. A nanomaterial-based catalyst is an example of a heterogeneous catalyst. Analogous events associated with substrate binding and product dissociation apply to homogeneous catalysts.

Although catalysts are not consumed by the reaction itself, they may be inhibited, deactivated, or destroyed by secondary processes. In heterogeneous catalysis, typical secondary processes include coking where the catalyst becomes covered by polymeric side products. Additionally, heterogeneous catalysts can dissolve into the solution in a solid–liquid system or sublimate in a solid–gas system.

Background

The production of most industrially important chemicals involves catalysis. Similarly, most biochemically significant processes are catalysed. Research into catalysis is a major field in applied science and involves many areas of chemistry, notably organometallic chemistry and materials science. Catalysis is relevant to many aspects of environmental science, e.g. the catalytic converter in automobiles and the dynamics of the ozone hole. Catalytic reactions are preferred in environmentally friendly green chemistry due to the reduced amount of waste generated, as opposed to stoichiometric reactions in which all reactants are consumed and more side products are formed. Many transition metals and transition metal complexes are used in catalysis as well. Catalysts called enzymes are important in biology.

A catalyst works by providing an alternative reaction pathway to the reaction product. The rate of the reaction is increased as this alternative route has a lower activation energy than the reaction route not mediated by the catalyst. The disproportionation of hydrogen peroxide creates water and oxygen, as shown below.

$$2\,H_2O_2 \rightarrow 2\,H_2O + O_2$$

This reaction is preferable in the sense that the reaction products are more stable than the starting material, though the uncatalysed reaction is slow. In fact, the decomposition of hydrogen peroxide is so slow that hydrogen peroxide solutions are commercially available. This reaction is strongly affected by catalysts such as manganese dioxide, or the enzyme peroxidase in organisms. Upon the addition of a small amount of manganese dioxide, the hydrogen peroxide reacts rapidly. This effect is readily seen by the effervescence of oxygen. The manganese dioxide is not consumed in the reaction, and thus may be recovered unchanged, and re-used indefinitely. Accordingly, manganese dioxide *catalyses* this reaction.

General Principles

Units

Catalytic activity is usually denoted by the symbol z and measured in mol/s, a unit which was called katal and defined the SI unit for catalytic activity since 1999. Catalytic activity is not a kind of reaction rate, but a property of the catalyst under certain conditions, in relation to a specific chemical reaction. Catalytic activity of one katal (Symbol 1 kat = 1 mol/s) of a catalyst means an amount of that catalyst (substance, in Mol) that leads to a net reaction of one Mol per second of the reactants to the resulting reagents or other outcome which was intended for this chemical reaction. A catalyst may and usually will have different catalytic activity for distinct reactions.

Typical Mechanism

Catalysts generally react with one or more reactants to form intermediates that subsequently give the final reaction product, in the process regenerating the catalyst. The following is a typical reaction scheme, where *C* represents the catalyst, X and Y are reactants, and Z is the product of the reaction of X and Y:

$$\mathrm{X} + C \rightarrow \mathrm{X}C \ (1)$$

$$\mathrm{Y} + \mathrm{X}C \rightarrow \mathrm{XY}C \ (2)$$

$$\mathrm{XY}C \rightarrow C\mathrm{Z} \ (3)$$

$$C\mathrm{Z} \rightarrow C + \mathrm{Z} \ (4)$$

Although the catalyst is consumed by reaction 1, it is subsequently produced by reaction 4, so it does not occur in the overall reaction equation:

$$\mathrm{X} + \mathrm{Y} \rightarrow \mathrm{Z}$$

As a catalyst is regenerated in a reaction, often only small amounts are needed to increase the rate of the reaction. In practice, however, catalysts are sometimes consumed in secondary processes.

The catalyst does usually appear in the rate equation. For example, if the rate-determining step in the above reaction scheme is the first stepX + C → XC, the catalyzed reaction will be second order with rate equation $v = k_{cat}[X][C]$, which is proportional to the catalyst concentration [C]. However [C] remains constant during the reaction so that the catalyzed reaction is pseudo-first order: $v = k_{obs}[X]$, where $k_{obs} = k_{cat}[C]$.

As an example of a detailed mechanism at the microscopic level, in 2008 Danish researchers first revealed the sequence of events when oxygen and hydrogen combine on the surface of titanium dioxide (TiO_2, or *titania*) to produce water. With a time-lapse series of scanning tunneling microscopy images, they determined the molecules undergo adsorption, dissociation and diffusion before reacting. The intermediate reaction states were: HO_2, H_2O_2, then H_3O_2 and the final reaction product (water molecule dimers), after which the water molecule desorbs from the catalyst surface.

Reaction Energetics

Generic potential energy diagram showing the effect of a catalyst in a hypothetical exothermic chemical reaction X + Y to give Z. The presence of the catalyst opens a different reaction pathway (shown in red) with a lower activation energy. The final result and the overall thermodynamics are the same.

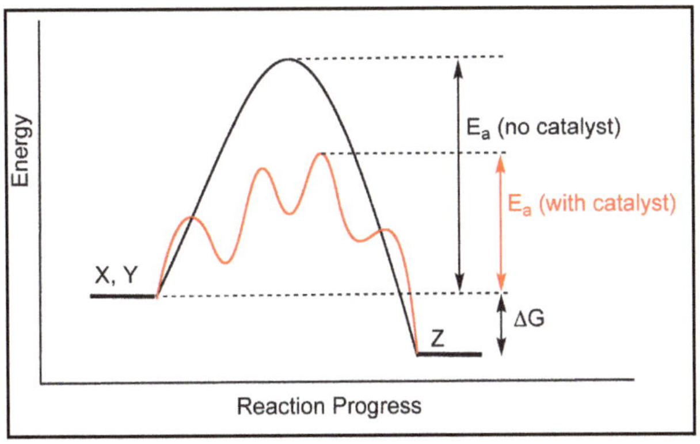

Catalysts work by providing an (alternative) mechanism involving a different transition state and lower activation energy. Consequently, more molecular collisions have the energy needed to reach the transition state. Hence, catalysts can enable reactions that would otherwise be blocked or slowed by a kinetic barrier. The catalyst may increase reaction rate or selectivity, or enable the reaction at lower temperatures. This effect can be illustrated with an energy profile diagram.

In the catalyzed elementary reaction, catalysts do not change the extent of a reaction: they have **no** effect on the chemical equilibrium of a reaction because the rate of both the forward and the reverse reaction are both affected. The second law of thermodynamics describes why a catalyst does not change the chemical equilibrium of a reaction. Suppose there was such a catalyst that shifted an equilibrium. Introducing the catalyst to the system would result in a reaction to move to the new equilibrium, producing energy. Production of energy is a necessary result since reactions are spontaneous only if Gibbs free energy is produced, and if there is no energy barrier, there is no need for a catalyst. Then, removing the catalyst would also result in reaction, producing energy; i.e. the addition and its reverse process, removal, would both produce energy. Thus, a catalyst that could change the equilibrium would be a perpetual motion machine, a contradiction to the laws of thermodynamics.

If a catalyst does change the equilibrium, then it must be consumed as the reaction proceeds, and thus it is also a reactant. Illustrative is the base-catalysed hydrolysis of esters, where the produced carboxylic acid immediately reacts with the base catalyst and thus the reaction equilibrium is shifted towards hydrolysis.

The SI derived unit for measuring the catalytic activity of a catalyst is the katal, which is moles per second. The productivity of a catalyst can be described by the turn over number (or TON) and the catalytic activity by the *turn over frequency* (TOF), which is the TON per time unit. The biochemical equivalent is the enzyme unit.

The catalyst stabilizes the transition state more than it stabilizes the starting material. It decreases the kinetic barrier by decreasing the *difference* in energy between starting material and transition state. It does not change the energy difference between starting materials and products (thermodynamic barrier), or the available energy (this is provided by the environment as heat or light).

Materials

The chemical nature of catalysts is as diverse as catalysis itself, although some generalizations can be made. Proton acids are probably the most widely used catalysts, especially for the many reactions involving water, including hydrolysis and its reverse. Multifunctional solids often are catalytically active, e.g. zeolites, alumina, higher-order oxides, graphitic carbon, nanoparticles, nanodots, and facets of bulk materials. Transition metals are often used to catalyze redox reactions (oxidation, hydrogenation). Examples are nickel, such as Raney nickel for hydrogenation, and vanadium(V) oxide for oxidation of sulfur dioxide into sulfur trioxide by the so-called contact process. Many catalytic processes, especially those used in organic synthesis, require "late transition metals", such as palladium, platinum, gold, ruthenium, rhodium, or iridium.

Some so-called catalysts are really precatalysts. Precatalysts convert to catalysts in the reaction. For example, Wilkinson's catalyst $RhCl(PPh_3)_3$ loses one triphenylphosphine ligand before entering the true catalytic cycle. Precatalysts are easier to store but are easily activated in situ. Because of this preactivation step, many catalytic reactions involve an induction period.

Chemical species that improve catalytic activity are called co-catalysts (cocatalysts) or promotors in cooperative catalysis.

Types:

Catalysts can be heterogeneous or homogeneous, depending on whether a catalyst exists in the same phase as the substrate. Biocatalysts (enzymes) are often seen as a separate group.

Heterogeneous Catalysts

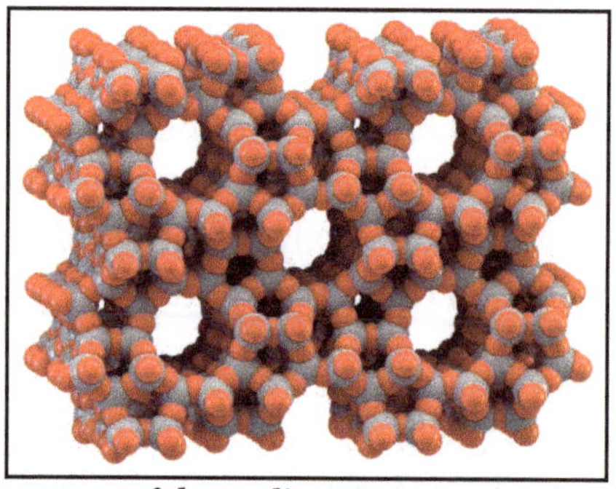

The microporous molecular structure of the zeolite ZSM-5 is exploited in catalysts used in refineries.

Zeolites are extruded as pellets for easy handling in catalytic reactors.

Heterogeneous catalysts act in a different phase than the reactants. Most heterogeneous catalysts are solids that act on substrates in a liquid or gaseous reaction mixture. Diverse mechanisms for reactions on surfaces are known, depending on how the adsorption takes place (Langmuir-Hinshelwood, Eley-Rideal, and Mars-van Krevelen). The total surface area of solid has an important effect on the reaction rate. The smaller the catalyst particle size, the larger the surface area for a given mass of particles.

A heterogeneous catalyst has active sites, which are the atoms or crystal faces where the reaction actually occurs. Depending on the mechanism, the active site may be either a planar exposed metal surface, a crystal edge with imperfect metal valence or a complicated combination of the two. Thus, not only most of the volume, but also most of the surface of a heterogeneous catalyst may be catalytically inactive. Finding out the nature of the active site requires technically challenging research. Thus, empirical research for finding out new metal combinations for catalysis continues.

For example, in the Haber process, finely divided iron serves as a catalyst for the synthesis of ammonia from nitrogen and hydrogen. The reacting gases adsorb onto active sites on the iron particles. Once physically adsorbed, the reagents undergo chemisorption that results in dissociation into adsorbed atomic species, and new bonds between the resulting fragments form in part due to their close proximity. In this way the particularly strong triple bond in nitrogen is broken, which would be extremely uncommon in the gas phase due to its high activation energy. Thus, the activation energy of the overall reaction is lowered, and the rate of reaction increases. Another place where a heterogeneous catalyst is applied is in the oxidation of sulfur dioxide on vanadium(V) oxide for the production of sulfuric acid.

Heterogeneous catalysts are typically "supported," which means that the catalyst is dispersed on a second material that enhances the effectiveness or minimizes their cost. Supports prevent or reduce agglomeration and sintering of the small catalyst particles, exposing more surface area, thus catalysts have a higher specific activity (per gram) on a support. Sometimes the support is merely a surface on which the catalyst is spread to increase the surface area. More often, the support and the catalyst interact, affecting the catalytic reaction. Supports are porous materials with a high surface area, most commonly alumina, zeolites or various kinds of activated carbon. Specialized supports include silicon dioxide, titanium dioxide, calcium carbonate, and barium sulfate.

Electrocatalysts

In the context of electrochemistry, specifically in fuel cell engineering, various metal-containing catalysts are used to enhance the rates of the half reactions that comprise the fuel cell. One common type of fuel cell electrocatalyst is based upon nanoparticles of platinum that are supported on slightly larger carbon particles. When in contact with one of the electrodes in a fuel cell, this platinum increases the rate of oxygen reduction either to water, or to hydroxide or hydrogen peroxide.

Homogeneous Catalysts

Homogeneous catalysts function in the same phase as the reactants, but the mechanistic principles invoked in heterogeneous catalysis are generally applicable. Typically homogeneous catalysts are dissolved in a solvent with the substrates. One example of homogeneous catalysis involves the influence of H^+ on the esterification of carboxylic acids, such as the formation of methyl acetate from acetic acid and methanol. For inorganic chemists, homogeneous catalysis is often synonymous with organometallic catalysts.

Organocatalysis

Whereas transition metals sometimes attract most of the attention in the study of catalysis, small organic molecules without metals can also exhibit catalytic properties, as is apparent from the fact that many enzymes lack transition metals. Typically, organic catalysts require a higher loading (amount of catalyst per unit amount of reactant, expressed in mol% amount of substance) than transition metal(-ion)-based catalysts, but these catalysts are usually commercially available in bulk, helping to reduce costs. In the early 2000s, these organocatalysts were considered "new generation" and are competitive to traditional metal(-ion)-containing catalysts. Organocatalysts are supposed to operate akin to metal-free enzymes utilizing, e.g., non-covalent interactions such as hydrogen bonding. The discipline organocatalysis is divided in the application of covalent (e.g., proline, DMAP) and non-covalent (e.g., thiourea organocatalysis) organocatalysts referring to the preferred catalyst-substrate binding and interaction, respectively.

Enzymes and Biocatalysts

In biology, enzymes are protein-based catalysts in metabolism and catabolism. Most biocatalysts are enzymes, but other non-protein-based classes of biomolecules also exhibit catalytic properties including ribozymes, and synthetic deoxyribozymes.

Biocatalysts can be thought of as intermediate between homogeneous and heterogeneous catalysts, although strictly speaking soluble enzymes are homogeneous catalysts and membrane-bound enzymes are heterogeneous. Several factors affect the activity of enzymes (and other catalysts) including temperature, pH, concentration of enzyme, substrate, and products. A particularly important reagent in enzymatic reactions is water, which is the product of many bond-forming reactions and a reactant in many bond-breaking processes.

In biocatalysis, enzymes are employed to prepare many commodity chemicals including high-fructose corn syrup and acrylamide.

Some monoclonal antibodies whose binding target is a stable molecule which resembles the transition state of a chemical reaction can function as weak catalysts for that chemical reaction by lowering its activation energy. Such catalytic antibodies are sometimes called "abzymes".

Nanocatalysts

Nanocatalysts are nanomaterials with catalytic activities. They have been extensively explored for wide range of applications. Among them, the nanocatalysts with enzyme mimicking activities are collectively called as nanozymes.

Tandem Catalysis

In tandem catalysis two or more different catalysts are coupled in a one-pot reaction.

Autocatalysis

In autocatalysis the catalyst *is* a product of the overall reaction, in contrast to all other types of catalysis considered in this article. The simplest example of autocatalysis is a reaction of type A + B → 2 B, in one or in several steps. The overall reaction is just A → B, so that B is a product. But since B is also a reactant, it may be present in the rate equation and affect the reaction rate. As the reaction proceeds, the concentration of B increases and can accelerate the reaction as a catalyst. In effect, the reaction accelerates itself or is autocatalyzed.

A real example is the hydrolysis of an ester such as aspirin to a carboxylic acid and an alcohol. In the absence of added acid catalysts, the carboxylic acid product catalyzes the hydrolysis.

Significance

Estimates are that 90% of all commercially produced chemical products involve catalysts at some stage in the process of their manufacture. In 2005, catalytic processes generated about $900 billion in products worldwide. Catalysis is so pervasive that subareas are not readily classified. Some areas of particular concentration are surveyed below.

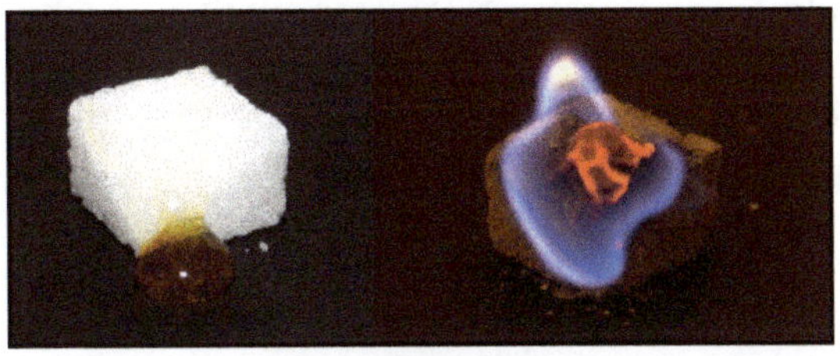

Left: Partially caramelised cube sugar, Right: burning cube sugar with ash as catalyst.

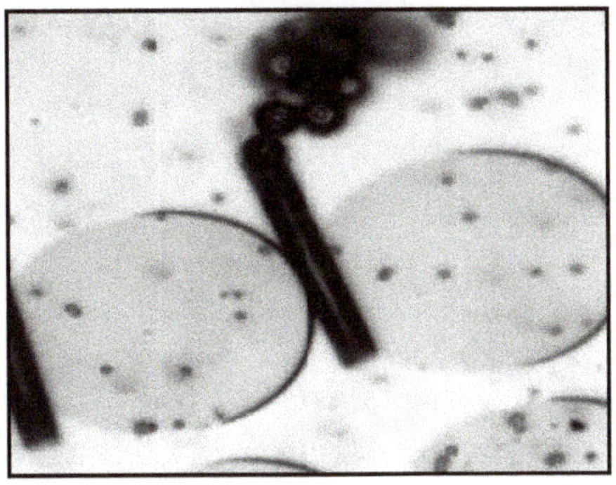

A Ti-Cr-Pt tube (~40 μm long) releases oxygen bubbles when immersed in hydrogen peroxide (via catalytic decomposition), forming a micropump.

Energy Processing

Petroleum refining makes intensive use of catalysis for alkylation, catalytic cracking (breaking long-chain hydrocarbons into smaller pieces), naphtha reforming and steam reforming (conversion of hydrocarbons into synthesis gas). Even the exhaust from the burning of fossil fuels is treated via catalysis: Catalytic converters, typically composed of platinum and rhodium, break down some of the more harmful byproducts of automobile exhaust.

$$2\,CO + 2\,NO \rightarrow 2\,CO_2 + N_2$$

With regard to synthetic fuels, an old but still important process is the Fischer-Tropsch synthesis of hydrocarbons from synthesis gas, which itself is processed via water-gas shift reactions, catalysed by iron. Biodiesel and related biofuels require processing via both inorganic and biocatalysts.

Fuel cells rely on catalysts for both the anodic and cathodic reactions.

Catalytic heaters generate flameless heat from a supply of combustible fuel.

Bulk Chemicals

Some of the largest-scale chemicals are produced via catalytic oxidation, often using oxygen. Examples include nitric acid (from ammonia), sulfuric acid (from sulfur dioxide to sulfur trioxide by the chamber process), terephthalic acid from p-xylene, and acrylonitrile from propane and ammonia.

Many other chemical products are generated by large-scale reduction, often via hydrogenation. The largest-scale example is ammonia, which is prepared via the Haber process from nitrogen. Methanol is prepared from carbon monoxide.

Bulk polymers derived from ethylene and propylene are often prepared via Ziegler-Natta catalysis. Polyesters, polyamides, and isocyanates are derived via acid-base catalysis.

Most carbonylation processes require metal catalysts, examples include the Monsanto acetic acid process and hydroformylation.

Fine Chemicals

Many fine chemicals are prepared via catalysis; methods include those of heavy industry as well as more specialized processes that would be prohibitively expensive on a large scale. Examples include the Heck reaction, and Friedel-Crafts reactions.

Because most bioactive compounds are chiral, many pharmaceuticals are produced by enantioselective catalysis (catalytic asymmetric synthesis).

Food Processing

One of the most obvious applications of catalysis is the hydrogenation (reaction with hydrogen gas) of fats using nickel catalyst to produce margarine. Many other foodstuffs are prepared via biocatalysis.

Environment

Catalysis impacts the environment by increasing the efficiency of industrial processes, but catalysis also plays a direct role in the environment. A notable example is the catalytic role of chlorine free radicals in the breakdown of ozone. These radicals are formed by the action of ultraviolet radiation on chlorofluorocarbons (CFCs).

$$Cl^{\cdot} + O_3 \rightarrow ClO^{\cdot} + O_2$$

$$ClO^{\cdot} + O^{\cdot} \rightarrow Cl^{\cdot} + O_2$$

History

In a general sense, anything that increases the rate of a process is a "catalyst". The concept of catalysis was invented by chemist Elizabeth Fulhame and described in a 1794 book, based on her novel work in oxidation-reduction experiments. The term *catalysis* was later used by Jöns Jakob Berzelius in 1835 to describe reactions that are accelerated by substances that remain unchanged after the reaction. Fulhame, who predated Berzelius, did work with water as opposed to metals in her reduction experiments. Other 18th century chemists who worked in catalysis were Eilhard Mitscherlich who referred to it as *contact* processes, and Johann Wolfgang Döbereiner who spoke of *contact action*. He developed Döbereiner's lamp, a lighter based on hydrogen and a platinum sponge, which became a commercial success in the 1820s that lives on today. Humphry Davy discovered the use of platinum in catalysis. In the 1880s, Wilhelm Ostwald at Leipzig University started a systematic investigation into reactions that were catalyzed by the presence of acids and bases, and found that chemical reactions occur at finite rates and that these rates can be used to determine the strengths of acids and bases. For this work, Ostwald was awarded the 1909 Nobel Prize in Chemistry.

Inhibitors, Poisons, and Promoters

Substances that reduce the action of catalysts are called catalyst inhibitors if reversible, and catalyst poisons if irreversible. Promoters are substances that increase the catalytic activity, even though they are not catalysts by themselves.

Inhibitors are sometimes referred to as "negative catalysts" since they decrease the reaction rate. However the term inhibitor is preferred since they do not work by introducing a reaction path with higher activation energy; this would not reduce the rate since the reaction would continue to occur by the non-catalyzed path. Instead they act either by deactivating catalysts, or by removing reaction intermediates such as free radicals.

The inhibitor may modify selectivity in addition to rate. For instance, in the reduction of alkynes to alkenes, a palladium (Pd) catalyst partly "poisoned" with lead(II) acetate ($Pb(CH_3CO_2)_2$) can be used. Without the deactivation of the catalyst, the alkene produced would be further reduced to alkane.

The inhibitor can produce this effect by, e.g., selectively poisoning only certain types of active sites. Another mechanism is the modification of surface geometry. For instance, in hydrogenation operations, large planes of metal surface function as sites of hydrogenolysis catalysis while sites catalyzing hydrogenation of unsaturates are smaller. Thus, a poison that covers surface randomly will tend to reduce the number of uncontaminated large planes but leave proportionally more smaller sites free, thus changing the hydrogenation vs. hydrogenolysis selectivity. Many other mechanisms are also possible.

Promoters can cover up surface to prevent production of a mat of coke, or even actively remove such material (e.g., rhenium on platinum in platforming). They can aid the dispersion of the catalytic material or bind to reagents.

Current Market

The global demand on catalysts in 2010 was estimated at approximately 29.5 billion USD. With the rapid recovery in automotive and chemical industry overall, the global catalyst market is expected to experience fast growth in the next years.

Materials Science

The interdisciplinarity field of materials science, also commonly termed materials science and engineering, involves the discovery and design of new materials, with an emphasis on solids. The intellectual origins of materials science stem from the Enlightenment, when researchers began to use analytical thinking from chemistry, physics, and engineering to understand ancient, phenomenological observations in metallurgy and mineralogy. Materials science still incorporates elements of physics, chemistry, and engineering. As such, the field was long considered by academic institutions as a sub-field of these related fields. Beginning in the 1940s, materials science began to be more widely recognized as a specific and distinct field of science and engineering, and major technical universities around the world created dedicated schools of the study.

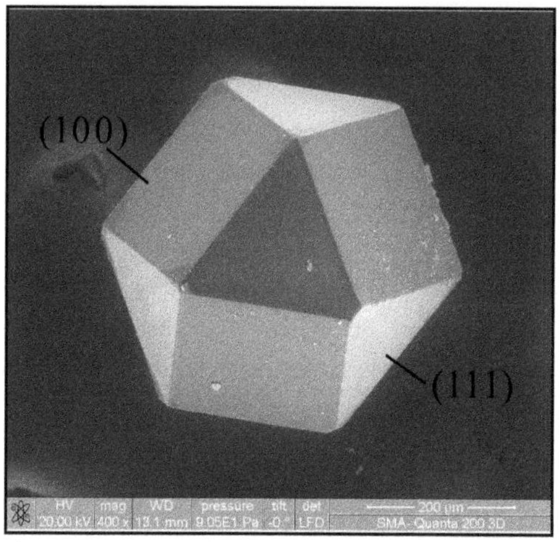

A diamond cuboctahedron showing seven crystallographic planes, imaged with scanning electron microscopy.

Materials science is a syncretic discipline hybridizing metallurgy, ceramics, solid-state physics, and chemistry. It is the first example of a new academic discipline emerging by fusion rather than fission.

Many of the most pressing scientific problems humans currently face are due to the limits of the materials that are available. Thus, breakthroughs in materials science are likely to affect the future of technology significantly.

Materials scientists emphasize understanding how the history of a material (its *processing*) influences its structure, and thus the material's properties and performance. The understanding of processing-structure-properties relationships is called the § materials paradigm. This paradigm is used to advance understanding in a variety of research areas, including nanotechnology, biomaterials, and metallurgy. Materials science is also an important part of forensic engineering and failure analysis - investigating materials, products, structures or components which fail or which do not operate or function as intended, causing personal injury or damage to property. Such investigations are key to understanding, for example, the causes of various aviation accidents and incidents.

History

The material of choice of a given era is often a defining point. Phrases such as Stone Age, Bronze Age, Iron Age, and Steel Age are great examples. Originally deriving from the manufacture of ceramics and its putative derivative metallurgy, materials science is one of the oldest forms of engineering and applied science. Modern materials science evolved directly from metallurgy, which itself evolved from mining and (likely) ceramics and the use of fire. A major breakthrough in the understanding of materials occurred in the late 19th century, when the American scientist Josiah Willard Gibbs demonstrated that the thermodynamic properties related to atomic structure in various phases are related to the physical properties of a material. Important elements of modern materials science are a product of the space race: the understanding and engineering of the metallic alloys, and silica and carbon materials, used in building space vehicles enabling the exploration of space. Materials science has driven, and been driven by, the development of revolutionary technologies such as rubbers, plastics, semiconductors, and biomaterials.

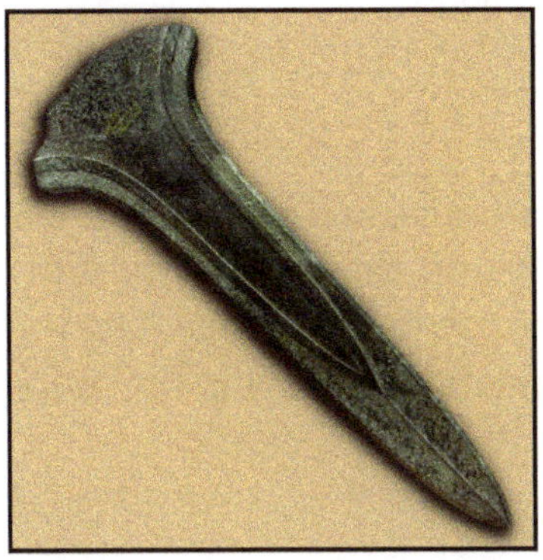

A late Bronze Age sword or dagger blade.

Before the 1960s (and in some cases decades after), many *materials science* departments were named *metallurgy* departments, reflecting the 19th and early 20th century emphasis on metals. The growth of materials science in the United States was catalyzed in part by the Advanced Research Projects Agency, which funded a series of university-hosted laboratories in the early 1960s "to expand the national program of basic research and training in the materials sciences." The field has since broadened to include every class of materials, including ceramics, polymers, semiconductors, magnetic materials, medical implant materials, biological materials, and nanomaterials.

Fundamentals

The materials paradigm represented in the form of a tetrahedron.

A material is defined as a substance (most often a solid, but other condensed phases can be included) that is intended to be used for certain applications. There are a myriad of materials around us—they can be found in anything from buildings to spacecraft. Materials can generally be divided into two classes: crystalline and non-crystalline. The traditional examples of materials are metals, semiconductors, ceramics and polymers. New and advanced materials that are being developed include nanomaterials and biomaterials, etc.

The basis of materials science involves studying the structure of materials, and relating them to their properties. Once a materials scientist knows about this structure-property correlation, they can then go on to study the relative performance of a material in a given application. The major determinants of the structure of a material and thus of its properties are its constituent chemical elements and the way in which it has been processed into its final form. These characteristics, taken together and related through the laws of thermodynamics and kinetics, govern a material's microstructure, and thus its properties.

Structure

As mentioned above, structure is one of the most important components of the field of materials science. Materials science examines the structure of materials from the atomic scale, all the way up to the macro scale. Characterization is the way materials scientists examine the structure of a material. This involves methods such as diffraction with X-rays, electrons, or neutrons, and various forms of spectroscopy and chemical analysis such as Raman spectroscopy, energy-dispersive spectroscopy (EDS), chromatography, thermal analysis, electron microscope analysis, etc. Structure is studied at various levels, as detailed below.

Atomic Structure

This deals with the atoms of the materials, and how they are arranged to give molecules, crystals, etc. Much of the electrical, magnetic and chemical properties of materials arise from this level of structure. The length scales involved are in angstroms. The way in which the atoms and molecules are bonded and arranged is fundamental to studying the properties and behavior of any material.

Nanostructure

Nanostructure deals with objects and structures that are in the 1—100 nm range. In many materials, atoms or molecules agglomerate together to form objects at the nanoscale. This causes many interesting electrical, magnetic, optical, and mechanical properties.

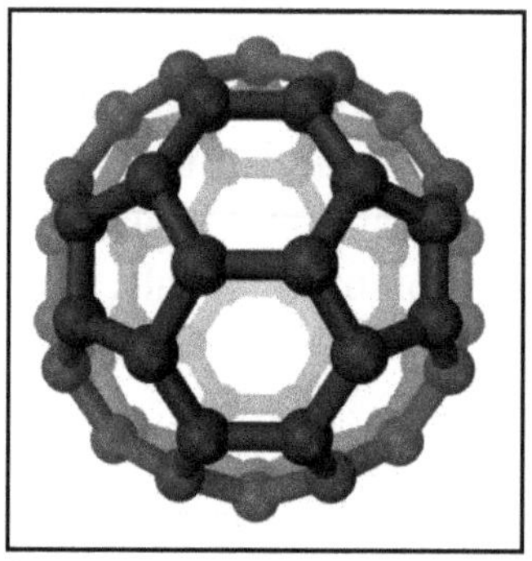

Buckminsterfullerene nanostructure.

In describing nanostructures it is necessary to differentiate between the number of dimensions on the nanoscale. Nanotextured surfaces have *one dimension* on the nanoscale, i.e., only the thickness of the surface of an object is between 0.1 and 100 nm. Nanotubes have *two dimensions* on the nanoscale, i.e., the diameter of the tube is between 0.1 and 100 nm; its length could be much greater. Finally, spherical nanoparticles have *three dimensions* on the nanoscale, i.e., the particle is between 0.1 and 100 nm in each spatial dimension. The terms nanoparticles and ultrafine particles (UFP) often are used synonymously although UFP can reach into the micrometre range. The term 'nanostructure' is often used when referring to magnetic technology. Nanoscale structure in biology is often called ultrastructure.

Materials which atoms and molecules form constituents in the nanoscale (i.e., they form nanostructure) are called nanomaterials. Nanomaterials are subject of intense research in the materials science community due to the unique properties that they exhibit.

Microstructure

Microstructure is defined as the structure of a prepared surface or thin foil of material as revealed by a microscope above 25× magnification. It deals with objects from 100 nm to a few cm. The microstructure of a material (which can be broadly classified into metallic, polymeric, ceramic and composite) can strongly influence physical properties such as strength, toughness, ductility, hardness, corrosion resistance, high/low temperature behavior, wear resistance, and so on. Most of the traditional materials (such as metals and ceramics) are microstructured.

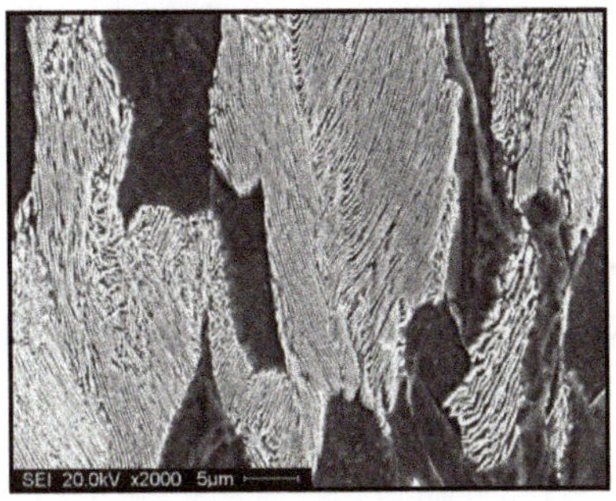

Microstructure of pearlite.

The manufacture of a perfect crystal of a material is physically impossible. For example, a crystalline material will contain defects such as precipitates, grain boundaries (Hall–Petch relationship), interstitial atoms, vacancies or substitutional atoms. The microstructure of materials reveals these defects, so that they can be studied.

Macro Structure

Macro structure is the appearance of a material in the scale millimeters to meters—it is the structure of the material as seen with the naked eye.

Crystallography

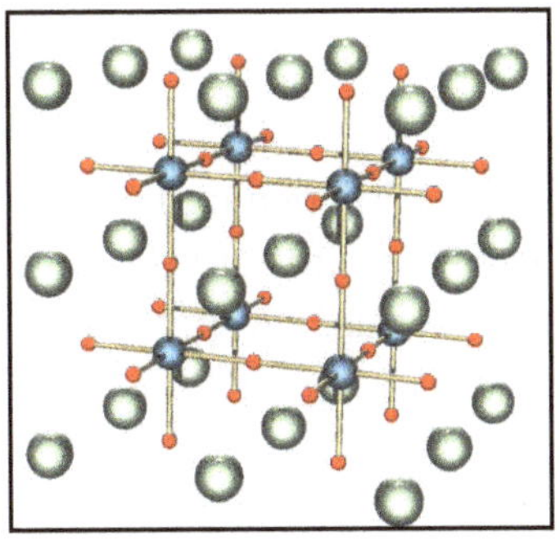

Crystal structure of a perovskite with a chemical formula ABX_3.

Crystallography is the science that examines the arrangement of atoms in crystalline solids. Crystallography is a useful tool for materials scientists. In single crystals, the effects of the crystalline arrangement of atoms is often easy to see macroscopically, because the natural shapes of crystals reflect the atomic structure. Further, physical properties are often controlled by crystalline defects. The understanding of crystal structures is an important prerequisite for understanding crystallographic defects. Mostly, materials do not occur as a single crystal, but in polycrystalline form, i.e., as an aggregate of small crystals with different orientations. Because of this, the powder diffraction method, which uses diffraction patterns of polycrystalline samples with a large number of crystals, plays an important role in structural determination. Most materials have a crystalline structure, but some important materials do not exhibit regular crystal structure. Polymers display varying degrees of crystallinity, and many are completely noncrystalline. Glass, some ceramics, and many natural materials are amorphous, not possessing any long-range order in their atomic arrangements. The study of polymers combines elements of chemical and statistical thermodynamics to give thermodynamic and mechanical, descriptions of physical properties.

Bonding

To obtain a full understanding of the material structure and how it relates to its properties, the materials scientist must study how the different atoms, ions and molecules are arranged and bonded to each other. This involves the study and use of quantum chemistry or quantum physics. Solid-state physics, solid-state chemistry and physical chemistry are also involved in the study of bonding and structure.

Properties

Materials exhibit myriad properties, including the following:

- Mechanical properties,
- Chemical properties,
- Electrical properties,
- Thermal properties,
- Optical properties,
- Magnetic properties.

The properties of a material determine its usability and hence its engineering application.

Synthesis and Processing

Synthesis and processing involves the creation of a material with the desired micro-nanostructure. From an engineering standpoint, a material cannot be used in industry if no economical production method for it has been developed. Thus, the processing of materials is vital to the field of materials science.

Different materials require different processing or synthesis methods. For example, the processing of metals has historically been very important and is studied under the branch of materials science named *physical metallurgy*. Also, chemical and physical methods are also used to synthesize other materials such as polymers, ceramics, thin films, etc. As of the early 21st century, new methods are being developed to synthesize nanomaterials such as graphene.

Thermodynamics

Thermodynamics is concerned with heat and temperature and their relation to energy and work. It defines macroscopic variables, such as internal energy, entropy, and pressure, that partly describe a body of matter or radiation. It states that the behavior of those variables is subject to general constraints, that are common to all materials, not the peculiar properties of particular materials. These general constraints are expressed in the four laws of thermodynamics. Thermodynamics describes the bulk behavior of the body, not the microscopic behaviors of the very large numbers of its microscopic constituents, such as molecules. The behavior of these microscopic particles is described by, and the laws of thermodynamics are derived from, statistical mechanics.

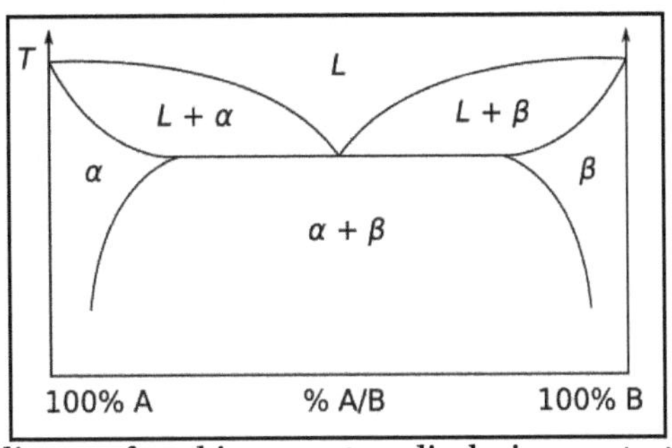

A phase diagram for a binary system displaying a eutectic point.

The study of thermodynamics is fundamental to materials science. It forms the foundation to treat general phenomena in materials science and engineering, including chemical reactions, magnetism, polarizability, and elasticity. It also helps in the understanding of phase diagrams and phase equilibrium.

Kinetics

Chemical kinetics is the study of the rates at which systems that are out of equilibrium change under the influence of various forces. When applied to materials science, it deals with how a material changes with time (moves from non-equilibrium to equilibrium state) due to application of a certain field. It details the rate of various processes evolving in materials including shape, size, composition and structure. Diffusion is important in the study of kinetics as this is the most common mechanism by which materials undergo change.

Kinetics is essential in processing of materials because, among other things, it details how the microstructure changes with application of heat.

In research

Materials science has received much attention from researchers. In most universities, many departments ranging from physics to chemistry to chemical engineering, along with materials science departments, are involved in materials research. Research in materials science is vibrant and consists of many avenues. The following list is in no way exhaustive. It serves only to highlight certain important research areas.

Nanomaterials

Nanomaterials describe, in principle, materials of which a single unit is sized (in at least one dimension) between 1 and 1000 nanometers (10^{-9} meter) but is usually 1—100 nm.

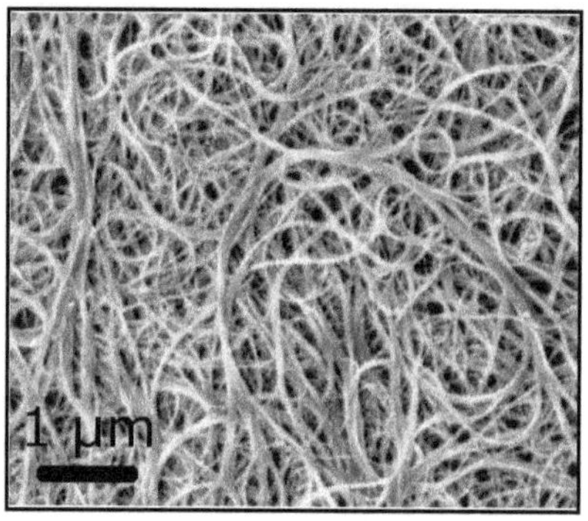

A scanning electron microscopy image of carbon nanotubes bundles.

Nanomaterials research takes a materials science-based approach to nanotechnology, leveraging advances in materials metrology and synthesis which have been developed in support of microfabrication research. Materials with structure at the nanoscale often have unique optical, electronic, or mechanical properties.

The field of nanomaterials is loosely organized, like the traditional field of chemistry, into organic (carbon-based) nanomaterials such as fullerenes, and inorganic nanomaterials based on other elements, such as silicon. Examples of nanomaterials include fullerenes, carbon nanotubes, nanocrystals, etc.

Biomaterials

A biomaterial is any matter, surface, or construct that interacts with biological systems. As a science, *bio materials* is about fifty years old. The study of biomaterials is called *bio materials science.* It has experienced steady and strong growth over its history, with many companies investing large amounts of money into developing new products. Biomaterials science encompasses elements of medicine, biology, chemistry, tissue engineering, and materials science.

The iridescent nacre inside a nautilus shell.

Biomaterials can be derived either from nature or synthesized in a laboratory using a variety of chemical approaches using metallic components, polymers, ceramics, or composite materials.

They are often used and/or adapted for a medical application, and thus comprises whole or part of a living structure or biomedical device which performs, augments, or replaces a natural function. Such functions may be benign, like being used for a heart valve, or may be bioactive with a more interactive functionality such as hydroxylapatite coated hip implants. Biomaterials are also used everyday in dental applications, surgery, and drug delivery. For example, a construct with impregnated pharmaceutical products can be placed into the body, which permits the prolonged release of a drug over an extended period of time. A biomaterial may also be an autograft, allograft or xenograft used as an organ transplant material.

Electronic, Optical, and Magnetic

Semiconductors, metals, and ceramics are used today to form highly complex systems, such as integrated electronic circuits, optoelectronic devices, and magnetic and optical mass storage media. These materials form the basis of our modern computing world, and hence research into these materials is of vital importance.

Negative index metamaterial.

Semiconductors are a traditional example of these types of materials. They are materials that have properties that are intermediate between conductors and insulators. Their electrical conductivities are very sensitive to impurity concentrations, and this allows for the use of doping to achieve desirable electronic properties. Hence, semiconductors form the basis of the traditional computer.

This field also includes new areas of research such as superconducting materials, spintronics, metamaterials, etc. The study of these materials involves knowledge of materials science and solid-state physics or condensed matter physics.

Computational Science and Theory

With the increase in computing power, simulating the behavior of materials has become possible. This enables materials scientists to discover properties of materials formerly unknown, as well as to design new materials. Up until now, new materials were found by time consuming trial and error processes. But, now it is hoped that computational methods could drastically reduce that time, and allow tailoring materials properties. This involves simulating materials at all length scales, using methods such as density functional theory, molecular dynamics, etc.

In Industry

Radical materials advances can drive the creation of new products or even new industries, but stable industries also employ materials scientists to make incremental improvements and troubleshoot issues with currently used materials. Industrial applications of materials science include materials design, cost-benefit tradeoffs in industrial production of materials, processing methods (casting, rolling, welding, ion implantation, crystal growth, thin-film deposition, sintering, glass-blowing, etc.), and analytic methods (characterization methods such as electron microscopy, X-ray diffraction, calorimetry, nuclear microscopy (HEFIB), Rutherford backscattering, neutron diffraction, small-angle X-ray scattering (SAXS), etc.).

Besides material characterization, the material scientist or engineer also deals with extracting materials and convering them into useful forms. Thus ingot casting, foundry methods, blast furnace extraction, and electrolytic extraction are all part of the required knowledge of a materials engineer. Often the presence, absence, or variation of minute quantities of secondary elements and compounds in a bulk material will greatly affect the final properties of the materials produced. For example, steels are classified based on 1/10 and 1/100 weight percentages of the carbon and other alloying elements they contain. Thus, the extracting and purifying methods used to extract iron in a blast furnace can affect the quality of steel that is produced.

Ceramics and Glasses

Another application of material science is the structures of ceramics and glass, typically associated with the most brittle materials. Bonding in ceramics and glasses uses covalent and ionic-covalent types with SiO_2 (silica or sand) as a fundamental building block. Ceramics are as soft as clay or as hard as stone and concrete. Usually, they are crystalline in form. Most glasses contain a metal oxide fused with silica. At high temperatures used to prepare glass, the material is a viscous liquid. The structure of glass forms into an amorphous state upon cooling. Windowpanes and eyeglasses are important examples. Fibers of glass are also available. Scratch resistant Corning Gorilla Glass is a well-known example of the application of materials science to drastically improve the properties of common components. Diamond and carbon in its graphite form are considered to be ceramics.

Si_3N_4 ceramic bearing parts.

Engineering ceramics are known for their stiffness and stability under high temperatures, compression and electrical stress. Alumina, silicon carbide, and tungsten carbide are made from a fine powder of their constituents in a process of sintering with a binder. Hot pressing provides higher density material. Chemical vapor deposition can place a film of a ceramic on another material. Cermets are ceramic particles containing some metals. The wear resistance of tools is derived from cemented carbides with the metal phase of cobalt and nickel typically added to modify properties.

Composites

Filaments are commonly used for reinforcement in composite materials.

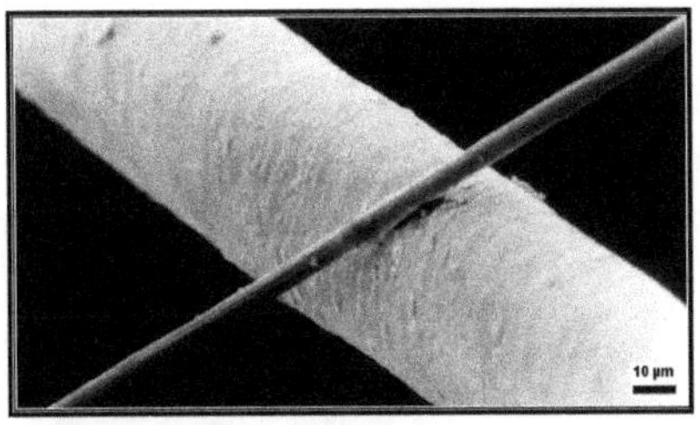

A 6 μm diameter carbon filament (running from bottom left to top right) siting atop the much larger human hair.

Another application of materials science in industry is making composite materials. These are structured materials composed of two or more macroscopic phases. Applications range from structural elements such as steel-reinforced concrete, to the thermal insulating tiles which play a key and integral role in NASA's Space Shuttle thermal protection system which is used to protect the surface of the shuttle from the heat of re-entry into the Earth's atmosphere. One example is reinforced Carbon-Carbon (RCC), the light gray material which withstands re-entry temperatures up to 1,510 °C (2,750 °F) and protects the Space Shuttle's wing leading edges and nose cap. RCC is a laminated composite material made from graphite rayon cloth and impregnated with a phenolic resin. After curing at high temperature in an autoclave, the laminate is pyrolized to convert the resin to carbon, impregnated with furfural alcohol in a vacuum chamber, and cured-pyrolized to convert the furfural alcohol to carbon. To provide oxidation resistance for reuse ability, the outer layers of the RCC are converted to silicon carbide.

Other examples can be seen in the "plastic" casings of television sets, cell-phones and so on. These plastic casings are usually a composite material made up of a thermoplastic matrix such as acrylonitrile butadiene styrene (ABS) in which calcium carbonate chalk, talc, glass fibers or carbon fibers have been added for added strength, bulk, or electrostatic dispersion. These additions may be termed reinforcing fibers, or dispersants, depending on their purpose.

Polymers

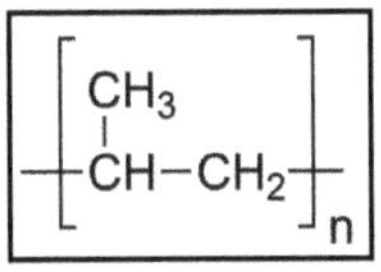

The repeating unit of the polymer polypropylene.

Polymers are chemical compounds made up of a large number of identical components linked together like chains. They are an important part of materials science. Polymers are the raw materials (the resins) used to make what are commonly called plastics and rubber. Plastics and rubber are really the final product, created after one or more polymers or additives have been added to a resin during processing, which is then shaped into a final form. Plastics which have been around, and which are in current widespread use, include polyethylene, polypropylene, polyvinyl chloride (PVC), polystyrene, nylons, polyesters, acrylics, polyurethanes, and polycarbonates and also rubbers which have been around are natural rubber, styrene butadiene rubber, chloroprene, and butadiene rubber. Plastics are generally classified as *commodity, specialty* and *engineering* plastics.

Polyvinyl chloride (PVC) is widely used, inexpensive, and annual production quantities are large. It lends itself to an vast array of applications, from artificial leather to electrical insulation and cabling, packaging, and containers. Its fabrication and processing are simple and well-established. The versatility of PVC is due to the wide range of plasticisers and other additives that it accepts. The term "additives" in polymer science refers to the chemicals and compounds added to the polymer base to modify its material properties.

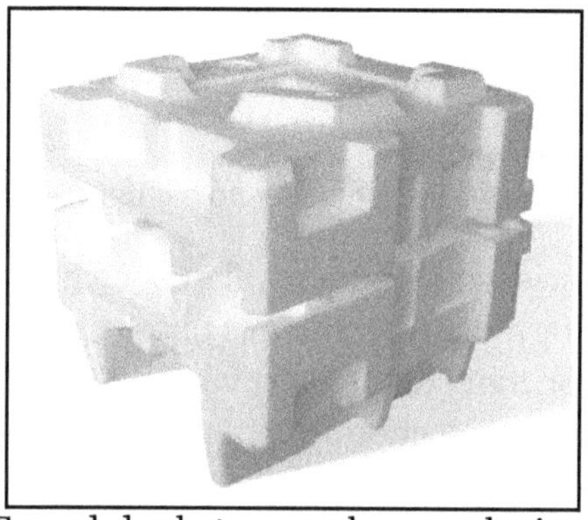

Expanded polystyrene polymer packaging.

Polycarbonate would be normally considered an engineering plastic (other examples include PEEK, ABS). Such plastics are valued for their superior strengths and other special material properties. They are usually not used for disposable applications, unlike commodity plastics.

Specialty plastics are materials with unique characteristics, such as ultra-high strength, electrical conductivity, electro-fluorescence, high thermal stability, etc.

The dividing lines between the various types of plastics is not based on material but rather on their properties and applications. For example, polyethylene (PE) is a cheap, low friction polymer commonly used to make disposable bags for shopping and trash, and is considered a commodity plastic, whereas medium-density polyethylene (MDPE) is used for underground gas and water pipes, and another variety called ultra-high-molecular-weight polyethylene (UHMWPE) is an engineering plastic which is used extensively as the glide rails for industrial equipment and the low-friction socket in implanted hip joints.

Metal Alloys

The study of metal alloys is a significant part of materials science. Of all the metallic alloys in use today, the alloys of iron (steel, stainless steel, cast iron, tool steel, alloy steels) make up the largest proportion both by quantity and commercial value. Iron alloyed with various proportions of carbon gives low, mid and high carbon steels. An iron carbon alloy is only considered steel if the carbon level is between 0.01% and 2.00%. For the steels, the hardness and tensile strength of the steel is related to the amount of carbon present, with increasing carbon levels also leading to lower ductility and toughness. Heat treatment processes such as quenching and tempering can significantly change these properties however. Cast Iron is defined as an iron–carbon alloy with more than 2.00% but less than 6.67% carbon. Stainless steel is defined as a regular steel alloy with greater than 10% by weight alloying content of Chromium. Nickel and Molybdenum are typically also found in stainless steels.

Other significant metallic alloys are those of aluminium, titanium, copper and magnesium. Copper alloys have been known for a long time (since the Bronze Age), while the alloys of the other three metals have been relatively recently developed. Due to the chemical reactivity of these metals, the electrolytic extraction processes required were only developed relatively recently. The alloys of aluminium, titanium and magnesium are also known and valued for their high strength-to-weight ratios and, in the case of magnesium, their ability to provide electromagnetic shielding. These materials are ideal for situations where high strength-to-weight ratios are more important than bulk cost, such as in the aerospace industry and certain automotive engineering applications.

Wire rope made from steel alloy.

Semiconductors

The study of semiconductors is a significant part of materials science. A semiconductor is a material that has a resistivity between a metal and insulator. It's electronic properties can be greatly altered through intentionally introducing impurities, or doping. From these semiconductor materials, things such as diodes, transistors, light-emitting diodes (LEDs), and analog and digital electric circuits can be built, making them materials of interest in industry. Semiconductor devices have replaced thermionic devices (vacuum tubes) in most applications. Semiconductor devices are manufactured both as single discrete devices and as integrated circuits (ICs), which consist of a number—from a few to millions—of devices manufactured and interconnected on a single semiconductor substrate.

Of all the semiconductors in use today, silicon makes up the largest portion both by quantity and commercial value. Monocrystalline silicon is used to produce wafers used in the semiconductor and electronics industry. Second to silicon, gallium arsenide (GaAs) is the second most popular semiconductor used. Due to its higher electron mobility and saturation velocity compared to silicon, its a material of choice for high speed electronics applications. These superior properties are compelling reasons to use GaAs circuitry in mobile phones, satellite communications, microwave point-to-point links and higher frequency radar systems. Other semiconductor materials include germanium, silicon carbide, and gallium nitride and have various applications.

Relation to Other Fields

Materials science evolved—starting from the 1960s—because it was recognized that to create, discover and design new materials, one had to approach it in a unified manner. Thus, materials science and engineering emerged at the intersection of various fields such as metallurgy, solid state physics, chemistry, chemical engineering, mechanical engineering and electrical engineering.

The field is inherently interdisciplinary, and the materials scientists/engineers must be aware and make use of the methods of the physicist, chemist and engineer. The field thus, maintains close relationships with these fields. Also, many physicists, chemists and engineers also find themselves working in materials science.

The overlap between physics and materials science has led to the offshoot field of *materials physics*, which is concerned with the physical properties of materials. The approach is generally more macroscopic and applied than in condensed matter physics.

The field of materials science and engineering is important both from a scientific perspective, as well as from an engineering one. When discovering new materials, one encounters new phenomena that may not have been observed before. Hence, there is lot of science to be discovered when working with materials. Materials science also provides test for theories in condensed matter physics.

Materials are of the utmost importance for engineers, as the usage of the appropriate materials is crucial when designing systems. As a result, materials science is an increasingly important part of an engineer's education.

Emerging Technologies in Materials Science

Emerging technology	Status	Potentially marginalized technologies	Potential applications
Aerogel	Hypothetical, experiments, diffusion, early uses	Traditional insulation, glass	Improved insulation, insulative glass if it can be made clear, sleeves for oil pipelines, aerospace, high-heat & extreme cold applications
Amorphous metal	Experiments	Kevlar	Armor
Conductive polymers	Research, experiments, prototypes	Conductors	Lighter and cheaper wires, antistatic materials, organic solar cells
Femtotechnology, picotechnology	Hypothetical	Present nuclear	New materials; nuclear weapons, power
Fullerene	Experiments, diffusion	Synthetic diamond and carbon nanotubes (e.g., Buckypaper)	Programmable matter

Graphene	Hypothetical, experiments, diffusion, early uses	Silicon-based integrated circuit	Components with higher strength to weight ratios, transistors that operate at higher frequency, lower cost of display screens in mobile devices, storing hydrogen for fuel cell powered cars, filtration systems, longer-lasting and faster-charging batteries, sensors to diagnose diseases
High-temperature superconductivity	Cryogenic receiver front-end (CRFE) RF and microwave filter systems for mobile phone base stations; prototypes in dry ice; Hypothetical and experiments for higher temperatures	Copper wire, semiconductor integral circuits	No loss conductors, frictionless bearings, magnetic levitation, lossless high-capacity accumulators, electric cars, heat-free integral circuits and processors
LiTraCon	Experiments, already used to make Europe Gate	Glass	Building skyscrapers, towers, and sculptures like Europe Gate
Metamaterials	Hypothetical, experiments, diffusion	Classical optics	Microscopes, cameras, metamaterial cloaking, cloaking devices
Metal foam	Research, commercialization	Hulls	Space colonies, floating cities
Multi-function structures	Hypothetical, experiments, some prototypes, few commercial	Composite materials mostly	Wide range, e.g., self health monitoring, self healing material, morphing, ...
Nanomaterials: carbon nanotubes	Hypothetical, experiments, diffusion, early uses	Structural steel and aluminium	Stronger, lighter materials, space elevator
Programmable matter	Hypothetical, experiments	Coatings, catalysts	Wide range, e.g., claytronics, synthetic biology
Quantum dots	Research, experiments, prototypes	LCD, LED	Quantum dot laser, future use as programmable matter in display technologies (TV, projection), optical data communications (high-speed data transmission), medicine (laser scalpel)
Silicene	Hypothetical, research	Field-effect transistors	

Superalloy	Research, diffusion	Aluminum, titanium, composite materials	Aircraft jet engines
Synthetic diamond	early uses (drill bits, jewelry)	Silicon transistors	Electronics

Pigment

A pigment is a material that changes the color of reflected or transmitted light as the result of wavelength-selective absorption. This physical process differs from fluorescence, phosphorescence, and other forms of luminescence, in which a material emits light.

Natural ultramarine pigment in powdered form.

Many materials selectively absorb certain wavelengths of light. Materials that humans have chosen and developed for use as pigments usually have special properties that make them ideal for coloring other materials. A pigment must have a high tinting strength relative to the materials it colors. It must be stable in solid form at ambient temperatures.

Synthetic ultramarine pigment is chemically identical to natural ultramarine.

For industrial applications, as well as in the arts, permanence and stability are desirable properties. Pigments that are not permanent are called fugitive. Fugitive pigments fade over time, or with exposure to light, while some eventually blacken.

Pigments are used for coloring paint, ink, plastic, fabric, cosmetics, food, and other materials. Most pigments used in manufacturing and the visual arts are dry colorants, usually ground into a fine powder. This powder is added to a binder (or vehicle), a relatively neutral or colorless material that suspends the pigment and gives the paint its adhesion.

A distinction is usually made between a pigment, which is insoluble in its vehicle (resulting in a suspension), and a dye, which either is itself a liquid or is soluble in its vehicle (resulting in a solution). A colorant can act as either a pigment or a dye depending on the vehicle involved. In some cases, a pigment can be manufactured from a dye by precipitating a soluble dye with a metallic salt. The resulting pigment is called a lake pigment. The term biological pigment is used for all colored substances independent of their solubility.

In 2006, around 7.4 million tons of inorganic, organic and special pigments were marketed worldwide. Asia has the highest rate on a quantity basis followed by Europe and North America. By 2020, revenues will have risen to approx. US$34.2 billion. The global demand on pigments was roughly US$20.5 billion in 2009, around 1.5-2% up from the previous year. It is predicted to increase in a stable growth rate in the coming years. The worldwide sales are said to increase up to US$24.5 billion in 2015, and reach US$27.5 billion in 2018.

Physical Basis

Pigments appear the colors they are because they selectively reflect and absorb certain wavelengths of visible light. White light is a roughly equal mixture of the entire spectrum of visible light with a wavelength in a range from about 375 or 400 nanometers to about 760 or 780 nm. When this light encounters a pigment, parts of the spectrum are absorbed by the molecules or ions of the pigment. In organic pigments such as diazo or phthalocyanine compounds the light is absorbed by the conjugated systems of double bonds in the molecule. Some of the inorganic pigments such as vermilion (mercury sulfide) or cadmium yellow (cadmium sulfide) absorb light by transferring an electron from the negative ion (S^{2-}) to the positive ion (Hg^{2+} or Cd^{2+}). Such compounds are designated as charge-transfer complexes, with broad absorption bands that subtract most of the colors of the incident white light. The other wavelengths or parts of the spectrum are reflected or scattered. The new reflected light spectrum creates the appearance of a color. Pigments, unlike fluorescent or phosphorescent substances, can only subtract wavelengths from the source light, never add new ones.

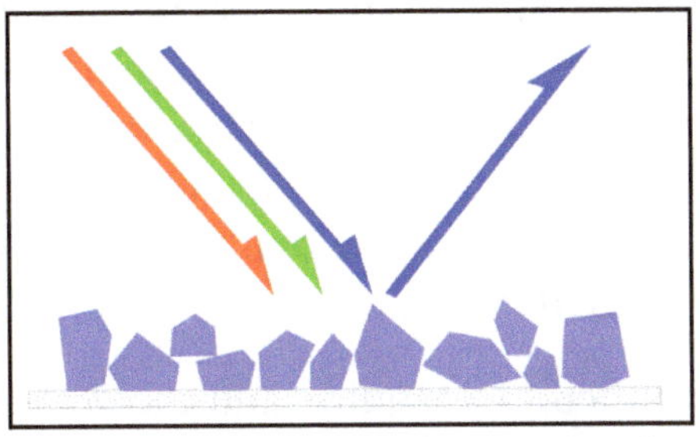

A wide variety of wavelengths (colors) encounter a pigment. This pigment absorbs red and green light, but reflects blue, creating the color blue.

The appearance of pigments is intimately connected to the color of the source light. Sunlight has a high color temperature, and a fairly uniform spectrum, and is considered a standard for white light. Artificial light sources tend to have great peaks in some parts of their spectrum, and deep valleys in others. Viewed under these conditions, pigments will appear different colors.

Color spaces used to represent colors numerically must specify their light source. Lab color measurements, unless otherwise noted, assume that the measurement was taken under a D65 light source, or "Daylight 6500 K", which is roughly the color temperature of sunlight.

Other properties of a color, such as its saturation or lightness, may be determined by the other substances that accompany pigments. Binders and fillers added to pure pigment chemicals also have their own reflection and absorption patterns, which can affect the final spectrum. Likewise, in pigment/binder mixtures, individual rays of light may not encounter pigment molecules, and may be reflected as is. These stray rays of source light contribute to a slightly less saturated color. Pure pigment allows very little white light to escape, producing a highly saturated color. A small quantity of pigment mixed with a lot of white binder, however, will appear desaturated and pale, due to the high quantity of escaping white light.

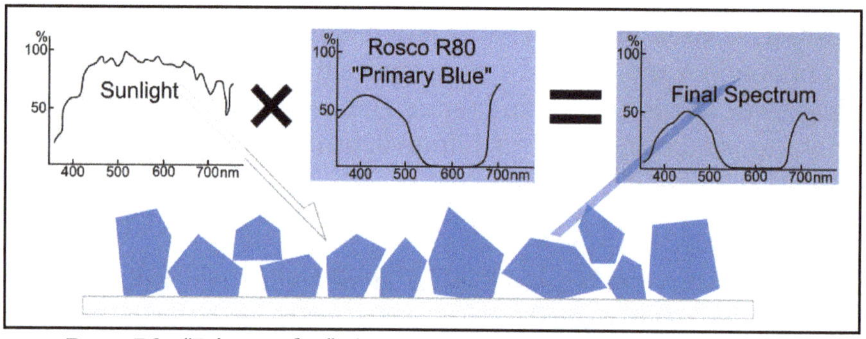

Sunlight encounters Rosco R80 "Primary Blue" pigment. The product of the source spectrum and the reflectance spectrum of the pigment results in the final spectrum, and the appearance of blue.

History

Naturally occurring pigments such as ochres and iron oxides have been used as colorants since prehistoric times. Archaeologists have uncovered evidence that early humans used paint for aesthetic purposes such as body decoration. Pigments and paint grinding equipment believed to be between 350,000 and 400,000 years old have been reported in a cave at Twin Rivers, near Lusaka, Zambia.

Before the Industrial Revolution, the range of color available for art and decorative uses was technically limited. Most of the pigments in use were earth and mineral pigments, or pigments of biological origin. Pigments from unusual sources such as botanical materials, animal waste, insects, and mollusks were harvested and traded over long distances. Some colors were costly or impossible to obtain, given the range of pigments that were available. Blue and purple came to be associated with royalty because of their rarity.

Biological pigments were often difficult to acquire, and the details of their production were kept secret by the manufacturers. Tyrian Purple is a pigment made from the mucus of one of several species of Murex snail. Production of Tyrian Purple for use as a fabric dye began as early as 1200 BCE by the Phoenicians, and was continued by the Greeks and Romans until 1453 CE, with the fall of Constantinople. The pigment was expensive and complex to produce, and items colored with it became associated with power and wealth. Greek historian Theopompus, writing in the 4th century BCE, reported that "purple for dyes fetched its weight in silver at Colophon [in Asia Minor]."

Mineral pigments were also traded over long distances. The only way to achieve a deep rich blue was by using a semi-precious stone, lapis lazuli, to produce a pigment known as ultramarine, and the best sources of lapis were remote. Flemish painter Jan van Eyck, working in the 15th century, did not ordinarily include blue in his paintings. To have one's portrait commissioned and painted with ultramarine blue was considered a great luxury. If a patron wanted blue, they were obliged to pay extra. When Van Eyck used lapis, he never blended it with other colors. Instead he applied it in pure form, almost as a decorative glaze. The prohibitive price of lapis lazuli forced artists to seek less expensive replacement pigments, both mineral (azurite, smalt) and biological (indigo).

Miracle of the Slave by Tintoretto (c. 1548). The son of a master dyer, Tintoretto used Carmine Red Lake pigment, derived from the cochineal insect, to achieve dramatic color effects.

Spain's conquest of a New World empire in the 16th century introduced new pigments and colors to peoples on both sides of the Atlantic. Carmine, a dye and pigment derived from a parasitic insect found in Central and South America, attained great status and value in Europe. Produced from harvested, dried, and crushed cochineal insects, carmine could be, and still is, used in fabric dye, food dye, body paint, or in its solid lake form, almost any kind of paint or cosmetic.

Natives of Peru had been producing cochineal dyes for textiles since at least 700 CE, but Europeans had never seen the color before. When the Spanish invaded the Aztec empire in what is now Mexico, they were quick to exploit the color for new trade opportunities. Carmine became the region's second most valuable export next to silver. Pigments produced from the cochineal insect gave the Catholic cardinals their vibrant robes and the English "Redcoats" their distinctive uniforms. The true source of the pigment, an insect, was kept secret until the 18th century, when biologists discovered the source.

Girl with a Pearl Earring by Johannes Vermeer (c. 1665).

While Carmine was popular in Europe, blue remained an exclusive color, associated with wealth and status. The 17th-century Dutch master Johannes Vermeer often made lavish use of lapis lazuli, along with Carmine and Indian yellow, in his vibrant paintings.

Development of Synthetic Pigments

The earliest known pigments were natural minerals. Natural iron oxides give a range of colors and are found in many Paleolithic and Neolithic cave paintings. Two examples include Red Ochre, anhydrous Fe_2O_3, and the hydrated Yellow Ochre ($Fe_2O_3.H_2O$). Charcoal, or carbon black, has also been used as a black pigment since prehistoric times.

Two of the first synthetic pigments were white lead (basic lead carbonate, $(PbCO_3)_2Pb(OH)_2$) and blue frit (Egyptian Blue). White lead is made by combining lead with vinegar (acetic acid, CH_3COOH) in the presence of CO_2. Blue frit is calcium copper silicate and was made from glass colored with a copper ore, such as malachite. These pigments were used as early as the second millennium BCE Later premodern additions to the range of synthetic pigments included vermillion, verdigris and lead-tin-yellow.

The Industrial and Scientific Revolutions brought a huge expansion in the range of synthetic pigments, pigments that are manufactured or refined from naturally occurring materials, available both for manufacturing and artistic expression. Because of the expense of Lapis Lazuli, much effort went into finding a less costly blue pigment.

Prussian Blue was the first modern synthetic pigment, discovered by accident in 1704. By the early 19th century, synthetic and metallic blue pigments had been added to the range of blues, including French ultramarine, a synthetic form of lapis lazuli, and the various forms of Cobalt and Cerulean Blue. In the early 20th century, organic chemistry added Phthalo Blue, a synthetic, organometallic pigment with overwhelming tinting power.

Self Portrait by Paul Cézanne. Working in the late 19th century, Cézanne had a palette of colors that earlier generations of artists could only have dreamed of.

Discoveries in color science created new industries and drove changes in fashion and taste. The discovery in 1856 of mauveine, the first aniline dye, was a forerunner for the development of hundreds of synthetic dyes and pigments like azo and diazo compounds which are the source of a wide spectrum of colors. Mauveine was discovered by an 18-year-old chemist named William Henry Perkin, who went on to exploit his discovery in industry and become wealthy. His success attracted a generation of followers, as young scientists went into organic chemistry to pursue riches. Within a few years, chemists had synthesized a substitute for madder in the production of Alizarin Crimson. By the closing decades of the 19th century, textiles, paints, and other commodities in colors such as red, crimson, blue, and purple had become affordable.

Development of chemical pigments and dyes helped bring new industrial prosperity to Germany and other countries in northern Europe, but it brought dissolution and decline elsewhere. In Spain's former New World empire, the production of cochineal colors employed thousands of low-paid workers. The Spanish monopoly on cochineal production had been worth a fortune until the early 19th century, when the Mexican War of Independence and other market changes disrupted production. Organic chemistry delivered the final blow for the cochineal color industry. When chemists created inexpensive substitutes for carmine, an industry and a way of life went into steep decline.

New Sources for Historic Pigments

Before the Industrial Revolution, many pigments were known by the location where they were produced. Pigments based on minerals and clays often bore the name of the city or region where they were mined. Raw Sienna and Burnt Sienna came from Siena, Italy, while Raw Umber and Burnt Umber came from Umbria. These pigments were among the easiest to synthesize, and chemists created modern colors based on the originals that were more consistent than colors mined from the original ore bodies. But the place names remained.

The Milkmaid by Johannes Vermeer (c. 1658). Vermeer was lavish in his choice of expensive pigments, including lead-tin-yellow, natural ultramarine and madder lake, as shown in this vibrant painting.

Historically and culturally, many famous natural pigments have been replaced with synthetic pigments, while retaining historic names. In some cases, the original color name has shifted in meaning, as a historic name has been applied to a popular modern color. By convention, a contemporary mixture of pigments that replaces a historical pigment is indicated by calling the resulting color a hue, but manufacturers are not always careful in maintaining this distinction. The following examples illustrate the shifting nature of historic pigment names:

- Indian Yellow was once produced by collecting the urine of cattle that had been fed only mango leaves. Dutch and Flemish painters of the 17th and 18th centuries favored it for its luminescent qualities, and often used it to represent sunlight. In the novel *Girl with a Pearl Earring*, Vermeer's patron remarks that Vermeer used "cow piss" to paint his wife. Since mango leaves are nutritionally inadequate for cattle, the practice of harvesting Indian Yellow was eventually declared to be inhumane. Modern hues of Indian Yellow are made from synthetic pigments.

- Ultramarine, originally the semi-precious stone lapis lazuli, has been replaced by an inexpensive modern synthetic pigment, French Ultramarine, manufactured from aluminium silicate with sulfur impurities. At the same time, Royal Blue, another name once given to tints produced from lapis lazuli, has evolved to signify a much lighter and brighter color, and is usually mixed from Phthalo Blue and titanium dioxide, or from inexpensive synthetic blue dyes. Since synthetic ultramarine is chemically identical with lapis lazuli, the "hue" designation is not used. French Blue, yet another historic name for ultramarine, was adopted by the textile and apparel industry as a color name in the 1990s, and was applied to a shade of blue that has nothing in common with the historic pigment ultramarine.
- Vermilion, a toxic mercury compound favored for its deep red-orange color by old master painters such as Titian, has been replaced in painters' palettes by various modern pigments, including cadmium reds. Although genuine Vermilion paint can still be purchased for fine arts and art conservation applications, few manufacturers make it, because of legal liability issues. Few artists buy it, because it has been superseded by modern pigments that are both less expensive and less toxic, as well as less reactive with other pigments. As a result, genuine Vermilion is almost unavailable. Modern vermilion colors are properly designated as Vermilion Hue to distinguish them from genuine Vermilion.

Titian used the historic pigment Vermilion to create the reds in the great fresco of Assunta, completed c. 1518.

Manufacturing and Industrial Standards

Pigments for sale at a market stall in Goa, India.

Before the development of synthetic pigments, and the refinement of techniques for extracting mineral pigments, batches of color were often inconsistent. With the development of a modern color industry, manufacturers and professionals have cooperated to create international standards for identifying, producing, measuring, and testing colors.

First published in 1905, the Munsell color system became the foundation for a series of color models, providing objective methods for the measurement of color. The Munsell system describes a color in three dimensions, hue, value (lightness), and chroma (color purity), where chroma is the difference from gray at a given hue and value.

By the middle years of the 20th century, standardized methods for pigment chemistry were available, part of an international movement to create such standards in industry. The International Organization for Standardization (ISO) develops technical standards for the manufacture of pigments and dyes. ISO standards define various industrial and chemical properties, and how to test for them. The principal ISO standards that relate to all pigments are as follows:

- ISO-787 General methods of test for pigments and extenders.
- ISO-8780 Methods of dispersion for assessment of dispersion characteristics.

Other ISO standards pertain to particular classes or categories of pigments, based on their chemical composition, such as ultramarine pigments, titanium dioxide, iron oxide pigments, and so forth.

Many manufacturers of paints, inks, textiles, plastics, and colors have voluntarily adopted the Colour Index International (CII) as a standard for identifying the pigments that they use in manufacturing particular colors. First published in 1925, and now published jointly on the web by the Society of Dyers and Colourists (United Kingdom) and the American Association of Textile Chemists and Colorists (USA), this index is recognized internationally as the authoritative reference on colorants. It encompasses more than 27,000 products under more than 13,000 generic color index names.

In the CII schema, each pigment has a generic index number that identifies it chemically, regardless of proprietary and historic names. For example, Phthalocyanine Blue BN has been known by a variety of generic and proprietary names since its discovery in the 1930s. In much of Europe, phthalocyanine blue is better known as Helio Blue, or by a proprietary name such as Winsor Blue. An American paint manufacturer, Grumbacher, registered an alternate spelling (Thanos Blue) as a trademark. Colour Index International resolves all these conflicting historic, generic, and proprietary names so that manufacturers and consumers can identify the pigment (or dye) used in a particular color product. In the CII, all phthalocyanine blue pigments are designated by a generic color index number as either PB15 or PB16, short for pigment blue 15 and pigment blue 16; these two numbers reflect slight variations in molecular structure that produce a slightly more greenish or reddish blue.

Scientific and Technical Issues

Selection of a pigment for a particular application is determined by cost, and by the physical properties and attributes of the pigment itself. For example, a pigment that is used to color glass must have very high heat stability in order to survive the manufacturing process; but, suspended in the glass vehicle, its resistance to alkali or acidic materials is not an issue. In artistic paint, heat stability is less important, while lightfastness and toxicity are greater concerns.

The following are some of the attributes of pigments that determine their suitability for particular manufacturing processes and applications:

- Lightfastness and sensitivity for damage from ultra violet light,
- Heat stability,
- Toxicity,
- Tinting strength,
- Staining,
- Dispersion,
- Opacity or transparency,
- Resistance to alkalis and acids,
- Reactions and interactions between pigments.

Swatches

Swatches are used to communicate colors accurately. For different media like printing, computers, plastics, and textiles, different type of swatches are used. Generally, the medium which offers the broadest gamut of color shades is widely used across different media.

Printed Swatches

There are many reference standards providing printed swatches of color shades. PANTONE, RAL, Munsell etc. are widely used standards of color communication across different media like printing, plastics, and textiles.

Plastic Swatches

Companies manufacturing color masterbatches and pigments for plastics offer plastic swatches in injection molded color chips. These color chips are supplied to the designer or customer to choose and select the color for their specific plastic products.

Plastic swatches are available in various special effects like pearl, metallic, fluorescent, sparkle, mosaic etc. However, these effects are difficult to replicate on other media like print and computer display. wherein they have created plastic swatches on website by 3D modelling to including various special effects.

Computer Swatches

Pure pigments reflect light in a very specific way that cannot be precisely duplicated by the discrete light emitters in a computer display. However, by making careful measurements of pigments, close approximations can be made. The Munsell Color System provides a good conceptual explanation of what is missing. Munsell devised a system that provides an objective measure of color in three dimensions: hue, value (or lightness), and chroma. Computer displays in general are unable to show the true chroma of many pigments, but the hue and lightness can be reproduced with relative accuracy. However, when the gamma of a computer display deviates from the reference value, the hue is also systematically biased.

The following approximations assume a display device at gamma 2.2, using the sRGB color space. The further a display device deviates from these standards, the less accurate these swatches will be. Swatches are based on the average measurements of several lots of single-pigment watercolor paints, converted from Lab color space to sRGB color space for viewing on a computer display. Different brands and lots of the same pigment may vary in color. Furthermore, pigments have inherently complex reflectance spectra that will render their color appearance greatly different depending on the spectrum of the source illumination; a property called metamerism. Averaged measurements of pigment samples will only yield approximations of their true appearance under a specific source of illumination. Computer display systems use a technique called chromatic adaptation transforms to emulate the correlated color temperature of illumination sources, and cannot perfectly reproduce the intricate spectral combinations originally seen. In many cases, the perceived color of a pigment falls outside of the gamut of computer displays and a method called gamut mapping is used to approximate the true appearance. Gamut mapping trades off any one of lightness, hue, or saturation accuracy to render the color on screen, depending on the priority chosen in the conversion's ICC rendering intent.

#990024	PR106 - #E34234	#FFB02E
Tyrian red	Vermilion (genuine)	Indian yellow

PB29 - #003BAF	PB27 - #0B3E66
Ultramarine blue	Prussian blue

Biological Pigments

In biology, a pigment is any colored material of plant or animal cells. Many biological structures, such as skin, eyes, fur, and hair contain pigments (such as melanin). Animal skin coloration often comes about through specialized cells called chromatophores, which animals such as the octopus and chameleon can control to vary the animal's color. Many conditions affect the levels or nature of pigments in plant, animal, some protista, or fungus cells. For instance, the disorder called albinism affects the level of melanin production in animals.

Pigmentation in organisms serves many biological purposes, including camouflage, mimicry, aposematism (warning), sexual selection and other forms of signalling, photosynthesis (in plants), as well as basic physical purposes such as protection from sunburn.

Pigment color differs from structural color in that pigment color is the same for all viewing angles, whereas structural color is the result of selective reflection or iridescence, usually because of multilayer structures. For example, butterfly wings typically contain structural color, although many butterflies have cells that contain pigment as well.

Pigments by Elemental Composition

Transition metal compounds. From left to right, aqueous solutions of: Co(NO
3)
2 (red); K
2Cr
2O
7 (orange); K
2CrO
4 (yellow); NiCl
2 (turquoise); CuSO
4 (blue); KMnO
4 (purple).

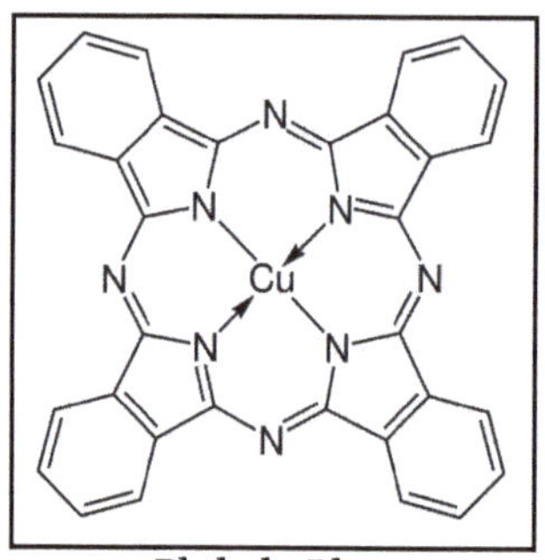

Phthalo Blue.

Metal-based Pigments

- Cadmium pigments: cadmium yellow, cadmium red, cadmium green, cadmium orange, cadmium sulfoselenide.
- Chromium pigments: chrome yellow and chrome green.
- Cobalt pigments: cobalt violet, cobalt blue, cerulean blue, aureolin (cobalt yellow).
- Copper pigments: Azurite, Han purple, Han blue, Egyptian blue, Malachite, Paris green, Phthalocyanine Blue BN, Phthalocyanine Green G, verdigris, viridian.
- Iron oxide pigments: sanguine, caput mortuum, oxide red, red ochre, Venetian red, Prussian blue.
- Lead pigments: lead white, cremnitz white, Naples yellow, red lead, lead-tin-yellow.
- Manganese pigments: manganese violet.
- Mercury pigments: vermilion.

- Titanium pigments: titanium yellow, titanium beige, titanium white, titanium black.
- Zinc pigments: zinc white, zinc ferrite.

Other Inorganic Pigments

- Carbon pigments: carbon black (including vine blac, lamp black), ivory black (bone char).
- Clay earth pigments (iron oxides): yellow ochre, raw sienna, burnt sienna, raw umber, burnt umber.
- Ultramarine pigments: ultramarine, ultramarine green shade.

Biological and Organic

- Biological origins: alizarin (synthesized), alizarin crimson (synthesized), gamboge, cochineal red, rose madder, indigo, Indian yellow, Tyrian purple.
- Non biological organic: quinacridone, magenta, phthalo green, phthalo blue, pigment red 170, diarylide yellow.

References

- Lindlar, H. and Dubuis, R. (2016). "Palladium Catalyst for Partial Reduction of Acetylenes". Org. Synth. doi:10.15227/orgsyn.046.0089. ; Coll. Vol., 5, p. 880
- Wei, Hui; Wang, Erkang (2013-06-21). "Nanomaterials with enzyme-like characteristics (nanozymes): next-generation artificial enzymes". Chemical Society Reviews. 42 (14). doi:10.1039/C3CS35486E. ISSN 1460-4744
- "Sto AG, Cabot Create Aerogel Insulation". Construction Digital. 15 November 2011. Retrieved 28, January 2020
- Nelson, D. L. and Cox, M. M. (2000) Lehninger, Principles of Biochemistry 3rd Ed. Worth Publishing: New York. ISBN 1-57259-153-6
- Rayner-Canham, Marelene; Rayner-Canham, Geoffrey William (January 2001). Women in Chemistry: Their Changing Roles from Alchemical Times to the Mid-Twentieth Century. American Chemical Society. ISBN 9780841235229

Permissions

All chapters in this book are published with permission under the Creative Commons Attribution Share Alike License or equivalent. Every chapter published in this book has been scrutinized by our experts. Their significance has been extensively debated. The topics covered herein carry significant information for a comprehensive understanding. They may even be implemented as practical applications or may be referred to as a beginning point for further studies.

We would like to thank the editorial team for lending their expertise to make the book truly unique. They have played a crucial role in the development of this book. Without their invaluable contributions this book wouldn't have been possible. They have made vital efforts to compile up to date information on the varied aspects of this subject to make this book a valuable addition to the collection of many professionals and students.

This book was conceptualized with the vision of imparting up-to-date and integrated information in this field. To ensure the same, a matchless editorial board was set up. Every individual on the board went through rigorous rounds of assessment to prove their worth. After which they invested a large part of their time researching and compiling the most relevant data for our readers.

The editorial board has been involved in producing this book since its inception. They have spent rigorous hours researching and exploring the diverse topics which have resulted in the successful publishing of this book. They have passed on their knowledge of decades through this book. To expedite this challenging task, the publisher supported the team at every step. A small team of assistant editors was also appointed to further simplify the editing procedure and attain best results for the readers.

Apart from the editorial board, the designing team has also invested a significant amount of their time in understanding the subject and creating the most relevant covers. They scrutinized every image to scout for the most suitable representation of the subject and create an appropriate cover for the book.

The publishing team has been an ardent support to the editorial, designing and production team. Their endless efforts to recruit the best for this project, has resulted in the accomplishment of this book. They are a veteran in the field of academics and their pool of knowledge is as vast as their experience in printing. Their expertise and guidance has proved useful at every step. Their uncompromising quality standards have made this book an exceptional effort. Their encouragement from time to time has been an inspiration for everyone.

The publisher and the editorial board hope that this book will prove to be a valuable piece of knowledge for students, practitioners and scholars across the globe.

Index

CPSIA information can be obtained
at www.ICGtesting.com
Printed in the USA
LVHW062013100122
708212LV00002B/216

9 781639 875535